BUGS IN THE SYSTEM

BUGS IN
THE SYSTEM

Insects and Their Impact on Human Affairs

May R. Berenbaum

HELIX BOOKS

Addison-Wesley Publishing Company

Reading, Massachusetts   Menlo Park, California   New York
Don Mills, Ontario   Wokingham, England   Amsterdam   Bonn
Sydney   Singapore   Tokyo   Madrid   San Juan
Paris   Seoul   Milan   Mexico City   Taipei

Many of the designations used by manufacturers and sellers to distinguish their products are claimed as trademarks. Where those designations appear in this book and Addison-Wesley was aware of a trademark claim, the designations have been printed in initial capital letters (The Dragonfly).

Library of Congress Cataloging-in-Publication Data
Berenbaum, M. (May)
    Bugs in the system : insects and their impact on human affairs / May R. Berenbaum.
      p. cm.
    Includes bibliographical references (p.   ) and index.
    ISBN 0-201-62499-0
    1. Insects.   2. Insects—Ecology.   3. Insect pests.   4. Beneficial insects.   I. Title.
    QL463.B46   1995
    595.7—dc20   94-14032
                                                    CIP

Jacket design by Lynne Reed
Text design by David Kelley
Set in 11-point Minion by Compset, Inc.

1 2 3 4 5 6 7 8 9 10-MA-979695
First printing, November 1994

Dedicated with all my heart to

Hannah Leskosky

Contents

Acknowledgments

FIRST OF ALL, I have to say that if, after reading this book, you are in any way displeased with it, I alone am responsible for all of the errors, misstatements, oversights, and bad jokes that may have contributed to your reaction. If, however, you like the book, I have to share the credit with quite a number of very fine people. As Cinderella's fairy godmother turned a pumpkin and some mice into a coach and six horses, Jack Repcheck, at Addison-Wesley, granted my fondest wish, to turn a jumbled set of course notes into a book for the general public. Unlike that fairy godmother, though, Jack couldn't rely on magic—he took a big chance when he picked me up out of the editorial page of *The New York Times* and, because he liked an op-ed piece I had written about insects, wrote me a letter asking whether or not I would like to write a book sometime. He also displayed incredible patience when faced with my inability to meet deadlines and was irrepressibly encouraging throughout the whole process.

Assembling the information was a mighty undertaking, in which I was assisted by innumerable people. My colleagues in the Department of Entomology and in many other units on the campus have proved to be inexhaustible sources of information, and the librarians of the many branches of the UIUC library were invaluable partners in this enterprise, tolerating a remarkable amount of rule-bending on my behalf. When I personally couldn't go to the library, Candace Pontbriant and Marjorie Montez faithfully and reliably sought out articles, no matter how dingy or dusty the journal. I have also benefited from the collective wisdom of the 220 or so students who have taken Entomology 105, "Insects and People," over the years and traded their expertise in every imaginable discipline with me for a little bit of entomological knowledge. Dot Houchens and Mary Wisniewski provided expert, patient, and good-humored secretarial assistance, without which the task of assembling a manuscript would have been far more intimidating.

Many people with knowledge far deeper than mine in many areas of entomology also contributed immeasurably to this book by reading parts of the manuscript and keeping the text accurate. My colleague Dr. Gene Robinson, in the Department of Entomology at the University of Illinois, read Chapters 3 and 4 and checked them for accuracy; Dr. Susan Paskewicz, of the Department of Entomology at the University of Wisconsin at Madison did the same for

Chapter 7. Dr. Ellis MacLeod, who knows more about insects than any two other people I know put together, was kind enough to read the book in its entirety and saved me from many a potentially embarrassing faux pas. Attorney-at-Law James Martinkus, our family lawyer and friend, checked the section of Chapter 10 dealing with insects and the law and proved that clichés about lawyers are just as untrue as are clichés about entomologists. Much of the illustrative material was taken from the department archives—the turn-of-the-century lantern slide collection as well as original artwork by A. Paterno, who created thousands of images for the teaching collection a half-century ago.

Then there are the people whose contributions took on many forms. My parents Morris and Adrienne Berenbaum served as my personal article-clipping service for over 20 years; no reference to insects, no matter how obscure, was overlooked. My mother's library science training held her in good stead, in that every article was meticulously documented as to source, date, and page number. Their confidence in me was and continues to be my inspiration. More recently, Helen Leskosky has happily taken on the task of clipping articles for her daughter-in-law, for which I am very grateful; I am also grateful that she makes it seem that she has always believed every family ought to have at least one entomologist in it. My husband Richard Leskosky read the manuscript in its entirety, even before it was presentable enough to show to anyone else; in addition to providing me with gentle suggestions for grammatical improvement, he also provided an abundance of obscure and fascinating information as well as catchy phrases and witty remarks. While not all of the witty remarks were incorporated into the book, all of them cheered me to no end and deepened my great affection for him.

There is one other person I have to mention, whose contributions were less tangible. She didn't read a single word of text—actually, at this point in her life, her tastes run more to *The Cat in the Hat* than to entomology books. She didn't provide me with any useful references nor did she type or provide illustrations or otherwise assist with preparing the book. Truth be told, Hannah doesn't even like insects all that much at the moment, although I'm optimistic that will change with time. All she did was make me happier than I ever imagined possible and this book is for her, in the hope that someday she'll be proud of her mommy the entomologist.

Preface

THE VAST MAJORITY of people consider it a high priority to minimize the extent of their interaction with the insect world. Homes are sealed, sprayed, and kept meticulously clean so as to reduce the probability that they will be invaded by insects; similarly, bodies are bathed, hair is shampooed, and clothing regularly washed in order to eliminate any unwanted contact with six-legged life forms. In the overwhelmingly vast majority of daily conversations, insects are conspicuous in their absence; those rare conversations in which insects feature prominently are generally carried out in guarded tones, often with a touch of embarrassment. After all, no one likes to admit, even to close personal friends, to being stung, bitten, infested, invaded, or otherwise bested by the loathsome insects that manage to get around the safeguards.

It is indeed a laudable goal to try to distance oneself from the insect world, but it is, alas, an impossible one. There is no other life form on the planet whose lives are as inextricably bound up with our own as are members of the class Insecta. For one thing, they intrude by force of sheer number. Of the world's species, almost 80% of them are insects—in other words, four out of every five creatures have six legs at some point during their lifetime. Over 800,000 species of insects are known to science, and there's really no way of telling how many there are altogether; estimates taking into account species yet undescribed and awaiting discovery range from 2 million to upwards of 30 million. As individuals, they collectively outweigh every other form of life on the planet as well. The total number of individual insects on earth at any given moment has been estimated at 10 quintillion (or 10,000,000,000,000,000,000), a number that's not that unreasonable considering that some termite colonies house over a million individuals and locust swarms can contain up to a billion individuals.

In view of these enormous numbers, it's not altogether surprising that insects can be found just about everywhere (and certainly everywhere that humans have staked a claim). There are insects that live in Antarctica, in cracks in the snow, and in hot springs in Yellowstone, in water where temperatures approach the boiling point. Insects live in horse intestines, where the acidity levels, even in a horse without heartburn, are comparable to vinegar; they live in pools of petroleum in oil fields, in jars of formaldehyde in morgues, and in baptismal fonts in churches. They thrive as well at the tops of the highest mountains as they do in mines almost a mile below the earth's surface, and they are equally at home in the driest of deserts and in the most humid of rain forests. About the only place on earth where they are not well represented is in the ocean, but even there a few hardy species have set up residence—ocean skaters that can be found gliding on the water surface a mile or more from shore, or lice that live in the nostrils of sea lions and stay relatively dry as they accompany their hosts on deep dives underwater.

So wherever humans have broken ground, whatever frontiers humans have explored, they have discovered that they are latecomers, following in the six-legged footsteps of insects. Whatever resources humans have wanted to garner as their own, insects have had a prior claim on. Thus it is that they are our chief competitors, exacting their toll in the form of destruction of crops, domesticated animals, stored products, timber, rangeland, and even human life (since many insects view humans as nothing more than a meal and in the course of feeding can transmit an enormous variety of debilitating and even deadly diseases).

But, because insects are in many cases the chief architects of terrestrial ecosystems, they are also our principal partners in making a living on earth. About a third of our diet (and a higher percentage for vegetarians) is the direct result of insect pollination; insect-pollination services in the U.S. amount to more than 9 billion dollars every year. Without insects, there would be no oranges in Florida, no cotton in Mississippi, no cheese in Wisconsin, no peaches in Georgia, and no potatoes in Idaho. By eating dung, carrion, and other ordure spurned by more discriminating beasts, insects keep the earth's surface free and clear of debris. Moreover, aside from economic services, insects contribute economic products in magnitude unequalled by any other group of organisms. Entire economies have revolved around insect products. Aztecs paid tribute in the form of dead bodies of scale insects, which, due to the pigment they produce, were worth more than gold; fortunes, even lives, were made and lost in the silk trade.

Because the interactions are so profound, all-encompassing, far-reaching, most people are completely unaware of the extent to which life and culture are shaped by insects. Insects have been present on every battlefield of every war and have determined the outcome of those battles more often than have bullets or bombs. Alexander the Great, Napoleon, and other brilliant military strategists were more often defeated by arthropods than by their opponents and would have been well-advised to have studied their habits more closely. Were it not for insects, there may never have been certain major social innovations, like the rise of a middle class in Europe, or scientific advances, like the germ theory of disease or the theory of natural selection. There may never have even been a science of genetics or a field of computer science without insects. Moreover, the world would have been a drab and colorless place, literally and figuratively. To remove all references to insects from English literature would be to gut the works of Chaucer, Shakespeare, Tennyson, and Keats, and to expunge all insect images rendered by artists would be to tamper with the genius of Van Gogh and Dali. Like it or not, insects are part of where we have come from, what we are now, and what we will be. It seems to me that's a pretty good reason for getting better acquainted with them.

BUGS IN THE SYSTEM

Chapter 1

CLASSIFICATION AND NOMENCLATURE ("A ROSE-CHAFER BY ANY OTHER NAME . . ")

History of classification

THROUGHOUT THE WORLD and throughout history, people have had a penchant for naming things. Even in the Bible, one of the first tasks assigned to Adam and Eve was naming all the other creatures in the Garden of Eden. Taxonomy is the science of naming, classifying, and identifying organisms. Several compelling reasons underlie the universal urge to identify, name, and classify things. Assigning a name tends to improve communication; for example, the statement "Hand me that thingamabob over there" isn't quite as clear to a listener as, say, "Hand me that socket wrench over there." Classification is the arrangement of things into groups sharing certain specified similarities. Thus, by knowing how something is classified, you immediately know something about it. If someone comes up to you and asks you (for whatever reason) to describe the flea he's holding clutched in his hand behind his back, you don't have to be a psychic to tell him that it is flattened side-to-side, wingless, and equipped with sucking mouthparts that it will use to imbibe the blood of some hapless warm-blooded vertebrate—because *all* fleas (members of the order Siphonaptera) exhibit these endearing traits. Finally, by determining the criteria by which groups are to be recognized, taxonomists greatly facilitate the chore of identifying hitherto unknown items. When in 1823 a German zoologist found a strange "turtle-like little animal" crawling in and around ant nests in the stumps of old oak trees, he was baffled by its appearance, equipped as it was with a "footless, naked belly," "fleshy tentacles" and other peculiar appurtenances; for want of a better idea, he suggested that he had discovered a new and "beautiful addition to the snail fauna of his own fatherland" (Berenbaum 1994); three-quarters of a century later, when the strange little beast was reared through to its adult stage, with three body segments, six legs, and two wings, it was instantly recognizable not as a snail but as an insect—specifically, as a fly (Fig. 1.1) . No matter how they start out in life, only flies (members of the order Diptera) end up as six-legged creatures with two wings.

So, in order for things to be classified, they must have some importance to those in charge of classifying things. Insects for a long time were not carefully classified, at least in part because until relatively recently they were not regarded as terribly important—annoying, yes, but, with a handful of exceptions, not of any consequence to the smooth and efficient functioning of society. Moreover, without the aid of a microscope, most insects are so small as to be distressingly similar in appearance, so distinguishing among them presented real difficulties.

One of the earliest attempts to classify insects was by Aristotle, in his monumental *Historia Animalium*. Aristotle classified insects according to whether they were winged or wingless; wingless insects were subsequently divided according to the number and type of wings they possessed. This emphasis on wings in particular, and locomotory appendages in general, proved to be remarkably durable; even today the names of major insect orders reflect wing characters. Aristotle's logical system prevailed despite the subsequent popularity of the writings of Pliny the Elder. Approximately four centuries after Aristotle's *Historia Animalium,* Pliny the Elder offered his interpretation of insect classification in the form of his magnum opus *Historia Naturalis.* Pliny wasn't so much interested in insects as he was in everything that existed; by the time he died in A.D. 79, he had authored at least thirty-seven volumes in his series.

Pliny the Elder was definitely not a detail man—not surprising in someone whose ambitious goal was to describe Nature in its entirety. Many of his "facts" were completely unsubstantiated (such as, for example, the notion that caterpillars origi-

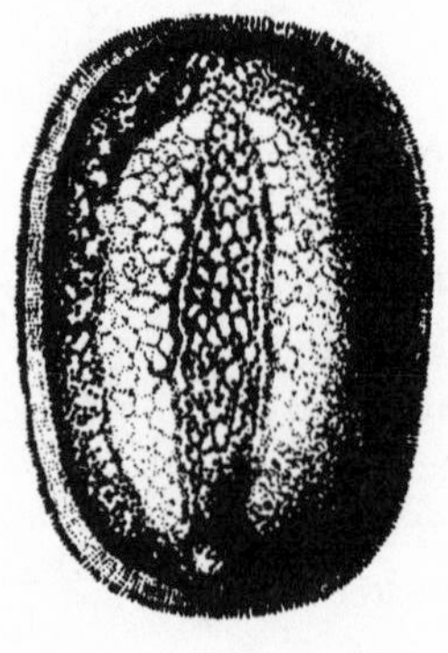

Figure 1.1
Larval stage of *Microdon,* a fly once thought to be a mollusc (original drawing by C.L. Metcalf).

nate from dew on radish leaves). Despite its inaccuracies, Pliny's *Historia Naturalis* was the authoritative source on natural history for the next 1,400 years. Medieval compilations borrowed heavily from his text and few innovations were made during the Middle Ages. For example, Bartholomaeus Anglicus (name notwithstanding, a Frenchman) compiled nineteen volumes around A.D. 1230 entitled *De Proprietatibus Rerum*. The work was intended to be a complete description of the universe. Book 12 is a discussion of air and includes an alphabetical list of flying things that inhabit the air, lumping flying insects indiscriminately in with birds. The bee, appearing along with birds as one of the "ornaments of the heavens," is described as "a little short beast with many feet. And though he might be classified among flying creatures, yet he uses his feet so much that he can reasonably be considered among ground going animals." Also considered a bird was the locust, "a worm engendered by a south wind" that "dies in a northwind." Book 18 discusses terrestrial animals, classified by their means of locomotion and their habitat. Thus, "Creeping beasts and worms pass from place to place by stretching of the body and then drawing it together; worms, adders and serpents move in this way. And they have different means of movement; some draw themselves by the mouth, like small worms, some draw themselves forward by the strength of their sides and the flexibility of their bodies as adders and serpents, and so on." Bartholomaeus Anglicus likely owes Aristotle for the idea of using means of motion as the basis for classifying living creatures.

The word "bug" dates back to this era and refers to a ghost or hobgoblin—something difficult to see and vaguely unpleasant (a term quite apt for most insects medieval people were likely to encounter). The word "insect," on the other hand, entered the English language only in 1601, when Philemon Holland published a translation of Pliny's *Historia Naturalis*. A year later, Ulysses Aldrovandus, an Italian, introduced a few taxonomic innovations of his own. Insects were divided according to habitat into Terrestria, or land-dwelling species, and Aquatica, the water-dwelling species. Each group was further divided according to the presence or absence of appendages (Pedata and Apoda, accordingly) and then subdivided further according to whether wings were present or absent (Alata and Aptera, respectively). Legs and wings were then tallied for finer taxonomic distinctions.

The introduction of devices that magnify optical images did wonders for the classification of insects. Indeed, insects were among the first objects of inspection once microscopes became readily available. The earliest recorded microscopical investigations, by Federico Cesi and Francesco Stelluti, were studies of a bee and a weevil, in 1625 and 1630, respectively (published, curiously, not in a scientific journal, but in Stelluti's translation of the first century *Satires of Persius*). With magnification, many of the anatomical features differentiating insect species were clearly visible for the first time and classification schemes based on morphological features, rather than habitat or means of motion, began to appear. The more detailed the observation, the more complex the name became. One species of butterfly, for example, was known as *Papilio media alis pronis praefertim interioribus maculis oblongis argenteis perbelle depictis.* The disadvantages of such a naming system are abundantly clear—by the time somebody rushed over and told you he'd seen one, it would be long gone.

For convenience, names were often shortened. In 1758, Carl Linné (or as he was called in scientific circles, Carolus Linnaeus—the tendency to Latinize names extended even to people) published a book called *Systema Naturæ*, in which he used a binomial, or two-name, system, consistently for the first time. The system so impressed people that it was universally adopted; no scientific names published before Linnaeus's time are considered valid and all subsequent names have conformed (and must continue to conform) to the Linnaean system.

Linnaeus was born in a small town in southern Sweden on May 23, 1707. As a young boy, he disliked school intensely, partly because of a series of uninspiring tutors and partly because he preferred puttering around his father's garden to studying. At the age of 19, his teachers decided he was not suited to the priesthood, and he further disappointed his parents by taking up the study of medicine. He went on to study natural history and medicine at the University of Uppsala, where he wrote a thesis on plant sexuality that was to become the basis of his botanical system of classification. At the time, the idea that plants were sexual organisms was vigorously decried by the church, and proponents of the theory were subject to discipline from the Vatican. Linnaeus didn't help matters much by drawing analogies between plant and human sexual practices in his writings (explaining, for example, that poppy and linden flowers were to be placed in the class he called Polyandria, from *poly,* or many, and *andros,* or men, because their sexual organs were effectively "twenty males or more in the same bed with the female"). The Catholic Church notwithstanding, Linnaeus was eminently successful not only as a taxonomist but as a physician—his practice included the Queen of Sweden as a patient.

That the binomial system works is evidenced by the fact that Linnaeus described only about 2,000 species of insects and today there are more than 750,000 with Linnaean names. The two-part, or binomial, name of a species consists of the genus (always capitalized) and the species (never capitalized). Because even today scientific names are rendered in Latin (or at least are Latinized), they are always written in italics, as are all foreign words in English text. Despite the general aversion people feel toward scientific names, they are exceedingly useful. For one thing, they're universally understood, so scientific exchanges can be carried out with precision; in contrast, common names, or vernacular names, for any given insect may vary in different parts of the country, and they certainly vary from country to country. *Helicoverpa zea* (Fig. 1.2), for example, a caterpillar with very eclectic feeding habits, is called the corn earworm in Illinois, the false tobacco budworm in North Carolina, the cotton bollworm in Arizona, and the tomato fruitworm in California. Secondly, the scientific name conveys information about the place of an organism in the hierarchy of things. *Helicoverpa zea* used to be called *Heliothis zea* until about thirty years ago, when a man named D.F. Hardwick realized that the species exhibited several anatomical features completely absent in species placed rightfully in the long-established genus *Heliothis* and accordingly invented a new generic name to convey its distinctiveness. Finally, if

Figure 1.2
Helicoverpa zea, the corn earworm, living up to its common name.

you're up on Greek or Latin, a scientific name can tell you a lot about an organism. The "*zea*" in *H. zea,* for example, means "corn," one of the caterpillar's favorite food plants.

Scientific names don't have to be off-putting. Some don't even differ dramatically from their classically derived common name equivalent—e.g., *Mantis religiosa,* the European praying mantis. Some are even shorter than their common name equivalents—*Ips pini* is positively pithy in comparison with "California five-spined engraver beetle." And some names even provide insights into the taxonomist's hobbies (e.g., *Dicrotendipes thanatogratus* (Fig. 1.3), a small fly, from *thanatos* meaning "dead" and *gratus* meaning "grateful," named by its discoverer J.H. Epler in honor of the venerable rock band the Grateful Dead) or personality (e.g., *Heerz lukenatcha,* a parasitic wasp described by inveterate punster P.M. Marsh).

Arthropod arrangements

TAXONOMISTS, THAT IS, people who classify things, have set up an organizational hierarchy for cataloging living organisms. The hierarchy runs from the largest category, the kingdom, progressively through the ranks of phylum, class, order, family, and genus, down to the smallest category, the species. Insects belong to the kingdom Animalia, the animal kingdom. Although many people are of the opinion that only warm and furry creatures are animals, the term is actually much broader, biologically speaking. It encompasses organisms that are multicellular (made up of more than one cell) and that eat food to

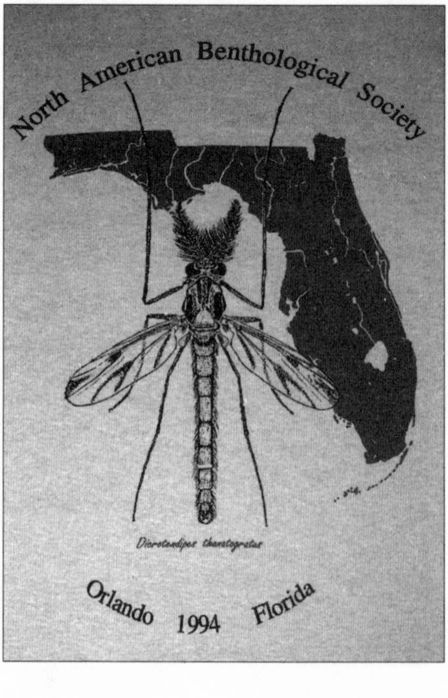

Figure 1.3
Dicrotendipes thanatogratus, on a t-shirt printed for the 1994 meetings of the North American Benthological Society (drawing by John Epler).

obtain nourishment (rather than, as plants do, manufacture their own from sunlight, carbon dioxide, and water). Although there are insects that are smaller in size than a one-celled protozoan (such as fairyflies, which are tiny wasps that spend their formative days inside the eggs of water beetles and other aquatic insects), even tiny insects are made of thousands, or even millions, of cells. Since insects can and certainly do eat a remarkable variety of things (a fact to which anyone who has shared an apartment with cockroaches can attest), their status as animals is secure.

Every kingdom is made up of a number of smaller units, called phyla (singular, phylum). The phylum Arthropoda, to which the insects belong, consists of all jointed-legged animals (*arthro* is Greek for "jointed" and *poda* for "foot") covered with an external, or exo-, skeleton. Aside from the jointed legs and the exoskeleton, another feature that characterizes arthropods is that their bodies are segmented. Arthropods are by no means unique in this respect; even humans and other vertebrates are segmented, but in arthropods, the segmentation is readily apparent, whereas in humans only a close look at the spinal column or vertebrate musculature reveals evidence of segmented construction. Needless to say, there's more to being an arthropod than being segmented. While there are other groups of segmented invertebrates (members of the phylum Annelida, the earthworms and leeches, come to mind), arthropods have done more with their segments than virtually any other group of animals.

Slice open an earthworm and it looks pretty much the same no matter where you slice it; segments are basically repeated identical units. In arthropods, segments become specialized for particular functions and often become consolidated in groups. This process of grouping adjacent segments together for particular functions, or tagmosis, allows for more efficient performance of such tasks as eating or detecting environmental stimuli. In bilaterally symmetrical organisms such as arthropods, these tasks are usually handled by appendages attached to a group of segments that collectively make up the head (the head usually being the first part of the body to encounter the environment, although as always with arthropods there are exceptions—witness crabs that walk sideways and the little tropical insects called webspinners that habitually walk backwards away from danger). Locomotion is usually carried out by appendages attached to segments that collectively make up the thorax, and reproduction and digestion, the province of segments comprising the abdomen.

Organizational plans differ with ecological demands, however, and classification of arthropods is usually based on differences in the specialization and organization of body segments. Although most people tend to regard anything with an excess of legs as an insect, in reality insects have plenty of multilegged company in the phylum Arthropoda. In fact, there is an entire subphylum, the Chelicerata, with nary an insect to claim as its own. Members of the subphylum Chelicerata have two major body regions: a cephalothorax (a fused head-and-chest arrangement with appendages for eating and locomotion) and an abdomen. The name "Chelicerata" refers to the presence of a pair of pincerlike appendages, called chelicerae, on the first segment behind the mouth opening. In addition to chelicerae, chelicerates have a pair of leglike appendages called pedipalps, and four pairs of walking legs attached to the cephalothorax.

Of the three classes of chelicerates, two are exclusively marine—the horseshoe crabs and the sea spiders. The third class, primarily a terrestrial one and therefore the most familiar, is the class Arachnida, the arachnids. Members of this class are as different from insects, taxonomically speaking, as humans are from snakes, fish, or birds. Among the arachnids are the scorpions (order Scorpiones), notable for their pincerlike pedipalps and long segmented abdomen tipped with a venomous sting. They're all carnivorous and use the sting to immobilize small, mostly insect, prey (although they are not averse to using the sting to teach blundering humans a painful lesson). Like scorpions, daddy longlegs (order Opiliones) have four pairs of legs but, as the name suggests, these legs are typically very long and slender; also distinctive is their apparent lack of a waistlike constriction between cephalothorax and abdomen. The daddy longlegs are probably scavengers, although there is some controversy as to what they eat when no one is looking. Mites and ticks (order Acari) are the most numerous of arachnids; while they too have no waist to speak of, they are easily distinguished from daddy longlegs and most other arachnids by their small size (the word *acari* actually means "tiny" in Greek) and apparent lack of segmentation. Ticks are exclusively parasitic on vertebrates, but mites lead a staggering array of lifestyles, ranging from bloodsucking to plant-feeding to scavenging. Finally, spiders (order Araneida) are the arachnids with a conspicuous constriction, or "waist," between the cephalothorax and abdomen. All spiders are predators, mostly on other arthropods, but a few of the larger species can take down small birds and mammals. They have from two to six structures on the abdomen, called spinnerets, used in spinning silk. Silk is spun for a variety of purposes; silken webs or snares entrap prey, silken pouches contain sperm during copulation, and silken blankets swaddle eggs, to cite a few.

Insects are among the members of the subphylum Mandibulata, whose members, in contrast with chelicerates, have a set of appendages called mandibles on the second segment past the mouth opening. Whereas chelicerates are pretty conservative in terms of number of legs, all more or less sticking to four (though on occasion sporting as many as six) pairs, mandibulate arthropods go to extremes in both directions—from none to hundreds. Members of the class Crustacea go in for legs in a big way; there are legs on every segment of head and thorax, which are occasionally fused, and in some crustaceans legs on every abdominal segment. All told, crustaceans can have anywhere from three to seventy pairs of legs. As well, they are unique among arthropods in having not one but two pairs of antennae; chelicerate arthropods have none and all other mandibulate arthropods have only a single pair. Familiar crustacean faces include barnacles, sowbugs, crabs, lobsters, shrimp, and crayfish; not so familiar are the tadpole shrimp, fairy shrimp, water fleas, and ostracods. Crustaceans are sufficiently different from the rest of the mandibulate arthropods (particularly in their predilection for aquatic habitats) that some people actually place them in their own subphylum.

Diplopods and chilopods are often lumped together in a superclass called Myriapoda, *myria* meaning "many." The name is appropriate since chilopods are otherwise known as centipedes and diplopods as millipedes. Both the scientific and common names convey the same idea—*cent* means "hundred" in Latin, *chilioi* means "thousand" in Greek, and *milli* means "thousand" in Latin. While it's not exactly true that centipedes have precisely 100 legs and millipedes ten times that number, it is true that they have a lot of legs, at least as adults (baby millipedes start life with, like their insect relatives, only three pairs of legs). Both groups have thirteen or more pairs of legs; the major difference between the groups is not in the number of legs but in how the legs are attached to the body. Millipedes have two pairs of legs per visible segment (hence *diplo* or "two") and centipedes have a single pair of legs per segment. The apparent doubling up of legs on millipede segments is due to the fact that each apparent segment is actually a fused double segment. While millipedes are inoffensive scavengers, centipedes are predaceous and have a set of poison glands in their jaws that they use to paralyze prey. Actually, millipedes are not totally inoffensive—while their eating habits don't involve venom, many have paired glands along the length of their bodies that, when a millipede feels threatened or offended in some way, ooze all kinds of noxious substances, including, in one species, hydrogen cyanide and in another a chemical that is structurally and functionally very similar to the active ingredient of Quaaludes, a powerful tranquilizer.

The group that lays claim to being the dominant life form on earth is the class Insecta. Three major body divisions—head, thorax, and abdomen—with three pairs of legs (no more) attached to the thorax, set them apart from all other arthropods. The body division is the trait to which insects owe their name: *Insecta* derives from the Latin for "cut into," as in "cut into pieces," a succinct way to describe segmentation. (*Entoma* is from the Greek for the same thing, whence cometh "entomology.") Insects are the only arthropods—in fact, the only invertebrates—with wings.

Recognizing insects is oftentimes easier in concept than in practice. Definitive characters are often so highly modified that they are no longer practicably recognizable; insect bodies have proved amazingly malleable to demands placed on them by a tremendous variety of lifestyles. Wings can be variously fringed, sheathlike, membranous, scaly, or hairy; these trademark anatomical appendages even disappear in some adult forms, as in the aptly named Siphonaptera, or fleas (from *siphon* for "tube" or "pipe" and *aptera* for "wingless"). Mouthparts are variously modified as syringes, stilettos, sponges, or shears, and legs can be oars, hooks, wrenches, vises, springs, or not there at all. Major anatomical features suffice to place most insects in some reasonable sort of order—actually, in about two dozen different orders, technically speaking. Ordinal names, characteristics, and prominent members are provided in Appendix 1, which should serve the function of a dramatis personae for the reader to consult when the cast of characters gets confusing. For the true insect aficionado, it should serve as a preview of coming attractions.

References

History of classification

Bardell, D., 1988. Why are the oldest records of microscopical observations in a work of literature? *Proc. R. Microsc. Soc.* 23: 365–67.

Beavis, I.C., 1988. *Insects and Other Invertebrates in Classical Antiquity.* Exeter, U.K.: University of Exeter Press.

Berenbaum, M., 1993. Buzzwords. Apis, Apis, Bobapis . . . *Am. Entomol.* 39: 4–5.

Berenbaum, M., 1994. Department of Ant-omology? *Am. Entomol.* 40: 70–71.

Blackwelder, R.E., 1967. *Taxonomy: A Text and Reference Book.* New York: Wiley.

Blunt, W., 1971. *The Compleat Naturalist: A Life of Linnaeus.* New York: Viking.

Bodenheimer, F.R.S., 1958. *The History of Biology: An Introduction.* London: Dawson.

Mitter, C., R.W. Poole, and M. Matthews, 1993. Biosystematics of the Heliothinae (Lepidoptera: Noctuidae). *Annu. Rev. Entomol.* 38: 207–225.

Arthropod arrangements

Borror, D., D. DeLong, and C. Triplehorn, 1976. *An Introduction to the Study of Insects.* New York: Holt, Rinehart and Winston.

Carrel, J. and T. Eisner, 1984. Spider sedation induced by defensive chemicals of millipede prey. *Proc. Nat. Acad. Sci.* 81: 806–810.

Metcalf, R.L. and R.A. Metcalf, l993. *Destructive and Useful Insects.* New York: McGraw-Hill.

Chapter **2**

PHYSIOLOGY

Good things in small packages

BY ANY STANDARD of measurement, insects can be said to be a very successful group. They owe their success at least in part to several physiological and anatomical innovations that give them a competitive edge over other animals. The majority of these unique features reflect the fact that, on a global scale, insects are small animals. Small size has been in large part responsible for the overwhelming success of the insects; at the same time, it imposes severe physiological and ecological restrictions. The range of size represented by living organisms on earth today encompasses some thirteen orders of magnitude—from a blue whale weighing in at 100 metric tons to a rotifer, or wheel animal, weighing less than 0.01 milligram. Insects range over only five orders of magnitude—from tropical scarabeid beetles (like the Atlas or Hercules beetles) weighing 30 grams (more than the average field mouse) to midges less than 0.1 milligram. So, eight orders of magnitude are missing in the class Insecta. Problems at the upper limit of life on earth involve support, transport, and overcoming inertia—clearly not problems for organisms like insects, which must deal with a whole different set of problems.

Most of the problems insects face can be traced back to one simple relationship—the surface area/volume ratio. As the size of an organism (or object) increases, its surface area increases in proportion to the square of length (since area is measured in two dimensions), and its weight (mass or volume) increases with the cube of length (since volume is a three-dimensional measure). The larger the number, the greater the discrepancy between its square and its cube; the difference between 3^2 and 3^3 (i.e., the difference between 9 and 27) is greater than the difference, for example, between 2^2 and 2^3 (i.e., the difference between 4 and 8). So, because squaring a number does not make it as big as cubing a number, as animals get larger they have proportionately less surface area (a squared measure) relative to their volume (a cubic measure) in comparison with smaller animals.

This relationship between surface area and volume can work to an insect's advantage. Since muscle strength is proportional to cross-sectional area, insect muscles appear to be very powerful, since they are moving a muscle mass (or volume) that is relatively small compared to the cross-sectional area. Hence, statements like "A grasshopper as large as a man could leap across a football field in a single jump" or "if a man were as strong as an ant, he could pull two boxcars without sweating." The problem with all of these statements is that they don't take into account the surface area/volume ratio. The truth of it is that, if a grasshopper were as big as a man, it could jump probably only about as far as a man can jump, since its muscles would have to move a much greater bulk or volume than a grasshopper is accustomed to moving (Fig. 2.1).

Virtually all the physical or anatomical traits that set insects apart from other organisms represent adjustments to the surface area/volume ratio. One of the first innovations, according to the

Figure 2.1
Turn of the century novelty postcard depicts a physiologically improbable scene.

fossil record, was a tough, hard external skeleton, or exoskeleton. In fact, all of today's organisms with exoskeletons are descended from fossil ancestors over 500 million years old. Today's exoskeletons are multilayers—layer upon layer of different material, each of which contributes to a unique property of the exoskeleton. The outside covering of an insect—the body wall that makes up the exoskeleton—is called the cuticle. The cuticle itself consists of several layers: the endocuticle (about 10 to 200 microns thick), the exocuticle (of variable thickness), and the epicuticle (only $\frac{1}{10}$ to $\frac{1}{10,000}$ as thick as the endocuticle and exocuticle together).

Although it is the thinnest layer, the epicuticle is probably the most important in providing one of the properties that makes an exoskeleton so valuable—it is responsible for waterproofing the insect. Water loss is a particularly acute problem for insects and other small animals in that the rate of water loss from a body is proportional to the surface area of that body, so insects, with a high surface area/volume ratio, are particularly at risk of desiccating under dry conditions. The wax component of the epicuticle is primarily responsible for guarding against water loss. Insect cuticular wax is highly water-repellent and acts as a seal to prevent loss of water through the skin. Preventing water loss was particularly important in facilitating the move onto dry land—arthropods were among the first colonists back about 500 million years ago, when new terrestrial habitats first opened up. Even today, terrestrial arthropods tend to have a more highly developed wax layer than do their aquatic relatives.

The remaining layers, the endocuticle and exocuticle, are responsible for conferring upon the exoskeleton the properties of strength and rigidity. The strength and rigidity result from a chemical reaction between the two major components of cuticle—chitin, a long chain molecule, or polymer, of repeating units of N-acetylglucosamine, a kind of sugar, and a group of proteins known collectively as sclerotin. The protein cross-links with the chitin (that is to say, it forms a series of crisscrossing chemical bonds) to make the cuticle rigid. Hardened cuticle is just the thing for terrestrial living because a hard, external covering made possible the arthropod exoskeleton. An exoskeleton is a great invention from the insect point of view; sturdy but lightweight, it provides protection for tiny bodies against bumps and collisions, it allows for extensive muscle attachment and leverage, it prevents drying out, and it makes a great set of wings.

The cuticular exoskeleton also put the "poda" (the feet, that is) in "Arthropoda." Insect legs are basically hollow cuticular cylinders.

Such an arrangement is nice for insects because, being so small, they have a limited amount of material to work with and an exoskeleton is more conservative of material than an endoskeleton would be. It is a general law of physics that hollow tubes have greater resistance to bending and to static loads (heavy weights) than do solid rods composed of an equivalent amount of material (that's why, for example, metal scaffolds and aluminum patio chairs are hollow inside rather than solid). The exoskeleton also made leg joints possible, and jointed legs greatly increased the speed and accuracy of movement. With these jointed appendages, insect bodies could get lifted off the ground to facilitate rapid movement; thus, arthropods, unlike earthworms, slugs, and other less well-endowed invertebrates, can run and jump rather than just slither and crawl. To illustrate the effectiveness of jointed legs in increasing speed, compare the top speed of onychophorans, sluglike animals intermediate in form between the arthropods and the earthworms (they have chitinous cuticle but no jointed legs), with that of your average spider or earwig; the best an onychophoran can manage is about 5 to 9 millimeters (1/5 to 3/8 inch) per second at a dead run while a wolf spider can hotfoot it all the way up to 250 millimeters (about 10 inches) per second, and an earwig can top 98 millimeters (about 4 inches) per second.

Despite its many appealing features, the external skeleton has one monumental drawback that actually is the reason insects are stuck at being smaller than most other living creatures. Nobody has yet devised an external skeleton that grows along with the organism; a rigid exoskeleton must be shed in order for an insect to increase in size. So all arthropods must molt or shed their skin in order to increase in size. The process is called ecdysis by technical types, from the Greek *ekdysis,* meaning "getting out." This molting wouldn't present problems except for one thing: when an insect molts, its fresh new cuticle takes a little while to harden. That's all right for a small insect, because the pull of gravity is not very great. But for a large insect, the force of gravity (which is proportional to mass) is so great that a soft, newly molted insect runs a serious risk of collapsing in on itself before its cuticle can harden.

The waterproof cuticle paved the way for the next innovation in arthropod evolution, the tracheal respiratory system, which first appears in the fossil record around 400 to 430 million years ago. Insects breathe not through lungs, as humans do, but rather through a system of branched tubes that connect to the outside through a series of small openings, called spiracles, on the insect body. Each main trunk divides, and each subsidiary trunk (or tracheole) continues to divide until a tiny tube delivers oxygen directly to each cell of the insect body. Tracheae were a great boon to insectkind because they provided a partial solution to the problem faced by any terrestrial organism—that is, how to obtain oxygen without losing water at the same time.

The challenge lies in the fact that a molecule of oxygen, with a molecular weight of 32, is larger than a molecule of water, with a molecular weight of 18. Any membrane or surface with holes large enough to admit oxygen is also going to be large enough to release water. Losing body water is a serious threat to the health and well-being of any would-be land-dwelling organism. Tracheae represent a compromise—they minimize the external surface susceptible to water loss by internalizing, or invaginating, that surface area. The only part of the insect body open to the outside, the spiracles, can be opened and closed by muscular contraction to prevent water loss in stress situations. Water loss is further minimized by the fact that the tracheal oxygen delivery system is lined with waterproof cuticle.

Oxygen does not diffuse very well in aqueous environments—it can only travel about 1 millimeter before slowing down to such an extent that it effectively stops. That's why organisms without a specialized oxygen delivery system are rarely more than 2 millimeters (less than 1/10 inch) across (and as a consequence are generally unfamiliar to the general public). Vertebrates use a slightly different tack in delivering oxygen to cells, tissues, and organs. Oxygen is carried through the circulatory system in blood cells, bound to a red pigment known as hemoglobin. The red blood cells in turn are circulated throughout the body in vessels, powered by the force of a beating heart. In some ways, the insect system is actually more efficient than the vertebrate system. Although the tracheae are long, they are filled with air, and oxygen moves more than 300,000 times faster in air than in a fluid such as water. The slowest part of oxygen delivery inside an insect is not from the spiracle to the tracheole but rather from the tracheole through the cell to the mitochondria (the cell's energy factory), a distance only 1/10,000 as long. Because oxygen is essential for burning metabolic fuel, having a watertight system for rapidly delivering oxygen in air to all the body cells allowed insects to pursue a high-energy, active terrestrial lifestyle.

One of the most energy-intensive things that insects do is fly, an activity made possible by the development of wings back about 300 million years ago. Wings are thought to have begun as airfoils, simple extensions of the cuticle from the thorax. Their locomotory value greatly increased once they developed joints at their base. Flight was greatly facilitated by the small size of insects and in turn flight allowed insects to cover far more terrain in less time than even six legs could cover. For example, top ground speed of running insects is about 10 centimeters per second (4 inches/second, or 0.23 miles per hour). In contrast, the top flight speeds are in the neighborhood of 36 miles per hour, as is the case for a large Australian dragonfly. Distances covered are far greater by wing rather than by foot. Horse flies, with flight speeds of 15 miles per hour, cover a range of over 60 miles, and monarch butterflies, in the course of their annual migrations south, can cover a thousand miles or more. Having wings allows insects to disperse to new habitats when the ones they occupy become unsuitable, to avoid their earthbound enemies more effectively, and to forage for food over a greater area.

Although wings provided a solution to several problems, they created a few new ones. In order for wings to be tough and strong, they consist only of cuticle; for the most part, they lack the epidermal tissues that underlie cuticle throughout the rest of the body.

Figure 2.2
Life cycle of the snowy tree cricket (*Oecanthus fultoni*), an example of gradual development (Department of Entomology, University of Illinois archives).

Dropping the epidermal tissue is an effective way to lighten the load without affecting wing strength or resiliency. The problem is that the epidermis is the tissue that synthesizes cuticle, so, once it degenerates, it cannot produce new cuticle. Insects therefore have only one set of wings—they can't be molted and replaced like the rest of the cuticle. The only adult insects that molt are thysanurans, or silverfish, and they can do so ad infinitum because they lack wings.

Insects came up with two solutions to this problem of unmoltable wings: incomplete or gradual metamorphosis, called by entomologists hemimetabolous development, and complete metamorphosis, called holometabolous development. In gradual metamorphosis, wings develop externally from buds throughout development and reach full size only at the final molt (Fig. 2.2). In complete metamorphosis, wings and some other adult tissues are quiescent throughout larval development in little islands of adult cells called imaginal discs; there is a transitional stage, called the pupa, in which these cells are activated and undifferentiated tissues begin to develop. When the adult emerges from the pupa, the wings, developed internally, are finally inflated (Fig. 2.3).

One characteristic of complete metamorphosis is that immature and adult stages of such insects can be completely different in appearance and habits, as is the case with grubs and beetles, maggots and flies, and caterpillars and butterflies, to name a few. This radical change in appearance and habits allows one species to take advantage of variability in the environment—to specialize on one food as a larva, for example, and another food as an adult. Larval mosquitoes, for example, feed on aquatic bacteria and decaying plant debris, a seemingly obscure bit of trivia of interest only to entomologists; on the other hand, adult female mosquitoes feed on the blood of vertebrates, a fact with which almost every person is distressingly familiar. Thus, insects with complete development can get along in a greater variety of habitats than can insects with gradual metamorphosis, and, accordingly, insects with complete metamorphosis planetwide outnumber insects with gradual metamorphosis by about four to one.

Although insects gained exceptional mobility when they acquired wings, they still have had to contend with physical forces in the air and on the ground that greatly affect that mobility. Due to their small size, insects face a number of unusual challenges in getting around in the world. A lot of physical phenomena look unfamiliar at high surface area/volume ratios (that is, in the world of insects). Kinetic energy, or the energy of movement, increases in

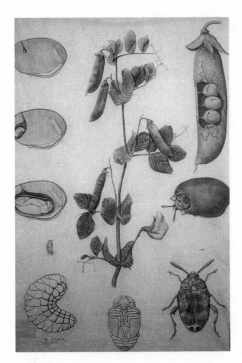

Figure 2.3
Life cycle of the pea weevil
(*Bruchus pisorum*), an example
of complete development
(Department of Entomology,
University of Illinois archives).

proportion not with the cube of length (like volume) but with the fifth power, so an insect 1/10 the size of a mouse hits the ground with only 1/100,000 the energy— which is why insects don't explode or fall apart when you drop them off a table. This physical relationship is no doubt the source of the expression "The bigger they are, the harder they fall." Flies can walk up walls be- cause the small gravitational force acting on them in proportion to their small mass is considerably less than the surface adhesion, a force related to surface area, acting on the spongy pads on their feet. Living around surface forces opens up new habitats for insects, too. Water striders skim around the water surface in part because the surface tension of the water is sufficiently strong to support them. Yet there is a down side to the surface area/volume ratio. Insects must struggle with forces larger animals can ignore. The same amount of energy that a flea expends to jump 30 centimeters into the air can propel a leopard to a height almost ten times greater— because the leopard, with a relatively small ratio of surface to vol- ume, isn't overcome by air resistance, a force, like friction, that acts in proportion to surface area.

The water surface is just one of thousands of habitats that in- sects can exploit due to their small size. Small size allows them ac- cess to nutrients that are otherwise too meager or too inaccessible to support life on a larger scale. Being small, insects can divide up their environments into far smaller livable pieces than can larger animals, so opportunities abound for successful establishment. However, there is a catch. According to insect expert Dan Janzen, "As organ- isms get smaller (elephant down to insect), their control over their individual environment decreases until they get so small (mites to viruses) that the individual's gambit is contained within a very small microhabitat and then their environmental control begins to increase again." Insects have little control of their environment; correspondingly, they must adapt to it, rather than adapt it to them. This ability to adapt to a staggering variety of environments has no doubt contributed in a fundamental way to the tremendous species diversity of insects.

Metamorphosis (quick-change artistry)

What is a butterfly? At best
He's but a caterpillar dressed.
—BENJAMIN FRANKLIN

MOLTING IS A COMPLICATED, time-consuming, and potentially dangerous procedure that all arthropods must endure. The process is essentially the same, physiologically, throughout the phylum. It's not clear exactly what first triggers the process; some people think that sensors built into the body wall send signals to the brain when an insect reaches a certain size. The molting process is known to be initiated in the brain. When the time is right, the brain releases a hormone (called by imaginative entomologists "brain hormone") into the blood or hemolymph, where it travels until it reaches a gland in the prothorax called the prothoracic gland. The prothoracic gland is then stimulated to produce another hormone, called ecdysone (sometimes called "prothoracic gland hormone"). Ecdysone is a steroid hormone, like the human steroid hormones estrogen and testosterone. Ecdysone in insects stimulates the epidermal cells that underlie the cuticle. During most of an insect's life, the epidermal cells are relatively inactive, but once stimulated by ecdysone they begin dividing, throwing themselves into folds, and rearranging.

The next step in molting is apolysis, the separation of the old cuticle from the newly activated epidermis. The old cuticle remains in place the entire time to protect the new cuticle as it forms. For the separation process, the epidermis secretes enzymes that dissolve away everything but the hardest parts of the old cuticle. This recycling allows insects to conserve on the amount of material that has to be synthesized to produce new cuticle. The new cuticle is systematically laid down as the old cuticle fades away. The first material to be laid down is cuticulin, which is a proteinlike substance that is extremely resistant to the action of the digestive enzymes, so it can prevent the newly forming cuticle from dissolving away as it's made. Next the endo- and exocuticle are formed, and finally the wax and cement layers (to complete the multilayer epicuticle). Wax gets ferried up to the surface of the new cuticle by way of pore canals.

At this point, the insect, equipped with a brand-new but not-yet-hardened cuticle, is trapped inside its old cuticle and now must shed its skin, which it does in any of a number of ways: gulping air, swallowing water, or by contracting its abdominal muscles to make use of the hydrostatic pressure of its blood. The old cuticle splits along lines of weakness where it has been dissolved away from underneath, and the insect wriggles free. Once out of its old cuticle, the insect must expand and harden its new cuticle. The process of sclerotization, or tanning, of the exocuticle—the cross-linking of proteins with chitin to provide rigidity—can take anywhere from an hour to several days to complete, depending on the species. During this so-called teneral period, the insect is soft and vulnerable, so molting is often practiced in concealed and well-protected places.

Just exactly what an insect turns into at each molt is regulated by another set of glands, the corpora allata. These glands produce a hormone called juvenile hormone—sort of an elixir of youth. As long as the corpora allata keep pumping out juvenile hormone, the insect will remain in a juvenile state. As the level in the hemolymph (or blood) drops, maturation proceeds. When, at the final molt, juvenile hormone is no longer produced, an insect finally becomes an adult.

Unlike the steroid hormones, which are found in both vertebrates and invertebrates, juvenile hormones seem to be uniquely arthropodan and as such have proved a useful target for humans intent on developing new ways to destroy insects. Chemists have devised compounds that look and act like juvenile hormone in the insect body and effectively prevent insects from reaching adulthood. Despite the appeal of the concept of eternal youth to humans, failure to reproduce is pretty much a losing situation for any organism from the perspective of posterity. Hydroprene, one such juvenile hormone mimic, is a component of many household preparations for cockroach control and is currently marketed as "birth control for roaches." Methoprene, another juvenile hormone analogue, is often incorporated into household flea control products, along with other insecticides, particularly pyrethroids; while the pyrethroids ensure rapid mortality of the adult-stage fleas, the methoprene works on the egg and larval stages to prevent population resurgences.

A developmental stage is called an instar. How many instars it takes to reach adulthood depends entirely on the species. Some flies complete the whole process in only four molts, while some beetle larvae (like wireworms, the immature stages of click beetles) can molt up to twenty-seven times before they finally pupate. In some species, the number of instars varies with the sex. In gypsy moths, for example, males complete development in five larval instars and females in six or seven. The amount and quality of food available can influence the number of molts as well.

The word "metamorphosis" simply means "change in form," and in general every molt is accompanied by a change in form—sometimes subtle, sometimes dramatic. Gradual metamorphosis doesn't necessarily look too gradual. In dragonflies, for example, the gills, the mouthparts, the gut, most of the abdominal musculature, parts of the head, and numerous other parts of the immature insect are completely rearranged into adult tissues. By the same token, in some of the more primitive insects with so-called complete metamorphosis (such as the dobsonfly), about the only conspicuous change is the acquisition of a set of wings. Some insects with gradual metamorphosis have stages that look for all the world like a pupal stage. Certain species of thrips and scale insects, for example, have pupalike "resting stages," which, in the case of male scale insects, provide for some pretty dramatic morphological changes. While male scale nymphs have no legs, no wings, no eyes, and no antennae, after a short while in this "resting stage," they emerge as fully equipped adults.

Some insects carry metamorphosis to an extreme and undergo major anatomical modification with each and every molt. This practice is known as hypermetamorphosis and is characteristic of some flies, bugs, and beetles. Blister beetles in the family Meloidae are masters of metamorphosis (Fig. 2.4). *Epicauta pennsylvanica,* the black blister beetle, starts out life as eggs underground. The eggs hatch to produce a long-legged active larva called a triungulin, which runs around looking for the egg case of a grasshopper. Once it locates an egg case, it chews its way in and molts into a stout, thick-legged form that eats grasshopper caviar until its next molt. The next molt produces a fat, C-shaped grub with tiny legs. After two more molts, a legless nonfeeding form, covered with thick dark

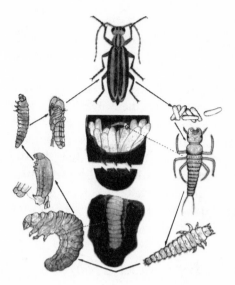

Figure 2.4
Life cycles of blister beetles
(Meloidae), an example of
hypermetamorphosis
(Department of Entomology,
University of Illinois archives).

exoskeleton, rests through the winter. Come spring, it molts again to produce another active, but legless, larva, which spends most of its time preparing a pupation chamber. It molts yet again to form a pupa and, after two weeks, molts into its final form, that of a shiny blue-black beetle. Although this sort of life cycle, even with so many changes of venue, may take only a few months to complete, such complexities mean that it may take an entomologist the better part of his or her own lifetime to figure out, and there are still thousands of species for which information about life cycles is woefully incomplete.

Under some circumstances, namely, when temperature, humidity, and the food supply are suitable, insects can develop continuously, but these conditions prevail in vanishingly few habitats. Life cycles of insects are varied in terms of number of generations, seasonal sequences, and the like, but what characterizes almost all life cycles are alternations of periods of active development with periods of developmental arrest coincident with habitat unsuitability. Environmental unsuitability can take on many forms: extremes of temperature, dryness, or food scarcity are but a few. The duration of the unsuitable period can be quite prolonged. The apple blossom weevil *Anthonomus pomosum* lives on apple buds and is basically inactive for ten to eleven months every year; some cone- and seed-feeding insects may have to wait years as pupae in a form of suspended animation, waiting for their host plants to reproduce so the next generation will have something to eat.

Such periods of inactivity in most circumstances fall within the definition of diapause. Diapause is a distinct physiological state, usually characterized by a cessation of development, a buildup in fat reserves, a reduction in body water, and an increased physiological resistance to extreme conditions. The physiological changes associated with diapause allow insects to survive unfavorable periods. For example, to withstand the extremely cold temperatures associated with winter, many insects accumulate large stores of glycerol, a chemical that not only looks a lot like polyethylene glycol but also shares its antifreeze properties. Thus, hemolymph containing large amounts of glycerol has a much lower freezing point than does hemolymph without such additives. Aside from producing compounds that lower the temperature at which their bodies freeze, some insects manufacture chemicals that protect them from damage caused by ice crystal formation inside cells and are actually capable of freezing up to 90% solid for much of the winter and then thawing out come spring without incident.

Reproduction (sex, bugs, and rock and roll)

> More courtship lives in carrion flies than Romeo.
> —WILLIAM SHAKESPEARE, *Romeo and Juliet*

IT ALMOST GOES without saying that, as a group, insects have been very successful in the area of sex and reproduction. The reproductive prowess of some species is legendary—queens in some termite colonies produce hundreds of thousands of eggs over the course of a lifetime and in some cultures are not too surprisingly regarded as fertility symbols.

Insects, of course, are not unique in their ability to reproduce—all animals do it. But insects are remarkable in the variety of ways in which they engage in reproduction. The vast majority of insects reproduce sexually, as opposed to asexually. The vast majority of successful organisms do. Asexual reproduction has its benefits—it guarantees a rapid increase in numbers and a close match between the organism and its environment, and it eliminates the often time-consuming and tedious process of finding and courting a mate. Asexual reproduction has probably been around for over 3 billion years, so it can't be an evolutionary dead end, either. But asexual organisms are at a singular disadvantage compared to sexual organisms. Biparental sexual reproduction involves the production of sex cells, or gametes, each of which contains half the genetic complement of an individual. Thus, two gametes must fuse to form a new individual. The zygote, the cell resulting from the fusion of two gametes, receives genetic information from two parents. The combination produces a genetically unique individual, and it is this genetic uniqueness of offspring that distinguishes sexual from asexual reproduction. Sexual reproduction generates variety, and variety is essential for survival if the environment is changeable, a characteristic that pertains to most environments (particularly the kinds of environments frequented by insects). So, over and above its other attractions, sex does provide evolutionary flexibility to a species.

Generally speaking, in sexual reproduction, there are two types of gametes. Sperm, conventionally described as the smaller, more mobile, and more numerous of the two gametes, is in insects produced by males, usually in the last larval instar, pupal stage, or in early adulthood. Insect sperm cells are in most cases unremarkable, mostly filamentous in appearance and equipped with a vaguely defined head containing the nucleus. Not all arthropod sperm is so simple in structure; crayfish spermatozoa, for example, are shaped like a five-pointed star. Size and number vary considerably and are in no way related to the size of the producer. A single sperm cell of the featherwing beetle (*Bambara*) is fully two-thirds the length of the beetle itself. Some backswimmers produce sperm cells 1.5 centimeters (3/5 inch) long; for perspective, it's worth noting that human sperm cells are a mere 0.7 centimeters in length. Sperm are produced by the testes, a pair of organs consisting of a series of follicles, or sex cell line production centers. From the testes, sperm travel through a series of tubes and ducts; stored in the seminal vesicle, they are discharged through the aedeagus, or penis, during copulation, accompanied by secretions from accessory glands.

The ovum, conventionally described as the immobile larger gamete, is produced in an ovary by a female insect. Each ovary consists of a variable number of follicles called ovarioles, which, unlike the testicular follicles, are free-floating rather than compacted into a single organ. The number of ovarioles differs with the species, and the range runs from sublime to ridiculous. While termite queens have over 2,000 ovarioles, a female tsetse fly has only one (tsetse flies produce only one offspring at a time, which, after hatching, stays inside the body of the female for most of its larval life, feeding on a milklike substance in a structure called a uterus). Females of many species have a structure known as a spermatheca, a shunt off the reproductive system used for storing sperm. In species that mate only once, sperm can be stored for months or, in the case of queen bees and ants, even years. A virgin queen bee mates only once, but during her nuptial (or mating) flight she may mate with a dozen or more males, acquiring millions of sperm to last several years. In one ant species, queens can store up to seven million sperm for ten years. Oogenesis, or egg production, usually occurs in the last larval instar, pupal stage, or adult (and seems to be influenced by the corpora allata, the same glands that regulate youthfulness in molting). An ovum released from an ovariole travels down the oviduct and is fertilized when sperm are released from the spermatheca.

Many females also have accessory glands, called colleterial glands, which provide materials that attach the egg to a surface or substrate. Eggs laid in the fall that must survive cold winter temperatures before hatching the following spring, such as those of the praying mantis, can be equipped with a layer of insulating foam. Other eggs come equipped with antipredator devices—green lacewing eggs, for example, are laid at the tip of an elongated stalk to keep them safe from bands of roving egg predators scouring leaves for an easy meal.

Sexual reproduction doesn't mean a thing unless the male and female gametes can get together; with biparental sexual reproduction, the two types of gametes are usually stored in two different bodies. Spermatozoa, being more mobile than ova, generally are responsible for doing most of the traveling. Sperm cells are basically aquatic and are built for swimming—they move through water by whipping their tails. Consequently, in aquatic arthropods like crustaceans, fertilization is generally external—males simply discharge their sperm directly into the water, generally following courtship procedures that increase the probability that females release their eggs more or less at the same time in the same general vicinity. On land, egg and sperm must get together inside the body of the female, via the process of internal fertilization. There are at least two ways that sperm are delivered into the body of the female. In indirect sperm transfer, the sperm are packaged and removed from the male genitalia. For example, male spiders spin a small sperm web, deposit sperm on it, and insert the whole package into the female genitalia with a pedipalp. Male silverfish package their sperm in a little bundle called a spermatophore, drop it on the ground, guide a female to the immediate vicinity, and strap her down with silk until she picks it up and inserts it into the appropriate orifice. The problems with indirect sperm transfer are manifold, not the least of which is that sperm are hardy, but they're not that hardy—they don't stand up too well to environmental unpleasantness such as rain or cold temperatures. Left exposed in the environment, they are also vulnerable to predators (including other males of the same species).

Direct sperm transfer is probably the most efficient delivery system of sperm to egg; in direct sperm transfer, sperm travel from the male reproductive system directly into the female reproductive system without passing "go" or collecting $200. In this way, the male insect has a guarantee that his sperm arrives intact at the appointed destination and fewer sperm cells are lost to environmental unpredictability. There are no guarantees, however, that once deposited in the female reproductive tract they'll stay there. Certain male damselflies, for example, have special flanges on their intromittent organs that are used to scoop out any sperm that may already have been deposited in the spermatheca of their mates.

The diversity with which internal fertilization is accomplished in the insect world is truly breathtaking. Insect genitalia are so variable that they are often used (almost to the exclusion of other characters) to identify insects to species. The oldest explanation for the remarkable diversification of genitalia is the lock-and-key hypothesis—the idea that if genitalia are sufficiently complex then only a member of the opposite sex of the same species will be able to gain access to the female reproductive tract and deliver sperm (thereby preventing inviable zygote formation). It has also been suggested that the elaborate cuticular variations in shape may serve to stimulate the couple sexually when contact is made. The most recent (and most popular) explanation is the so-called sexual selection hypothesis—that females discriminate among males competing for their attentions and genitalic structure is evidence of male acceptability. Females are choosier than males because their gametes are fewer in number and they invest substantially more time and energy in them, particularly after fertilization. Thus, since females have more at stake every time they mate (it is in all cases the female that gets pregnant and in the majority of cases the female that invests time and energy in care of offspring), females show "sales resistance" and males must display "salesmanship" in order to gain access to their gametes.

Sexual selection may also be responsible for sexual dimorphisms (gender-specific differences in body shapes) in structures other than genitalia. In many species, elaborate ornamentation is used for grasping the female during copulation. Dragonflies, for example, have a genital opening on the ninth abdominal segment, near the tip of the tail. However, the copulatory organs are on the abdomen seven segments up, two segments behind the thorax. The male has to deposit sperm from the tip of his tail into a storage sack behind his legs by bending his abdomen up and around. He then grabs a female with the tip of his tail right behind her head. She then bends her abdomen up and under the male's and places the tip of her abdomen up against the storage sack near the thorax. Once coupled in this peculiar fashion, they can fly around together for several days, either mating or trying to figure out how to get untangled. Still other nongenitalic contact organs are used by males to grasp other males in ritualistic combat over access to females. The large and often bizarre head ornaments of certain scarabeid beetles are used in some instances to pry an opponent off the back of a female while he is *in flagrante delicto.*

Then again, there are insects at the opposite extreme that have dispensed with traditional forms of copulation. The bed bug *Cimex lectularius,* a bloodsucking human parasite, may have the most unique method of insemination of all. Female bed bugs have no genital opening at all. As a result, the male makes his own—he punches through a particular weak section of her abdomen between the fifth and sixth segments with his large sharp hooked intromittent organ, and dumps sperm and accessory fluids into her abdominal cavity, bypassing the normal complex system of tubes and ducts devised to conduct sperm to the right place. The female has a structure called the organ of Berlese, which acts like a cushion to protect her against major collision damage. Much of the semen is actually digested by the female, but a few sperm do manage to make their way into the female reproductive tract. After the male withdraws, the opening he created eventually heals and closes over with a scar (you can find out how often a female has mated by counting the scars on her abdomen). This form of insemination is known appropriately enough as hypodermic or traumatic insemination.

In a few exceptional insects, the process of fertilization is greatly simplified by virtue of the fact that both male and female organs are housed in the same body. Some races of the cottony-cushion scale *Icerya purchasi* in California are hermaphroditic—individuals contain both male and female reproductive organs and they fertilize themselves (their internal anatomy is so configured that they can reach the appropriate orifices with the proper organs). The term "hermaphrodite" derives from the names of the Greek god Hermes, the messenger, and the Greek goddess of love, Aphrodite.

The process of sex determination in insects is generally similar to sex determination in mammals, but as always there are some unique insect twists. Chromosomes, which carry the genetic information in the cell, generally are present as matched pairs, with one exception, the pair known as the sex chromosomes (the others are called autosomes). As in mammals, an individual with a matched pair of sex chromosomes is a female; an individual with a pair that doesn't quite look identical (XY) is a male. In butterflies, moths, and caddisflies, though, the situation is reversed—the female is the so-called heterogametic sex (XY). In some strains of fruit flies in the genus *Drosophila,* sex determination is based not on the presence or absence of sex chromosomes but on the relative ratio of sex chromosomes to autosomes (sort of a majority rule form of gender selection).

In some species, males possess only a single sex chromosome. This X0 sex-determining mechanism is found in the bees, ants, and wasps (order Hymenoptera) and may have led to the rather remarkable form of parthenogenesis, or asexual reproduction, called haplodiploidy (arrhenotoky). Fertilized eggs of Hymenoptera turn into females, whereas unfertilized eggs turn into males. A virgin female, then, can produce an endless line of sons (and, in the case of pharaoh ants, actually mate with one of her own sons in order to produce females). One other peculiar aspect of reproduction in the Hymenoptera involves polyembryony, characteristic of several families of mostly tiny parasitic wasps. Normally, over the course of development a zygote begins to divide and the proliferating cells unite to form an embryo. At some point in species exhibiting polyembryony, each cell of the embryo becomes autonomous and begins to develop on its own. Thus, one egg can produce dozens of offspring; in some families, a single egg can yield over 1,500 progeny. This reproductive gimmick is one way to take advantage of favorable conditions, producing enormous numbers of offspring from a single egg-laying event.

Thelytoky is another form of parthenogenetic reproduction, in which diploid females produce diploid female offspring without mating. Many aphids are thelytokous for much of their lives. The advantages of thelytoky are abundant: no wasting time finding a mate, for example, or undergoing potentially life-threatening mating. Plus, thelytoky is fast—often inside the body of an aphid not only are her offspring clearly visible but inside the body of the offspring their offspring are also visible. Aphid numbers can thus build up explosively in a short time. Most aphids, however, alternate thelytokous periods with more conventional biparental sexual reproduction.

One other variation on reproduction found in the insects is temperature-dependent sex determination. Certain species of mosquitoes develop in vernal (spring) pools of water. If larvae are maintained in pools of water at a constant 28°C, every larva in the population will develop into a female mosquito. Apparently, male imaginal discs fail to develop at high temperatures. In some species of butterflies, males predominate at high temperatures and females at low temperatures.

So, even though sex may seem at times a hopelessly complicated process, everything is relative—compared to insects, humans have it easy.

Genetics (designer genes)

IMMEDIATELY APPARENT TO any entomologist, and even to the casual insect-watcher, is the fact that insects are extraordinarily variable. The same is true for most organisms, actually—no two individuals are exactly alike. The existence of variation is hardly a debatable point, but what generates and maintains variation, and the significance of variation within populations, are not so obvious. For centuries, typological thinking dominated the scientific community, that is, the idea that individuals are mere shadows of ideal "types" (*eidos,* according to Plato). The prevailing notion was that the essence of an organism is immutable and, hence, the observed variability in the world is of no consequence. This line of thinking, called essentialism, dominated hereditary thinking for centuries. In contrast to the essentialists, Charles Darwin formulated a theory that emphasized the variation rather than the type or essence. To the essentialists, the mean was real and the variations abstract; to Darwin the variations were real and the ideal type was an abstraction. His theory, advanced in 1859 in his book known as *On the Origin of Species,* was based partly on variations observed in domestic animals and in many different places visited during a five-year voyage around the southern end of South America. This theory involved several elements: that variation exists among individuals, that it is heritable, and that organisms produce more offspring than can possibly survive. Only those individuals with beneficial variations survive, and those beneficial variations that promote survival are passed along to offspring. The transmutation of species and the origin of new species is by accumulation of this heritable variation. The process by which those individuals with heritable traits conferring survival produce more offspring than do those individuals lacking such traits was called natural selection.

Darwin came to these conclusions by way of a long and circuitous passage. Charles Robert Darwin (1809–1882) was a recent graduate of Cambridge in 1831 when he was offered a position as naturalist on a five-year voyage around the southern tip of South America on the HMS *Beagle*. Darwin (at least in part due to a tendency toward seasickness) spent much of the trip on land, observing strange and wonderful plant and animal life, recording his observations, reading the books he'd brought along, and just thinking. One of the books he chose for his voyage was by a geologist named Lyell, who proposed that the earth was considerably older than it was fashionably believed to be at the time.

When Darwin returned from his trip, he cranked out several books on a number of assorted topics, including one on coral reefs and atolls, and one on South American geology. He spent about eight years researching and writing a four-volume treatise on barnacles. During this period, he also found time to read a book by Thomas Malthus, an economist who predicted in 1798 that uncontrolled population growth would lead to massive death and starvation since populations inevitably grow to outstrip the food supply.

In 1842, Darwin's life took a definite turn when he moved with his family to their country home in Kent. He lived the life of country squire and English gentleman (he had by that time married into the fabulously wealthy Wedgwood family), writing when the mood struck. As a country squire, he observed and studied the effects of variation under domestication—breeding dogs and pigeons were popular pursuits of British gentlemen of the day.

In 1859, his idyll was, if not shattered, then at least a little shaken when he received in the mail on June 18 an essay writted by a young naturalist named Alfred Russell Wallace. The essay described the theory that Darwin had been elaborating since his return from South America. In less than two weeks, Darwin had also completed a short essay and both papers were presented on July 1 to a meeting of the Linnaean Society (which neither Darwin nor Wallace attended—the papers were submitted by Charles Lyell and William Dalton Hooker). Although the papers received little notice, Darwin was prompted to put together the book on which he had been desultorily working, and, in November, *On the Origin of Species by Means of Natural Selection or the Preservation of Favoured Races in the Struggle for Life* was published.

Needless to say, the book was an immediate sensation. It was acclaimed by many scientists, who found that the theory of evolution by way of natural selection accounted for otherwise inexplicable patterns. It created a major furor in religious circles since it did not explicitly embrace church doctrine on a number of issues (not the least of which was the story of how humans came to be). Furor notwithstanding, it's instructive to note that the book in its first as well as subsequent editions was amply illustrated with examples of Darwin's observations of insects. Among the phenomena that he found supported his notions of evolution via natural selection were the myriad examples of cryptic (concealing) coloration ("when we see leaf-eating insects green"), homotypism (insects resembling "green or decayed leaves, dead twigs, bits of lichen, flowers, spines, excrement of birds"), and mimicry of toxic butterflies by edible species ("Now if a member of one of these persecuted and rare groups were to assume a dress so like that of a well-protected species that it continually deceived the practiced eye of an entomologist, it would often deceive predaceous birds and insects, and thus often escape destruction"). Wingless beetles on islands and blind insects in caves led him to suggest that "use . . . has strengthened and enlarged certain parts, and disuse diminished them; and that such modifications are inherited. . . . many animals possess structures which can be best explained by the effects of disuse." He continued on to suggest that "the wingless condition of so many Madeira beetles is mainly due to the action of natural selection, combined probably with disuse. For during many successive generations each individual beetle which flew least, either from its wings having been ever so little less perfectly developed or from indolent habit, will have had the best chance of surviving from not being blown out to sea; and, on the other hand, those beetles which most readily took to flight would oftenest have been blown to sea, and thus destroyed."

In the interest of equal time, it must also be said that occasionally his observations of insects left him without an explanation. He considered these cases "special difficulties of the theory of natural selection." The gradual acquisition, by accumulation of small variations, of luminous organs in fireflies was one problem, as well as the apparent lack of physical modification needed by fairyflies, tiny wasps smaller in size than a one-celled amoeba, to adopt an underwater lifestyle. His biggest problem, though, was the neuter, or sterile, caste in insect societies; neuters don't reproduce so they can't pass on beneficial variation and thus they posed "the acme of the difficulty; namely, the fact that the neuters of several ants differ, not only from the fertile females and males, but from each other, sometimes to an almost incredible degree." Darwin himself admitted, "It will indeed be thought that I have an overweening confidence in the principle of natural selection, when I do not admit that such wonderful and well-established facts at once annihilate the theory."

Darwin's theory had another big problem, namely, it did not provide a satisfactory mechanism for generating variation and transmitting that variation to offspring. Darwin proposed the mechanism of pangenesis, which he admitted was a "provisional hypothesis." According to this hypothesis, each cell in an adult sheds into the blood little copies of itself. These "gemmules" assemble in the reproductive organs to form sex cells. Upon fertilization and embryogenesis, the gemmules convert back to body cells. Pangenesis as a hypothesis was testable and, once tested, was found wanting. When Francis Galton transfused blood between white and black rabbits in an explicit test of the hypothesis, he failed to obtain spotted bunnies as the offspring. Had the hypothesis been correct, transfusing blood should have sufficed as a means of transferring gemmules from black to white rabbits and vice versa.

Unbeknownst to Darwin, and much of the rest of the world, an Austrian monk was conducting experiments that would eventually provide a workable mechanism for the generation and transmission of variation. Gregor Mendel was born in 1822; his early education was in a monastery and he continued his studies in mathematics and science at the University of Vienna. After failing the tests for a teaching certificate, Mendel returned to the monastery, where he remained for the rest of his life. He did not, however, abandon his interest in science. As a monk and eventually an abbot, he pursued a number of studies of plant hybridization in the monastery garden. He chose for his studies the garden pea, *Pisum sativum,* a plant that had been known for decades to produce so-called "true-breeding" varieties—varieties that produced progeny identical to the parents. After many different crosses, Mendel made several fundamental observations. First, he noted that plants transfer information from generation to generation, rather than physical properties or material; Mendel called these discrete pieces of information "factors." He also noted that every individual possesses two different factors for each trait, one inherited from each parent. The two different factors may be identical or they may take on alternative forms. Depending on the form a factor takes, it may or may not necessarily be visible or detectable. Finally, multiple cross-breeding experiments confirmed that the two different factors maintained their integrity in that they remained discrete, "uncontaminated" by other such factors.

From these observations, Mendel formulated two "laws." His first law, the Law of Segregation, states that the alternative forms of a particular factor remain discrete and segregate during the reproductive process independently. His second law, the Law of Independent Assortment, states that different pairs of alternate factors assort themselves independently—that pea shape (round or wrinkled), for example, is uninfluenced by pea color (green or yellow).

In 1900, Gregor Mendel's experiments with garden peas were rediscovered and the "gene" (a word coined by W. Johannsen in 1909) replaced Mendel's "factors" as the physical basis for heredity and carrier of inherited information. In 1902, Walter Sutton proposed the chromosomal theory of inheritance. Since chromosomes (structures of the cell nucleus) segregate during reproduction in the same fashion that Mendel observed factors segregating, then they must be the part of the cell that carries hereditary information. The number of chromosomes in any organism, however, is far less than the number of factors, a problem with the theory that proved to be troubling for several years.

Studies of the gene and of chromosomal inheritances underwent a giant leap forward when in 1909 Thomas Hunt Morgan proposed the use of a small nondescript fly called *Drosophila melanogaster,* the vinegar or pomace fly (incorrectly but widely known as the fruit fly), for genetic studies (Fig. 2.5). Easy to rear, the fly completes a generation every ten days; less than half the size of a house fly, it could be raised and housed in large numbers without requiring prohibitive amounts of space. Amazingly unfussy, it could thrive on a diet as cheap and easy to obtain as rotten bananas. It was, in short, the ideal laboratory animal. Thanks to *Drosophila,* by 1910 the world had a mechanism for generating variability—the mutation. T.H. Morgan described a mutant form of *Drosophila*—one not with normal red eyes, but instead with white eyes. Moreover, he determined its mode of inheritance; it was linked to one particular chromosome, which happened to be the X chromosome (a chromosome that males had only one of and females two and that was thought to determine sex), vindicating Sutton's notion that genes are located on chromosomes that assort independently during reproduction.

Things took off from there—many of the observations made by these early "geneticists" (a new discipline for the twentieth century) today form the basis for modern molecular biology. Many of the most fundamental genetic processes were first observed in fruit flies. Without the enormous number of visible mutations in the

Figure 2.5
Drosophila melanogaster, the vinegar or pomace fly known to generations of geneticists as the fruit fly (D. Voegtlin).

fruit fly, the discovery of many genetic phenomena, including sex-linked inheritance, cytoplasmic sterility, and transposable elements, may have gone undiscovered for years. With only a few specific exceptions, genetic processes that occur in fruit flies are identical to genetic processes in higher organisms, thus making the fruit fly an exceptionally good model species. Today's great genetic engineering feats—vegetables that resist decay, crops that are invulnerable to frost damage or insect attack, and bacteria that manufacture human hormones—as well as nascent gene transfer therapies to combat birth defects and congenital disorders in humans, effectively originated almost a century ago with rotten bananas and the flies that love them.

References

Good things in small packages

Alexander, R.M., 1985. The ideal and the feasible: physical constraints on evolution. *Biol. J. Linn. Soc.* 26: 345–358

Bonner, J.T., 1965. *Size and Cycle: An Essay on the Structure of Biology.* Princeton: Princeton University Press.

Gould, S.J., 1974. Size and shape. *Nat. Hist.* 83: 20–26.

Haldane, J.B.S., 1932. *On Being the Right Size and Other Essays.* J.M. Smith, ed. Reprint, Oxford: Oxford University Press, 1985.

Hinton, H., 1976. Enabling mechanisms. *Proc. XV Int. Cong. Entomol.* 1976: 71–83.

Janzen, D., l976. Why are there so many species of insects? *Proc. XV Int. Cong. Entomol.* l976: 84–94.

May, R., l978. The dynamics and diversity of insect faunas. *Symp. R. Entomol. Soc. Lond.* 9: l88–204.

Peters, R.H., 1983. *The Ecological Implications of Body Size.* Cambridge: Cambridge University Press.

Price, P., l984. *Insect Ecology.* New York: Wiley.

Schmidt-Nielsen, K., 1975. Scaling in biology: the consequences of size. *J. Exp. Zool.* 194: 287–308.

Schmidt-Nielsen, K., 1984. *Scaling: Why Is Animal Size So Important?* Cambridge: Cambridge University Press.

Southwood, T.R.E., 1978. The components of diversity. *Symp. R. Entomol. Soc. Lond.* 9: 19–40.

Vogel, S., 1988. *Life's Devices: The Physical World of Animals and Plants.* Princeton: Princeton University Press.

Metamorphosis (quick-change artistry)

Baust, J.S. and R. Rojas, 1985. Insect cold-hardiness: facts and fancy. *J. Insect Physiol.* 31: 755–760.

Blum, M., 1984. *Fundamentals of Insect Physiology.* New York: Wiley.

Chapman, R., 1972. *The Insects: Structure and Function.* New York: Elsevier.

Danilevskii, A.S., 1965. *Photoperiodism and Seasonal Development in Insects.* Edinburgh: Oliver and Boyd.

Powell, J., 1987. Records of prolonged diapause in Lepidoptera. *J. Res. Lep.* 25: 83–109.

Storey, K.B. and J.M. Storey, 1988. Freeze tolerance in animals. *Phys. Rev.* 68: 27–84.

Tauber, M., C. Tauber, and S. Masaki, 1986. *Seasonal Adaptations of Insects.* New York: Oxford University Press.

Reproduction (sex, bugs, and rock and roll)

Chapman, R., 1972. *The Insects: Structure and Function.* New York: Elsevier.

Eberhard, W.G., l985. *Sexual Selection and Animal Genitalia.* Cambridge: Harvard University Press.

Horsfall, W.R. and J. F. Anderson, 1961. Suppression of male characteristics of mosquitoes by thermal means. *Science* 133: 1830.

Schaller, F., 1971. Indirect sperm transfer by soil anthropods. *Ann. Rev. Entomol.* 16: 407–446.

Thornhill, R. and J. Alcock, 1983. *Evolution of Insect Mating Systems.* Cambridge: Harvard University Press.

Genetics (designer genes)

Darwin, C., 1859. *The Origin of Species.* Reprint, New York: Macmillan, 1962.

Rubin, G.M., 1988. *Drosophila melanogaster* as an experimental organism. *Science* 240: 1453–1459.

Sootin, H., 1959. *Gregor Mendel: Father of the Science of Genetics.* New York: Vanguard.

Strickberger, M.W., 1968. *Genetics.* New York: Macmillan.

Chapter 3

BEHAVIOR

Sensory physiology

> . . . every insect, ant, and golden bee, all so marvellously know their path,
> though they have not intelligence.
> —FYODOR DOSTOYEVSKI, *Brothers Karamozov*

LIKE ALL OTHER ANIMALS, insects behave. This is not to say they comport themselves in a manner consistent with good etiquette; rather, "behavior" is the ability of an organism to respond to environmental stimuli. The study of behavior is known as ethology, a word derived from the Latin for the "science of character." The term was used originally in the eighteenth and nineteenth centuries to refer to the sociological study of manners and customs in both human and animal societies. Until comparatively recently, the study of animal behavior was heavily influenced by anthropomorphism, the tendency to attribute to animals traits that are in all probability uniquely human. The concept of the "busy bee" (and, for that matter, the "killer bee") reflects this attitude. Modern ethologists, however, have made every attempt to divest themselves of human prejudices and instead concentrate on studying the responses of animals to stimuli in the context of their natural environment.

In order to respond to changes in its environment, an insect has to have some mechanism for evaluating its environment. Like most higher animals, insects perceive information about their environment through the nervous system. Most information is received by sensilla, or sensory organs. Sensilla are specialized cuticular structures that are equipped with nerve cells, or neurons. The structure of any given sensillum usually reflects its function; most sensilla are specialized for processing only certain types of sensory information. Sensilla that are specialized for touch are called mechanore-ceptors. Some mechanoreceptors process information about the environment; logically enough, these tend to be concentrated in parts of the body, such as the legs, that are often in contact with the ground. Other mechanoreceptors, called proprioceptors, monitor the position of various parts of the body with respect to each other, so such vital activities as walking, eating, or even copulating can be coordinated. Still others are hearing organs, which detect sound vibrations.

Detection of chemical stimuli is by chemoreceptors; olfactory receptors detect airborne chemical signals (or smells), and gustatory (or taste) receptors detect close-range chemical signals in solution. Chemoreceptors are found in abundance in some not-too-surprising places, from the human perspective—like humans, insects have structures analogous to taste buds in and around their mouths. But insect chemoreceptors also show up in places, such as, for example, on the legs, where humans are conspicuously poorly endowed. Some female butterflies deciding whether or not to lay an egg on a particular plant can "taste" the plant by drumming the leaf surface with their feet, which are studded with chemoreceptors. The feet of the red admiral butterfly (*Vanessa atalanta*) are about 200 times more sensitive to the presence of sugar than is the human tongue.

Some insects can detect changes in temperature with thermo-receptors. People are not the only ones to appreciate a nice, warm bed; *Cimex lectularius,* the bed bug, uses thermoreceptors to locate its warm-blooded vertebrate prey. A blood-feeding relative of the bed bug stridulates, or chirps, when it senses the presence of a warm-blooded body nearby. The U.S. Department of Defense actually considered issuing these bugs to soldiers in Vietnam to use as monitors of approaching Vietcong guerillas, although exactly how these bugs were to differentiate friend from foe was never clearly elucidated. Many blood-feeding parasites can also detect changes in humidity using hygroreceptors, or sensory hairs that detect changes in humidity; warm moist breath is a good indicator of the presence of a potential host nearby.

Photoreceptors allow insects not only to distinguish light from dark but also, depending on species, to distinguish shapes and images quite clearly. While some invertebrates have eyes that operate on the same principles as do vertebrate eyes (squids, for example), insects have hit upon their own solutions to seeing things. There are basically two types of insect photoreceptors, or eyes. Simple eyes, or ocelli, are quite variable in structure but generally consist of a corneal lens made out of cuticle, a crystalline cone to direct light rays, and visual cells, collectively forming a structure called a rhabdom, which respond to light and connect with nerves heading to the brain. Simple eyes are found in different places and in different numbers, depending on species. In caterpillars, several (called stemmata) are found on each side of the head; sawfly larvae have only a single ocellus on each side of the head. In contrast, true bugs have groups of three ocelli, usually on the front or top of the head.

Compound eyes consist of numerous individual light-processing units called ommatidia. Each ommatidium is topped by a cuticular hexagonal corneal lens; the lens is connected to the crystalline cone, which in turn is surrounded by visual cells forming a rhabdom. On the edge of each ommatidium are pigment cells. The number of ommatidia in a compound eye depends on the species and varies from less than a dozen to over 20,000 (as in adult dragonflies). Exactly what image an insect sees with such an oddly constructed eye has been subject to considerable discussion over the years. Current theory has it that insects perceive images as mosaics of light and dark spots (like a newspaper photograph). Popular conceptions of insect vision as single images multiplied thousands of times (as in low-budget insect fear films) are definitely off base. Nonetheless, they have given rise to an optical toy called "The Dragonfly", consisting of a multiprism dioptric lens that produces "24 upright images of anything viewed" (Van Cort Instruments, Northampton, MA).

Many, but not all, insects can see colors. In general, insect visual pigments are sensitive to a different array of colors than are human visual pigments. As a result, insects can see some colors (such as ultraviolet) in the short-wavelength end of the spectrum that humans cannot see. Bees and ants can also see polarized light, so they can tell the exact position of the sun even when it's behind a cloud.

Insects, then, are capable of detecting all sorts of environmental stimuli. They are also capable of responding in a tremendous variety of ways to such stimuli. Behavior falls into two major categories: innate and learned. Innate behavior—also called instinct—can be performed with no prior experience and has survival value in that it allows an organism to respond in an appropriate fashion to a particular situation the first time it is encountered. It is certainly true that not all innate behaviors require a brain—after all, one-celled protozoans can respond to environmental stimuli—but it helps.

The more neural circuits are involved, the more flexible the behavior, and the more finely attuned the response to the environment.

About the simplest form of behavior is a reflex, which does not necessarily even involve the brain. A stimulus is detected by a sensory neuron, which conveys a message to an intermediary neuron, which in turn conveys a message to a motor neuron, which initiates a response. Human actions such as sneezes, yawns, and the like involve such simple reflex arcs. Most are all-or-none sorts of things (try stopping yourself in the middle of a sneeze).

Slightly more sophisticated than a simple reflex is a kinesis, an undirected locomotory reaction. The frantic writhing of an apple maggot (the immature form of a picture-winged or true fruit fly) when it is unceremoniously dumped from the dark recesses of apple flesh onto a brightly illuminated picnic table is an example of a kinesis. A taxis is a directed movement either toward (as in positive taxis) or away from (as in negative taxis) a stimulus. The term comes from the Greek *tassein*, "to arrange." Tactic behavior requires a sensory system sensitive enough to determine the direction from which a stimulus comes and a locomotory system capable of correction for deviations from a true course.

Taxes come in a variety of forms. Phototaxis is orientation with respect to light. Most species of cockroaches that live in human habitations are negatively phototactic, that is, they run when the kitchen lights go on. Thigmotaxis is orientation to contact. The same household cockroaches like to press their bodies against as many surfaces as possible; hence, their tendency to scuttle in among the folds of paper grocery bags stored in drawers or under sinks. Chemotaxis, orientation to a chemical gradient, allows those cockroaches not only to locate even the tiniest unswept crumb in a kitchen but also to find members of the opposite sex, in order to mate and produce more cockroaches.

A transverse orientation is a behavior in which the body is aligned at a fixed angle relative to the source of a stimulus. The sun, moon, and stars provide stimuli to which many organisms display transverse orientations. Many dragonflies, for example, adjust the position in which they bask in the sun, maximizing their exposure when they need to warm up (as in the early morning) and minimizing their exposure when they are overheated and need to cool down (as after a vigorous flight in midday). The moon and stars serve as navigational guides for flying insects, just as they do for sailors at sea. The development by humans of long-lived point sources of light otherwise unprecedented in nature (such as candle flames and lightbulbs) offers a partial explanation behind the seemingly bizarre self-destructive tendencies of moths and many other nocturnal species around these light sources. Maintaining a fixed angle relative to a distant point source of light allows for relatively long-distance travel; however, maintaining the same fixed angle to a nearby source of light involves constant directional shifts after only short distances are traveled. Such shifts cause a flying insect to spiral inwards, eventually resulting in singed scales or worse.

Learning is an adaptive change in behavior resulting from experience (rather than from the process of maturation). Perhaps the simplest form of learning is habituation, the elimination of a response in the absence of reward or punishment. In other words, habituation is learning *not* to respond to a stimulus. Cockroaches are capable of detecting even very slight air movements and initiating escape behavior in response, but it's often of no benefit to respond to every stray air movement. After a while, if air movements present no danger, cockroaches will cease responding. In some apartments, they are so habituated to disturbance that slamming doors and screaming at them fails to evoke the desired response.

In contrast, probably the most sophisticated learning is insight learning, solving problems by reasoning (or learning without trial and error practice). Insight learning is usually regarded as the epitome of mental accomplishment, probably because people have assumed that it was a uniquely human attribute. However, tool use, a typical example of the application of insight learning, has been documented in ants and other social insects; some weaver ants, for example, use silk-producing aphids as bobbins, shuttling them back and forth between edges of a leaf to stitch them together. Such tool use may be explained away as a series of simple behaviors strung together to create the appearance of careful planning and rational thought, but it still gives one pause with respect to our place in the universe.

It's very difficult to remain objective about much insect behavior, not only because it often takes place in a context completely different from that of human (and, for that matter, vertebrate) behavior but also because making observations of insect behavior is operationally extremely challenging. Often, the types of environmental manipulations required for allowing careful and precise observation of insect behavior may end up totally disrupting the behavior. Completely understanding insect behavior necessitates perceiving the environment as they do, and since they see, hear, taste, touch, and smell things differently from the way humans do, much has to be left to the imagination.

Listening to sounds (cheep trills)

INSECTS PRODUCE SOUNDS for all kinds of reasons, among them to attract or find mates, to communicate danger to other members of their species, to establish territories, to locate offspring or to solicit parental care, and to advertise distastefulness to a potential predator. An amazing variety of sounds is produced in an amazing variety of ways, but basically there are two approaches to insect sound production. Many species have specially modified body parts that are stroked, slammed, or vibrated entirely independently of the environment in order to produce sound; other species must strike parts of their body against the ground or other hard surface in order to make sounds.

Acoustical signals have the decided advantage over other forms of communication of traveling a great distance (and thus covering a lot of territory) and of working under cover of darkness (when visual signals are less effective). The sounds that insects make to communicate tend to be toward the high end of the sound spectrum. This in part reflects the fact that wavelengths produced by living things generally are similar in magnitude to the size of the organism producing them. Humans, for example, produce sounds ranging from 80 to 800 Hz, or wavelengths from 4 to 0.4 meters. Because loudspeakers are most effective when their diameter is greater than 16% of the emitted wavelength, insects with sound-producing organs often go to great lengths to increase their size; mole crickets, for example, use the full length of their subterranean burrows to amplify the signals they produce.

Probably the most well-known of all insect sound producers are the crickets and their relatives in the order Orthoptera. Unlike many animal sounds, the chirp of a cricket really has the technical characteristics of a musical tone. The word "cricket" in fact derives from the sound—it means "little creaker," from the French *criquer* (in Dutch, they're called *krekel* and in Chinese *kwo kwo*). Crickets (as well as other noisy orthopterans) can produce sounds using only their bodies. Throughout the order, different body parts are modified for sound production. In crickets, the forewing is equipped with a thickened vein crossing near the base. Like a file, this vein is etched with from 50 to 300 ridges. On the upper side of the wing is a thick hardened area that serves as a scraper. A "singing" cricket is simply lifting its wings over its body and pulling the file on the underside of one wing over the scraper on the wing underneath. Adjacent parchmentlike portions of the wing vibrate and help to amplify the sound. The entire process is known as stridulation, from the Latin "to make a harsh sound." Most species of crickets in the family Gryllidae are right-winged—the right wing is raised and drawn over the left.

Not all crickets are singers. Wingless species (such as those living in bat caves) and wingless individuals of winged species (such as the immature nymphs) are mute. Despite the fact that they possess four fully functional wings, no female crickets sing. This sexual disparity is at least partly attributable to cricket sexual practices. The most conspicuous of all cricket songs is the so-called calling song, used by males for attracting a mate. Each species has a unique calling song, characterized by a particular frequency and pitch, although some species have regional dialects or "accents." Female crickets are able to recognize the song of males of their own species and can distinguish even very similar calls with great accuracy. Like humans, male and female crickets hear by means of a thin, flat membrane, or tympanum; unlike humans, their tympana are located on the lower part of their forelegs. The tympanum vibrates when it comes in contact with sound waves. Cordlike nerve cells then transmit the vibrations to the auditory nerve, which relays impulses to the brain. The tympanic organs of crickets are sensitive to a broad frequency range extending beyond 20,000 cycles per second—beyond the sensitivity of the human ear but conveniently within the range of cricket stridulation.

The fact that the calling song is genetically programmed and species-specific does not mean that it is entirely unaffected by environmental factors. In many species, the rate of stridulation is profoundly (and precisely) affected by temperature. This relationship is so precise that certain species of crickets can be used reliably as thermometers. In a classic paper published in 1897, titled "The cricket as thermometer," A. E. Dolbear determined that, for the snowy tree cricket, the number of chirps in fifteen seconds added to 37 will yield the temperature in degrees Fahrenheit. That crickets never caught on as household meteorological aids is due in part to the fact that Dolbear's formula applies only to a particular kind of cricket—each cricket species must be carefully calibrated before being used to assess the weather.

There is more to a cricket's repertoire than the calling song; even for crickets, courtship is a complicated business. After the female locates the male by following the calling song (some songs are audible over a distance of a mile), she touches him with her antennae. The male then begins a courtship song, which induces the female to mate with him. After mating is complete, males of some species break into yet another song (which some investigators refer to as the "triumphal song"). This last song may reinforce the mating bond and stop a female from seeking out another mate before she lays at least a few eggs.

Cricket song isn't always directed at females; fighting among males setting up territories is often accompanied by the aggressive or rivalry song. In China, crickets are bred for their aggressiveness and territorial combats are actually staged as sporting events (Fig. 3.1). Attending cricket matches (as it were) has been a popular pastime in China since the eighth century. The decline of the South Song Dynasty was attributed in part to the passion of Jia Sidao, the premier, for cricket fighting and his consequent neglect of affairs of state. Even today training and fighting crickets is big business. As soon as immature crickets are old enough to be sexed, males are caged in elaborate houses and solicitously fed vegetables, fruits, chopped fish, and even honey as a tonic (the voiceless females are unceremoniously sold to bird-fanciers as pet food). Champion crickets (*shou lip*) can pull in lifetime earnings of $20,000 or more

A CHINESE CRICKET ARENA

In the bowl is the cricket house with the lid on, at the upper it a food rack with a bean lying beside it, and at the upper ght a water trough (see, also, page 63).

Figure 3.1

(top) Accommodations for Chinese fighting crickets, containing cricket house with lid on, food rack, and water trough; (bottom) Fighting arena with two combatants (Department of Entomology, University of Illinois archives).

and become national heroes; matches are even televised in Shanghai.

In nature, stridulation is not without its risks; while singing certainly makes a male more conspicuous to a female, it also makes a male more conspicuous to potential predators. Certain parasitic flies can orient to the sound of cricket chirping. Upon locating a male cricket, the female fly then lays eggs on him, and the hatching maggots live as parasites inside his body, consuming his vital organs. A male, then, faces a tough decision—if he doesn't stridulate, he is unlikely to attract a mate, but if he does stridulate, he stands a significant risk of being detected by a parasitic fly and parasitized (talk about fatal attractions). So-called satellite males compromise and hang silently around the periphery of male cricket choruses, hoping to waylay a female on her way in or out.

Cicadas (members of the Homoptera) are also conspicuous chorusers. Only male cicadas are equipped with a pair of drum-like noisemaking structures. There's a pair of cuticular membranes stretched across cuticular rims on the first abdominal segment of cicadas that are attached to a series of muscles. When the muscles contract, the membrane vibrates and, since the chest cavity is mostly hollow, the vibrations are amplified.

Periodical cicadas, the Methuselahs of the insect world, spend either 13 or 17 years underground as nymphs, sucking on tree sap through the roots (Fig. 3.2). As if by prearranged signals, literally millions emerge over a relatively short period of time and molt into adulthood. Males engage in mass chorusing, and the volume can be louder than the din of a major metropolitan airport; one South African species checks in at over 100 decibels at full throttle. The idea behind the mass chorusing is not only to attract females but also to confuse predators. Since cicadas are absolutely defenseless as far as predators go (no venom, no nasty sprays, not even any dramatic escape behavior), they go for safety in numbers. In such a massive aggregation, any individual cicada enjoys a fairly low risk of being singled out by a predator and eaten.

Probably the most remarkable sound repertoire of any insect is possessed by the Passalidae, or patent leather beetles. These shiny brownish-black beetles live in family groups inside and around rotting logs in forests and woodlands. The eastern North American species *Odontotaenius disjunctus* possesses a fourteen-signal communication system, a greater repertoire of sounds than many birds. Adults rub the dorsal part of their fifth abdominal segment, which is rasped, against a hardened fold, etched like a file, in their membranous hind wing. Unlike crickets, both sexes stridulate. Courtship signaling can last twelve hours or longer (mating, the only silent part of the process, takes a mere ten to thirty minutes). Even larvae can make sounds, by rubbing their hind legs against a file on the upper joint of their middle legs.

Many insects must strike various body parts against a substrate in order to produce sounds. Stoneflies (order Plecoptera), for example, make noise with a hammerlike structure on their abdomen that is thumped against the ground. Booklice (order Psocoptera) make a ticking sound by knocking a knoblike device on the underside of their tail against a thin sheet of paper to cause it to vibrate (they can keep this up, five or six taps per second, for an hour or more). The female deathwatch beetle, *Testobium rufovillosum*, which makes tunnels in wood inside buildings, strikes its head against the sides of its burrows to make tapping sounds. This ticking was thought to be bad luck in a household: indeed, according to

Figure 3.2
The (long) life cycle of the periodical cicada (Riley, 1869). a. nymph, b. exuvia (cast skin), c. adult, d. oviposition damage, e. eggs.

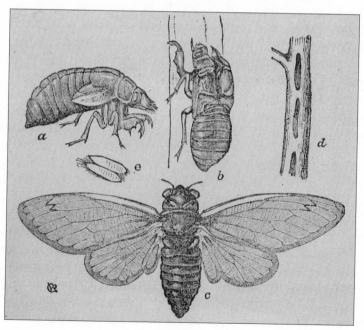

one nineteenth century author, "the fate of many a nervous and superstitious patient has been accelerated by listening in the silence and solitude of night, to this imagined knell of his approaching dissolution" (Cowan 1865).

Looking around (sights to behold)

The firefly wakens:
waken thou with me.
—A. TENNYSON
Now Sleeps the Crimson Petal
(1847)

VISUAL SIGNALS ABOUND in the process of mate finding. First of all, characteristic markings provide a visual indication of an appropriate mate. Sometimes the signals are visible to insects but invisible to potential predators. Sulfur butterflies (so-named for their bright yellow wing color), for example, have ultraviolet markings on their wings, which vary with the species. While ultraviolet is a color visible to many insects, it's invisible to many vertebrates, including people (with the exception of those individuals who have had their UV-filtering lens removed during cataract surgery). Often a complex behavioral repertoire accompanies a pretty face, as visual indicators of a potential mate. Complex rituals of actions and movements constitute courtship in many species. There's an added element of risk in some predaceous species like spiders—make the wrong move and your potential mate turns you into a meal.

One disadvantage of visual signals is that they are not very effective at night, when predation risks may be lower. Fireflies and a few other groups have gotten around this limitation by using light energy for signaling. A firefly is a beetle in the family Lampyridae (from the Latin for "shining fire"). In the abdomen of the adult beetle is a light-producing organ. Cells in this "lantern" contain a substance called luciferin. When oxygen, piped into the lantern via tracheoles, comes in contact with luciferin in the presence of the enzyme luciferase, it reacts chemically to release energy in the form of light. The light is intensified by a layer of reflector cells that act as mirrors to amplify the light signal. Light production by fireflies is far more efficient than by electric light bulbs; in firefly lanterns, almost 100% of the energy produced by the chemical reaction is released as light, whereas in an electric light bulb about 90% of the energy is lost in the form of heat. Producing light is a firefly family affair; many larvae, called glowworms, and even eggs, produce a continuous glow.

Adult fireflies control the flash pattern of their lanterns by regulating the oxygen supply to their light organs. Flashes are intermittent, almost never continuous, and each pattern is highly species-specific. Around dusk (depending on the species), males take to the air and fly over vegetation, flashing their species-specific coded signal. Females perched on vegetation reciprocate by flashing back the same signal in response after a short one- or two-second interval. After five to ten reciprocal signal exchanges, the male lands adjacent to the female, and copulation ensues.

Female fireflies in the genus *Photuris,* however, provide an interesting exception to the statement that every species of firefly has a unique species-specific flash pattern. These females are capable of returning the signal of any number of males of other species. It's not that they're promiscuous—rather, they are just hungry. Once they lure a male in, they eat him. These females are aptly known in the scientific literature as firefly "femmes fatales."

The luminescent properties of fireflies have proved to be an unlikely source of fun and profit for people of many descriptions. The aesthetic aspects of firefly illumination are contemplated throughout Japan, where for centuries fireflies have figured prominently in haiku, a genre of poetry. In contemporary times, firefly festivals, held in August when firefly viewing is optimal, attract 10,000 or more people every year. As for their more remunerative aspects, fireflies are big business in the scientific supply trade. Their lanterns are powered by a substance called adenosine triphosphate (ATP), the biological transporter of chemical energy in most living systems. In the absence of ATP, firefly lanterns can't flash, so the intensity of a lantern flash can be used as a means of monitoring the levels of ATP present in any kind of biological preparation. Because ATP powers so many different kinds of cellular reactions, a sensitive system for detecting levels of ATP is a useful biochemical tool. Thus it happens that many chemical supply companies are in the firefly business. Desiccated "tails" (actually, abdomens), whole fireflies, and purified luciferin and luciferase can be purchased for prices ranging from $40 for a gram (about $\frac{1}{28}$ ounce) of tails to $75 for a milligram ($\frac{1}{28,000}$ ounce) of purified luciferase enzyme.

Tastes and smells

Coming from every direction and apprised I know not how, here are forty lovers
eager to pay their respects to the marriageable bride born that morning . . .
Are there . . . effluvia of extreme sublety, absolutely imperceptible to ourselves and
yet capable of impressing a sense of smell better-endowed than our own?
—J.H. FABRE, (1907)

CHEMICALS GOVERN VIRTUALLY every aspect of insect life, from food
finding to defense to communication, even to burial of the dead.
Chemical signals offer a few advantages over other forms of signals,
in particular for relatively small organisms such as insects. First,
chemicals work in the dark and are effective around obstacles, conditions under which visual signals are ineffective. Chemical signals
also have the greatest range of flexibility in transmission distances.
Depending on structure, a chemical signal can be transmitted on
contact, at effectively no distance at all (as when a taste receptor is
stimulated), or as far as several miles. Moreover, chemical signals
can persist in time longer than most other forms of signals—a visual or acoustical signal, for example, is effectively instantaneous,
but a chemical signal, again depending upon the structure, can
linger for minutes, hours, or even days.

Virtually all organisms to some extent make use of chemical
signals, but the class Insecta raises chemical communication to an
art form. Chemicals are particularly important in mate location
and species recognition. The chemical messages involved in these
activities are known generally as pheromones. A pheromone is an
intraspecific chemical message—a chemical signal that conveys information from one member of a species to another. In the case of
sex pheromones, the information conveyed is usually on the order
of "Come and get me, big boy." Finding a mate is a needle-in-a-
haystack proposition for most insects, and chemical messages, with
their long transmission range and greater persistence, increase the
odds.

Pheromones by and large are produced in glands. Unlike en-
docrine glands, which secrete their contents inside the body, the
exocrine glands that produce pheromones release their products
outside the body. Insects are equipped with an enormous variety of
glands, associated with virtually every orifice. In most cases, pheromone glands are lined with cuticle, the same inert hardened pro-
tein-polysaccharide complex that forms the outer exoskeleton. This
inert cuticular lining protects the inner workings of the insect from
any potential harmful effects of the chemicals stored inside. In ad-

dition to serving as a manufacturing center and storage depot, glands also act as a delivery system. In order for a chemical signal to be of any use, it has to be dispersed, and insects have developed a diversity of mechanisms for spraying, oozing, or dripping glandular contents to get the message across.

The vast majority of known sex pheromones are produced by females for the purpose of attracting mates. These have been found in over 200 species in at least six orders (including cockroaches, scale insects, moths, butterflies, beetles, flies, and bees, among others). The very first pheromone ever identified from an insect was the female-produced sex pheromone of *Bombyx mori,* the Japanese silk moth. Bombykol, isolated in 1959, is a typical female-produced sex pheromone in that it is a small molecule, with only sixteen carbon atoms (most proteins, in contrast, have hundreds of carbon atoms). In addition to carbon atoms, sex pheromones also usually contain a few dozen hydrogen atoms and only a handful of oxygen atoms (bombykol has only one oxygen atom). Small size is associated with volatility (the ability to travel in a gaseous phase), which greatly increases the range of the signal. Another characteristic of female-produced sex pheromones is that they are often produced in tiny amounts and are active at extremely low concentrations. Females release as little as a billionth of a gram per hour. In order to isolate and purify bombykol, the abdomens of almost a million virgin females were extracted—and from these only 1.5 mg (only 1/20,000 ounce) of pure bombykol was extracted.

A chemical signal wouldn't work unless there was an equally sensitive receiver at the other end, and in most cases the chemore-

ceptors for female sex pheromones are located on the antennae of the males. The antennae of many moths are sexually dimorphic (that is, different in structure depending upon whether the owner is a male or female). Whereas female antennae tend to be thin and wirelike, the antennae of males are usually bushy or feathery and function like radar screens in that their large surface area is used to screen the skies, not

Figure 3.3
Antennae of luna moth (J. G. Sternburg).

for sound signals but for chemical signals (Fig. 3.3). The chemore-ceptors are highly specific and extremely sensitive—one molecule of pheromone is enough to trigger a chemoreceptor in the anten-nae of male silk moths. As little as 100 molecules of pheromone in a milliliter of air is enough to start a male off in pursuit of a chemical trail. Males generally orient to a chemical trail by following it up-wind toward its source (a behavior known as positive anemotaxis). The area occupied by scent molecules in concentrations sufficiently high to be detected by the receiver is known as the active space. The size and shape of the active space depend not only on the rate and amount of production by the female and the sensitivity of the male but also on general weather conditions. If the wind is blowing, the molecules will be scattered over a greater distance and the active space will be of a different shape.

Sex pheromones are involved not only in mate location but also in species recognition. Species-specific information can be conveyed in a number of ways in a chemical message. First, the chemical contents of the signal convey information. While a sex pheromone may consist of a single unique species-specific com-pound, more often it is a mixture of components that vary in con-centration. Species specificity can be conveyed by varying the relative composition of such a mixture. Even different populations of the same species can have different pheromone blends. For ex-ample, female European corn borers (*Ostrinia nubilalis*) in Iowa use a mixture of two components in a ratio of 97:3, but females of the same species in New York use a blend of the same two compo-nents in a ratio of 4:96. The timing of pheromone release is another factor that conveys information about species identity; different species can even use the same pheromone components, as long as they release them at different times of day.

Once mating has been accomplished, females of some species produce substances designed to discourage the advances of addi-tional suitors. Such compounds are called antiaphrodisiacs. Female *Pterostichus lucublandus,* a ground beetle, produce a "mace"—methacrylic acid is squirted at males that persist in courting a fe-male that has already mated. In a typical encounter, "a male, receiving the discharge, immediately stops running and tries to wipe the irritant from his face and antennae with his front legs. However, within 10 seconds, his movements become so uncoordi-nated that he may roll over on his back and be unable to regain footing, his legs become stiff and movement ceases. He remains in this deathlike coma for 1–3 hours but c[irc]a 15–30 minutes after the first signs of revival, recovery appears to be complete" (Kirk and Dupraz 1971).

In the majority of cases studied, females produce the long-distance attractants and it is up to the males to follow the trail. In some cases, a male can travel several miles in pursuit of a mate. Once he arrives, though, many males make their own contribution to chemical courtship. Males of many species produce sex pheromones, but, unlike the sex pheromones of females, in most cases the pheromones of males are effective only at relatively short range and serve not so much as attractants as excitants—something to put the female in the mood, to make her more likely to accept his advances. This sort of pheromone is accordingly known as an aphrodisiac. Males of at least sixty species, including flies, moths, butterflies, cockroaches, and bees, are known to produce aphrodisiacs.

It has been suggested that males produce aphrodisiacs so that the females can assess their quality as prospective mates. Often the substance serving as an aphrodisiac is a plant-derived poison that is stored in the body of the male; thus protected, he is unlikely to end up as a meal for a bird and is therefore likely to father strong, healthy offspring. In the case of the bella moths and their relatives, males sequester pyrrolizidine alkaloids from their host plants. In fact, without the alkaloids, their pheromone glands (called coremata) don't even develop properly—they end up twisted and useless, and males possessing such effronteries are invariably rejected by discriminating females.

One interesting variation on the idea of chemical courtship has shown up in a number of orders. In these cases, the signal is not olfactory but gustatory—the male appeals to the female's good taste. The food enticement can be a glandular secretion, as in the cockroach *Nauphoeta*, which has glands under its wings that are licked by the female as the male inserts his genitalia at her nether end. Many males offer a so-called nuptial or wedding gift to the female prior to mating. Female hangingflies (peculiar long-legged insects in the obscure order Mecoptera) have never been observed to hunt or catch prey in nature. Males catch prey, and when they begin to eat, they extrude glands on their abdomen. This musky odor attracts females, who then try to take the prey out of the male's grasp. While this struggle is going on, the tip of the male's abdomen is circling around in search of the female's. Once the connection is made, the male hands over the meal and settles down to copulate. The vast majority of species in which males present wedding gifts are predators; it's always a risk for a male to approach a carnivorous female, especially if he doesn't know when she had her last meal. With the nuptial gift system, she is distracted and, what's more, nice and well-nourished after the fact so she can fertilize lots of eggs with his sperm. So it's not exactly the spirit of unselfish gift-giving motivating these males.

Once the act has been consummated, many males also produce antiaphrodisiacs, with the intent of discouraging subsequent males from bothering his mate. Male heliconian butterflies equip their mates with a sort of chastity belt, a large "stink club" that serves two purposes: it blocks the female's genital opening and it smells so repulsive that other males immediately turn around and fly in the opposite direction without contemplating a sexual advance.

In some species, sex pheromone production is an equal opportunity affair. In these cases, the chemical signals serve as aggregation pheromones, bringing both sexes into the neighborhood. The pine bark beetle *Dendroctonus brevicomis* has an extraordinarily complex system of chemical signaling. Pine bark beetles are small, cylindrical beetles that carve out tunnels just beneath the bark of pine trees, crisscrossing the living, growing cambium layer. Some of these mines are feeding tunnels, made by adults who digest the soft woody tissue; others are egg tunnels or parent galleries, into which females sequester their eggs; and still others are larval tunnels, carved out by the growing grubs. These little beetles cause damage out of all proportion to their tiny size; their extensive tunneling can quickly starve a tree and cause it to die.

The assault begins when females attracted to volatile pine emissions arrive and release a pheromonal blend of exobrevicomin, a product synthesized by the beetle, and myrcene, a derivative of pine tree resin. These chemicals lure males, who upon arrival release another pheromone, frontalin, which attracts more males as well as females to the tree. Once things get too crowded for comfort (or the competition gets too intense), resident males begin to produce verbenone, which dissuades new males from landing.

The idea of this mass mating frenzy is that only large aggregations of pine bark beetles can overcome the defenses put up by the tree against the beetle. When attacked, a tree produces a copious flow of resin. This resin is life-threatening to the beetles, who stand to drown, or at least get mired, if their tunnels flood. If it is overwhelmed by beetles, however, a tree cannot react with its resin flow. So it's to every beetle's advantage to gather at a tree not only for mating but for egg-laying under the bark.

Some aggregation pheromones control not only the location of mating but also the timing of mating by controlling the rate of development. Cockroaches raised in groups grow faster than cockroaches raised in isolation due to the presence of an aggregation pheromone (one reason one almost never finds only a single cockroach in one's kitchen). Adult larder beetles (*Dermestes*) produce a fecal pheromone that synchronizes larval molts. Aggregation pheromones may be common among stored-product insects like cockroaches and larder beetles because, in nature, these species are associated with food resources that are unpredictable in time and space. Prior to the human custom of stockpiling food (which dates back only 10,000 years or so), large caches of seeds or grain were few and far between, restricted primarily to rodent nests. Aggregation pheromones provide a mechanism for widely dispersed individuals to find each other and produce offspring at an opportune time, to exploit an abundant resource.

Pheromonal communication is quite complex and sophisticated in many insects; chemical messages serve functions other than reproductive ones. Other types of pheromonal signals include alarm pheromones, which alert fellow members of the species to the presence of danger and evoke escape behaviors. These are common only in the social insects and in parthenogenetic (clonal) species like aphids. Many species of aphids, for example, produce farnesene when threatened, which when perceived by fellow aphids elicits the full complement of escape behaviors (although in an aphid that may amount to nothing more than falling off the branch). Egg-laying females of many species protect the environment in which they oviposit, or lay their eggs, from competitors by depositing an oviposition deterrent pheromone. Apple maggot flies (*Rhagoletis pomonella*) drag their ovipositor around the surface of an apple after laying their eggs; other females of the same species will avoid laying eggs in that apple for the next three to six days (the end result being that the first maggots to hatch can live out their larval lives in peaceful solitude, without having to worry where their next meal is coming from). In contrast with oviposition deterrent pheromones, which prevent members of the same species from sharing the same food resource, trail pheromones allow other members of the same species to find food. These are common in social insects but also in other species. Eastern tent caterpillars (*Malacosoma americanum*), for example, lay a chemical trail in their silk that fellow caterpillars can detect at concentrations of

0.00000000001 gram per square millimeter in order to locate branches with an abundant supply of leaves.

Because responses to pheromones are generally so automatic in insects, pheromones provide useful tools in insect control. Insects are relatively vulnerable to misinformation that is chemically conveyed and in human hands misinterpretation can be fatal. Insect sex pheromones have been used to that end in a number of ways: for monitoring (when female pheromones are used to bait a trap and draw in males in order to census the population, particularly in migratory species that may move into an area seasonally); for mating disruption (when female pheromones are released in such enormous concentrations that males in the field cannot locate calling females); and even for the outright taking of life (as when aggregation pheromones are incorporated into sticky traps for population control of stored-product pests). Pheromones provide an attractive alternative to synthetic chemicals for pest control (particularly of stored grain, processed food, or other edibles); they're biodegradable and hence leave little in the way of residue, they have practically no effects at all on organisms (including humans) other than the target species, and, because they're effective at extremely low concentrations, they are potentially cost-competitive with synthetic chemicals.

References

Sensory physiology

Alcock, J., 1975. *Animal Behavior.* Sunderland, MA: Sinauer.

Askew, R.R., 1971. *Parasitic Insects.* New York: Elsevier.

Atkins, M.D., 1980. *Introduction to Insect Behavior.* New York: Macmillan.

Blum, M., 1984. *Insect Physiology.* New York: Wiley.

Chapman, R., 1972. *The Insects: Structure and Function.* New York: Elsevier.

Cowan, F., 1865. *Curious Facts in the History of Insects.* Philadelphia: Lippincott.

Evans, H.E., 1968. *Life on a Little-Known Planet.* New York: Dutton.

Fraenkel, G.S. and D.L. Gunn, 1961. *The Orientation of Animals: Kineses, Taxes and Compass Reactions.* New York: Dover.

Roeder, K.D. and A. Treat, 1961. The detection and evasion of bats by moths. *Am. Scientist* 49: 135–148.

Listening to sounds (cheep trills)

Alexander, R. D., 1962. Evolutionary change in cricket acoustical communication. *Evolution* 16: 443–467.

Alexander, R.M., 1985. The ideal and the feasible: physical constraints on evolution. *Biol. J. Linn. Soc.* 26: 345–358.

Bentley, D. and R. R. Hoy, 1974. The neurobiology of cricket song. *Scientific American* 231: 34–44.

Busnel, R.G., 1964. *Acoustic Behavior of Animals.* New York: Elsevier.

Cade, W., 1975. Acoustically orienting parasitoids: fly phonotaxis to cricket song. *Science* 190: 1312–1313.

Chou, I., 1960. *A History of Chinese Entomology.* Wugong (Shaanxi): Entomotaxonomia.

Evans, H.E., 1968. *Life on a Little-Known Planet.* New York: Dutton.

Hoy, R.R., J. Hahn, and R. C. Paul, 1977. Hybrid cricket auditory behavior: evidence for genetic coupling in animal communication. *Science* 195: 82–84.

Ignatius, A., 1988. Even the winners crawl from the ring after these bouts. *Wall Street Journal,* 8 November 1988, pp. A1, A15.

Kalmring, K. and N.S. Eisner, eds., 1985. *Acoustical and Vibrational Communication in Insects.* New York: Paul Parey.

Laufer, B., 1927. *Insect-Musicians and Cricket Champions of China.* Field Museum of Natural History Dept. Anthropology Leaflet No. 22.

Riley, C.V., 1869. The periodical cicada. *First Ann. Rep. on the Noxious, Beneficial, and Other Insects of the State of Missouri.* Jefferson City (MO): Regan and Edwards, pp. 18–42.

Schuster, J.C., 1983. Acoustical signals of passalid beetles: complex repertoires. *Florida Entomol.* 66: 486–496.

Villet, M., 1987. Sound pressure levels of some African cicadas (Homptera: Cicadoidea). *J. Ent. Soc. S. Africa* 50: 269–273.

Looking around (sights to behold)

Lloyd J.E., 1965. Aggressive mimicry in *Photuris:* firefly femmes fatales. *Science* 149: 633–634.

Reitman, V., 1993. Scientists are abuzz over the decline of the gentle firefly. *Wall Street Journal,* 2 September 1993, pp. A1, A5.

Silberglied, R. and O. Taylor, 1978. Ultraviolet reflection and its behavioral role in the courtship of the sulfur butterflies *Colias eurytheme* and *Colias philodice. Behav. Ecol. Sociobiol.* 3: 203–243.

Tastes and smells

Boller, E.R., R. Schoni, and G.L. Bush, 1987. Oviposition deterring pheromone in *Rhagoletis cerasi*: biological activity of a pure single compound verified in semi-field test. *Entomol. Exp. Appl.* 45: 17–22.

Crump, D., R.M. Silverstein, H.J. Williams, and T.D. Fitzgerald, 1987. Identification of trail pheromone of larvae of eastern tent caterpillar *Malacosoma americanum* (Lepidoptera: Lasiocampidae). *J. Chem. Ecol.* 13: 397–402.

Harborne, J.B., 1982. *Introduction to Ecological Biochemistry.* London: Academic.

Kirk, V. and B. Dupraz, 1971. Discharge by a female ground beetle, *Pterostichus lucublandus* (Coleoptera: Carabidae), used as a defense against males. *Ann. Entomol. Soc. Am.* 65: 5113.

Klun, J.A. and S. Maini, 1979. Genetic basis of an insect chemical communication system: the European corn borer. *Environ. Entomol.* 8: 423–426.

McNeil, J.N., 1992. Evolutionary perspectives and insect pest control: an attractive blend for the deployment of semiochemicals in management programs. Pp 334–351 in *Insect Chemical Ecology: An Evolutionary Approach,* B.D. Roitberg and M.B. Isman, eds. New York: Chapman and Hall.

Nakai, Y. and Y. Tsubaki, 1986. Factors accelerating the development of German cockroach (*Blattella germanica*) nymphs reared in groups. *Japan. J. Appl. Entomol. Zool.* 30: 1–6.

Phelan, P.L. and T.C. Baker, 1987. Evolution of male pheromones in moths: reproductive isolation through sexual selection? *Science* 235: 205–207.

Prestwich, G.D. and G.J. Blomquist, eds. 1987. *Pheromone Biochemistry.* New York, Academic Press.

Roelofs, W.L. and R.L. Brown, 1982. Pheromones and evolutionary relationships of Tortricidae. *Ann. Rev. Ecol. Syst.* 13: 395–422.

Shorey, H.H., and J.J. McKelvey, 1977. *Chemical Control of Insect Behavior.* New York: Wiley-Interscience.

Teale, E.W., 1956. *The Fascinating Insect World of J. Henri Fabre.* New York: Fawcett Pub., Inc.

Wilson, E.O., 1970. Chemical communication within animal species. In: *Chemical Ecology.* E. Sondheimer and J.B. Simeone, eds. New York: Academic Press, pp. 133–155.

Chapter 4

SOCIAL LIVES

The joys of group living

> The thronging thousands to a passing view
> Seemed like an anthill's citizen.
> —PERCY BYSSHE SHELLEY

IT'S NO COINCIDENCE THAT "community" and "communication" both derive from the same Latin word, *communis,* "to exchange or share obligations"; communication is an essential part of group living. Sociality is living in a group of individuals of the same species that exchange information on a regular basis and cooperate in one way or another for reasons other than mating. Social behavior in the class Insecta covers a broad range of interactions. Not all groups of insects can be counted as social gatherings, however. Some insect gatherings are simply the result of many individuals responding in the same way to an environmental stimulus. This sort of gathering is called an aggregation. Caterpillars all hatching from an egg cluster may spend a few instars feeding together but just because they appear to get along together doesn't mean they are exchanging information or cooperating in any way. Moths swarming around a porch light at night are all responding to a common stimulus and are not really exhibiting true social behavior.

To organize and classify social interactions, three basic criteria are used. For one thing, social species exhibit cooperative care of offspring—members of a society cooperate collectively in the care of young. In addition, in social species there is reproductive division of labor—there are individuals in a society who mate and lay eggs and others who never have that opportunity. This is not to say that other individuals do not perform critical tasks asssociated with care of the young. Often, these tasks are divided among members of a colony, with some who care for young, and others who protect the colony from danger. Finally, in most of the highly developed societies, generations overlap so that offspring can grow up to assist their parents in care of the young.

When all three traits are exhibited by a species, it is said to be eusocial, or truly social (from *eu,* Greek for "true"). Many species exhibit only one or two of these traits and thus are regarded as subsocial. One of the remarkable attributes of eusocial animals is that they are the only ones that exhibit altruistic behavior (from the Latin *alter,* "other people"). An organism is altruistic when it acts in a way that increases the probability that other individuals will survive at its own expense. Humans display this sort of behavior often; for example, many people have rushed into burning buildings to rescue strangers trapped inside, or plunged into icy waters to save a drowning child.

Subsocial behavior is widespread and is known in many insect orders, including the cockroaches (some of which actually display maternal behavior), earwigs, webspinners, true bugs, aphids, and beetles. Subsocial behavior can also be found in noninsect arthropods; social spiders, for example, spin communal webs, and scorpions display maternal care. In contrast, eusocial behavior is practiced almost exclusively by two orders of insects: the Hymenoptera (bees, ants, and wasps), and the Isoptera (termites). In the Hymenoptera, social behavior is thought to have evolved at least ten different times. Recently, a few isolated cases of eusociality have been identified in the Homoptera (there are a few eusocial aphids) and the Thysanoptera (at least one species of thrips is eusocial).

The advantages of group living are numerous. One decided advantage is enhanced defense against enemies, a manifestation of the idea "in union there is strength," or, as expressed in an Ethiopian proverb, "When spider webs unite, they can halt a lion." Groups of animals that communicate with one another have a greater probability of detecting predators in the first place—there are more eyes, ears, or equivalent sensory structures on the alert. Many social species use pheromones—intraspecific chemical signals—to alert other members of a society to the presence or approach of danger. These alarm pheromones can evoke different responses depending on the content and composition of the secretion. For example, in species of *Atta* ants, the alarm pheromone is 4-methyl-3-heptanone; at low concentrations it stimulates attraction (so ant workers can come to the aid of their fellows), but at concentrations ten times higher it stimulates fast running with open mandibles, so an individual can escape an immediate and present danger. *Oecophylla* ants produce no fewer than thirty compounds from four glands as an alarm pheromone. Depending on the concentration and composition of the mixture, the responses of workers can include alerting, attraction, attraction and biting, and biting alone.

Yet another advantage of living in a group is that it reduces the probability that any individual can get singled out and eaten. Most predators have great difficulty selecting a victim to chase down and subdue when faced with a milling mass of prospective meals. Colonies of African driver ants (*Anomma wilverthi*) can contain over 22 million individuals that collectively weigh over 40 pounds—a formidable mass to deal with and one that is likely to send a predator off in search of a less challenging meal.

A predator hungry or perverse enough to take on the challenge of attacking a colony of social insects soon learns that sheer number is only part of the problem. Many social species have individuals specialized for defense of the colony. These individuals are known as soldiers and are generally well equipped morphologically for such activity. Among termite societies, soldiers are variously equipped with enlarged mandibles, enormous glands filled with acrid or toxic fluids, or heads shaped like stoppers, which they use to plug up holes in the nest to prevent entry by an enemy.

Another benefit of social life is that living in a group can improve success at raising young. Living in a group greatly facilitates the process of finding a mate. In some social species, breeding is synchronized by pheromones so that all members of a group are ready and willing to mate at the same time. Cooperative care of the young also increases the probability that the young will survive to reproduce. Reproductive females in ant or bee societies (queens) can lay over a million eggs in the course of their lifetime—raising all those offspring without a little assistance would no doubt be a daunting task.

Group living also improves the likelihood that an individual will find and secure food. Cooperative hunting allows members of a group to capture prey larger or faster than any single individual could successfully pursue. The subsocial burying, or sexton, beetles provide a case in point. Immature sexton beetles feed on the decaying flesh of dead birds and mammals, which is provided for them by their parents. Male and female sexton beetles work together to strip a carcass of its fur or feathers and bury it underground; although individually the beetles are less than an inch in length, together a couple can move and bury a 4-ounce mouse in a matter of minutes.

Group living also allows individuals to exploit a short-lived food source more efficiently. Colonies of honeypot ants (*Myrmecosystus*) have workers, called repletes, whose sole function is to store food. They are fed nectar collected from oak galls by foraging individuals and their abdomens become so enlarged that they can no longer walk around. Repletes are forced to hang suspended in the nest, where, when any other worker is hungry, they serve as honey dispensers. Communication among foragers is also important in exploiting food resources. Many ants and termites produce trail pheromones, which mark the path to a food source. Trail pheromones tend to be less volatile and more durable than either alarm or sex pheromones—which is why ants at a picnic will continue to walk a circuitous path around the place where a picnic basket rested long after the picnic basket has been removed to higher ant-free ground.

One additional benefit of group living is that, in a group, individuals are not so much at the mercy of the environment as they are individually. By collective effort, social insects are capable of astounding feats of engineering. Some termite mounds, for example, extend more than 15 feet below the soil, with galleries extending over 300 feet from the central nest. Underground nests of this size can house over a million individuals. One population of the mound-building ant *Formica exsectoides* consisted of seventy-three nests occupying a 10-acre spread (with 12 million workers in residence). In these massive structures, tiny residents are afforded protection against environmental disturbances such as temperature changes, wind shifts, or precipitation.

Group living is not without disadvantages. Protection from predation is somewhat offset by the fact that large aggregations are more conspicuous and are bigger targets for enemies. Many mammals have actually specialized in consuming social insects. Anteaters, scaly pangolins, and echidnas, for example, all spend much of their time in pursuit of ants and other nest-building social insects. They all have long sticky tongues for fishing ants out of a nest and either stiff hair or scales to protect themselves against the stings and venom of workers. Many other mammals, including humans, make a practice of raiding the stored food supplies of social insects. The honeyguide *Indicator indicator* is a honey-eating bird in Africa that loves to eat beeswax but lacks the strength or claws to break into a hive to get any. Instead, once a honeyguide finds a hive, it flies away to catch the attention of a honey badger (or a human), which pursues it back to the hive. When the hive is broken open by the more muscular mammal, the honeyguide eats its fill.

Another disadvantage is that individuals have to give up a lot to live in a group. Workers not only forego mating and egg-laying but also are called upon to give up their lives in defense of the colony. This altruistic behavior can take on very dramatic forms in the social insects—as in the case of the termite soldiers that engage in autothysis, or explosive defecation. In the face of danger, these soldiers expel defensive secretions with such force through the anus that their entire abdomen blows off. Thus, they perish, committing the ultimate sacrifice in defense of their nestmates.

Sociality takes many forms, depending on the organism. Strictly speaking and narrowly defined, a colony is a group of genetically identical individuals, usually physically joined to one another. As such, only lower invertebrates form true colonies. The Portugese man-o'-war, for example, a jellyfishlike marine creature, represents a true colony of genetically identical individuals physically joined to one another. The term is widely used, however, to refer to insect societies. Of these, the societies that may most closely fit the narrow definition are those of some very rare species of aphids that seem to display eusocial behavior. Many aphids reproduce parthenogenetically, or asexually; all offspring of such a single female should be genetically identical. Although it seems improbable, given the customary softbodied build and mild-mannered behavior of aphids in general, some of these aphid societies have soldier castes, called samurai after the warrior class of feudal Japan. The name refers not only to their aggressive behavior toward aphid predators but also to the sclerotized "helmet" ornamentation on their heads, which bears an uncanny resemblance to a samurai headdress.

Social structures

> For so work the honey bees,
> Creatures that by a rule in nature teach
> The act of order to a peopled kingdom.
> —WILLIAM SHAKESPEARE, *Henry V*

AMONG THE COMMON eusocial groups—the termites and the hymenopterans—societies differ in several fundamental ways. Termite colonies are distinctive in that they have both a king and a queen—a male and a female reproductive individual living with the colony. Despite the seemingly progressive equal opportunity approach to governance, the termites also have child labor, in which immature individuals, beginning about the second instar, begin to care for the very young and never do anything else. They don't even get to leave the colony, except for the few that graduate to become soldiers. Termites regularly engage in the process of anal trophallaxis—exchange of fluids from anus to mouth. This less than appetizing behavior serves a vital function in the lives of termites in that it is an effective way to transfer microbes that live in the guts of termites among the residents of a nest. By producing cellulases (enzymes that break down cellulose, the otherwise indigestible structural material in plant tissues), these microbes enable termites to use wood as a source of nutrition. Certain termites that are generally regarded as evolutionarily advanced, however, can manufacture their own cellulases.

All members of the order Isoptera are eusocial; in contrast, many species of Hymenoptera lead solitary lives. There is an entire suborder of Hymenoptera in which a social species is nowhere to be found; the Symphyta, or sawflies, feed on plants as larvae, and, although the larvae may be found in large aggregations, they exhibit no true social behavior. Even in the rest of the order, eusociality appears inconsistently. While all ants (members of the family Formicidae) are social, many bees and wasps are solitary or subsocial. This mixture of behaviors throughout the order suggests that eusociality may have arisen more than once in the evolutionary history of the Hymenoptera. There are many theories (and much discussion) about why eusociality evolved so many different times (and in so many different ways) in the Hymenoptera. One theory is based on the curious sex-determination characteristics of the order. As a result of the peculiar phenomenon of haplodiploidy, males are produced by unfertilized eggs and are haploid while females, which are diploid, arise from fertilized eggs. Females are therefore more closely related to their sisters, with whom they share 75% of their genes—half from their father (who, as a haploid individual, can pass along only one set of genes) and a quarter from their mother (who as a diploid can pass along one of two sets of genes)—than they would be to their own daughters (with whom they would share only half of their genes). Thus, when a female helps her mother to raise her own sisters, she is raising offspring who are genetically more similar to herself than would be her own daughters, and her genes are represented in the next generation in greater proportion than if she had had her own children.

According to another theory, the multiple origins of eusociality in the Hymenoptera are attributable to the development of a stinger, a modified ovipositor that can also inject venom. Venom not only allowed solitary hymenopterans to feed their offspring with paralyzed prey, but it also provided them with an effective defense against large enemies—a defense that is all the more effective when it is used by many individuals acting cooperatively. Despite the attractions of both theories, numerous exceptions and logical flaws prevent a resolution of the question.

The organization of hymenopteran societies can be illustrated by the society of the most familiar social insect, the honey bee, *Apis mellifera*. The honey bee society is made up of three types of individuals. The majority of members are workers, sterile females who, as the name implies, do most of the work around the hive or dwelling place. They gather nectar and pollen, they secrete beeswax, they build comb (hexagonal cells of wax), they feed the larvae, and in general keep the hive humming. Worker bees demonstrate polyethism, the division of labor by age. As a honey bee matures, it assumes different tasks, depending on its age and physiological state. During the first two or three days after emergence, a worker bee is assigned to cleaning detail, where her responsibilities include preparing the combs to receive eggs; later on, she may concentrate on removing dead or dying bees from the hive. She then progressses to the nurse contingent, the group that feeds and otherwise cares for the grubs. She assumes this position as the nurse glands in her head become active and secrete various nutritive secretions for growing bees. After a week or two, the wax glands in her abdomen develop rapidly and begin to secrete beeswax. The worker bee then participates in comb building and capping. Finally, two or three weeks after emerging from her cell, the worker is ready to leave the hive and forage for nectar and pollen, process nectar and pollen in the hive, and guard the hive entrance from intruders. Guard bees take stations near hive entrances with antennae poised to touch entering bees in order to ensure that they are colony members rather than passersby out to rob honey, and soldier bees aggressively defend the colony against intruders (Fig. 4.1).

Although honey bee behavior does change predictably with age, there are many factors involved in determining when and if any individual undertakes any particular task. Certain individuals appear to have a genetic predisposition for particular tasks; internal physio-logical stimuli as well as external environmental conditions can also be influential in determining how a worker spends her time.

(Despite the seeming heavy schedule, Martin Lindauer, a fa-mous bee biologist, has discovered that the image of the "busy bee" is something of a myth. He carefully observed thousands of indi-vidual bees in their daily routine and came to the conclusion that, while they do perform many different duties during the course of their adult life, they spend most of their time—up to 70%—hang-ing around the hive, just resting.)

The queen bee differs in appearance from the thousands of workers in the hive—she is longer (particularly in the abdomen), has shorter wings, and lacks the abdominal stripes that mark the workers. The queen is the only reproductively functional female in the whole hive. Her job is to lay eggs. This she does amazingly well; a productive queen can lay up to 1,000 eggs a day. A queen is simply a worker bee that received a special food called royal jelly from nurse bees throughout the course of her larval development. Royal jelly is a mixture of hormones, fatty acids, vitamins, acetylcholine, and nucleic acid bases that promotes extremely rapid growth and morphological change, at least in part by stimulating bee grubs to eat more food. Unlike the stinger of the worker bees, the stinger of a queen bee is unbarbed; although she almost never uses it against outsiders, she uses it effectively against potential rivals. There is

Figure 4.1
Guard bees at hive entrance (G. Robinson).

only one queen in each colony, largely because immediately after she cuts her way out of her pupation chamber a new queen searches out her royal sisters and immediately kills them.

Male honey bees are called drones. Although they are larger than worker bees, they are still considerably smaller than the queen. There aren't many of them in any given hive—only 200 or 300 compared to up to 30,000 workers. They don't do much, either. They don't, for example, build comb, forage for nectar or pollen, or feed the grubs. For that matter, they aren't even capable of feeding themselves for the first few days of their lives; worker bees must tend them as solicitously as they tend the legless, helpless grubs. They have no stingers and therefore cannot defend the colony against its enemies. About the only contribution to society that drones make is that they inseminate or fertilize the queen. After approximately twenty-four days of development (three days longer than a worker and over a week longer than a queen), drones emerge from their pupation chambers. They receive food from the workers for about a week, after which they begin to make short forays out of the hive to scope out the neighborhood. With time, these flights, which usually occur around the middle of the afternoon, get longer and take the drones further from the hive (up to two miles or more). Drones in search of a queen will opportunistically pursue almost anything that flies by, even birds or other insects. When they finally spot a virgin queen on a nuptial (mating) flight, they take off in pursuit, using her sex pheromone as a guide. A swarm can quickly form behind a queen in flight, often consisting of 100 or more anxious males; these swarms are called drone comets, due to their resemblance to the celestial bodies.

The queen can mate with half a dozen or more males on her nuptial flight, which may involve several forays, but she will never mate again after the completion of her nuptial flight and her return to the hive. Queens are known to store sperm for up to seven years. From the drones' point of view, mating is a mixed blessing. The act of copulation, while undoubtedly something for males to do to pass the time, is not only violent, it is actually lethal to the male—he "literally explodes [his] internal genitalia into the genital chamber of the queen and then quickly dies" (Wilson 1971). Drones are maintained in the colony throughout the summer months after the nuptial flight in the event that something happens to the queen mid-season and their services are again required to inseminate a replacement queen. At the end of the warm weather, however, drones are unceremoniously driven out of the hive to starve.

Queens are in a sense responsible for maintaining order in the hive and this they do chemically. Queen bees produce pheromones that suppress the reproductive development of the workers. This queen substance is transmitted through the hive by the constant grooming of the queen and the subsequent mutual food exchange among all the workers. It has been suggested that the reason why hives develop new queens and swarm is that, when populations build up to excessive numbers, the queen cannot produce sufficient quantities of queen substance to go around and keep all of the workers sexually suppressed. Workers will raise new queens under three circumstances: when the population gets so large that half or more of the resident workers depart for a new home with the old queen; when a queen proves inadequate or defective (in which case the new queen supersedes the old); or when an existing queen meets a premature end.

It's ironic (and a reflection of deep-seated western European prejudices) that for literally thousands of years people believed that honey bee society, like most of western European societies, was governed not by a queen but by a king. It wasn't until 1637 that Jan Swammerdam, by careful dissection, clearly and incontrovertibly established that the royal bee was indeed a queen and not a king.

The honey bee exemplifies social behavior at its finest. Bees have an extraordinarily effective antipredator tactic—they sting. The stingers, modified ovipositors that are no longer needed by workers for egg-laying, are barbed such that when they penetrate the skin of a vertebrate intruder (like a human), they cannot be withdrawn. The entire sting apparatus and venom sac, then, are pulled out of the bee's abdomen and left behind, leaving the worker minus major parts of her nether region and bound to die. The stinger continues to inject venom into the enemy even in the absence of the rest of the bee. This sort of self-sacrifice for the common good is a premier example of altruism. To make matters worse for the predators, at the same time the bee leaves her stinger behind, she releases both an alarm and recruitment pheromone. Thus, as she dies, a honey bee sends out a call for reinforcements and many more bees arrive at the scene to dispatch the predator. Bees often sting repeatedly at the same spot, where the alarm pheromone concentration is greatest. This phenomenon was remarked upon by Charles Butler in 1609, in his book about bees, *The Feminine Monarchie*:

> When you are stung, or any in the company, yea though a Bee have strike but your clothes, vou were best be packing as fast as you can for the other Bees smelling the ranke favour of the poison cast out

with the sting will come about you as thick as haile; so that fitly and lively did he expresse the multitude and fierceness of his enemies that said They came about me like Bees.

The alarm substance is isoamyl acetate, a seven-carbon molecule that evaporates quickly and can be detected over a considerable distance by the bees. Beekeepers, who must constantly invade and disturb a hive for business purposes, avoid the alarm pheromone problem by using a smoker, a bellowslike contraption in which paper, straw, or other combustible material is burned in order to generate clouds of smoke. When the smoke is pumped into a hive, the bees cannot detect alarm pheromone. Moreover, smoke has a narcotic effect on bees and, far from panicking, they gorge themselves on honey the whole time the smoker is in operation (one theory advanced to account for this peculiar behavior is that smoky conditions in nature may indicate a forest fire, in which event bees move out and carry their honey stores with them to their new home).

Enhanced reproductive success is an obvious outcome of bee cooperation. With all the workers cooperating, a queen can raise literally thousands of offspring. Exactly what's in it for the workers, who forego sex and reproduction to aid their royal sister, has been subject to speculation for years. Bees present a special challenge to haplodiploidy as an explanation for eusocial behavior because queen bees mate with many drones on their nuptial flight—so it's unlikely that any two workers are going to share the same father and thus share 75% of their genetic material.

House hunting and home maintenance are cooperative projects in a honey bee colony. Scouts venture out in the spring, just before the colony produces a new queen. When the scouts find a spot, they come back to the swarm and perform a dance, which indicates the location of the potential homesite, whereupon the old queen joins an entourage out to the new spot (this process of colony division is called swarming, Fig. 4.2). What honey bees—originally cavity nesters partial to hollowed-out trees and the like before humans provided them with hive boxes—look for in a homesite is a bit of a mystery but a few things seem to be especially important: the size of the cavity, the height off the ground, and proper protection from the elements, such as rain, wind, or dust (as opposed to proximity to good schools or convenient shopping, criteria that increase the value of real estate to human house hunters).

The limiting factor in scouting for a new home is the queen, who is built for egg-laying and not for flying; she often has trouble making it all the way to the new site. Once there, bees do a lot to

Figure 4.2
Capturing a wild swarm
in Israel (G. Robinson).

make things comfortable for themselves. They can both heat and air-condition their homes. In cold weather, when temperatures drop, bees form a tight cluster; the colder the temperature, the tighter the cluster. The workers use stored honey as fuel and shiver to generate metabolic heat. The bees on the outside of the cluster, with bodies amply endowed with hair, insulate the cluster while the innermost bees generate heat. Individuals in a cluster continually rotate position and alternate as heat producers and heat retainers. In winter, the temperature in a properly managed hive never falls below 63°F, or 17°C, even when the outside temperature drops to −28°C. In summer, when there's a danger of overheating in the hive, worker bees bring in drops of water and fan their wings to generate a breeze and take advantage of evaporative cooling for air-conditioning.

Social living facilitates food finding for *Apis mellifera* in a big way. Honey bees live on food resources that are not terribly abundant and that changes in availability over the season—nectar and pollen. Sociality makes it possible for them to thrive on such a diffuse food source while maintaining large populations year-round (Fig. 4.3). Honey bees have a very sophisticated means of communication to convey information to fellow foragers as to the location of an ample supply of nectar or pollen—they dance. The dance language of the honey bee allows the colony to direct large numbers of foragers to a temporarily rich food resource.

There are basically three types of dances that foraging honey bees use to communicate information to hivemates. The round dance is used when nectar sources are about 10 yards away or closer. Odors clinging to the dancing bee's body appear to be important in allowing other foragers to go out and pinpoint the location of nearby nectar sources. A C-shaped, or sickle, dance indicates that nectar or pollen sources are between 10 and 100 yards away. The waggle dance is conducted when a nectar source is more than 100 yards distant. The waggle dance conveys precise information about food location in two ways. Basically, the waggle dance is a figure eight in shape. The number of times a dance circuit is completed on the comb along with the number of times a bee waggles its abdomen while on the straightaway part of the dance informs fellow bees as to the distance of the food source (a rough simulation of the amount of energy needed to reach the source). The angle in the hive at which the dance is performed, relative to vertical (or to the pull of gravity), corresponds to the angle between the sun and the food source. If a dance is performed 15 degrees east of vertical, then the nectar source is 15 degrees east of the sun. Since bees can see polarized light, they can gain information about the sun's location when foraging even under a fairly heavy cloud cover.

Figure 4.3
Honey bee worker out foraging (J. G. Sternburg).

When bees visit flowers, they often get covered with pollen grains, which adhere to the branched hairs all over their bodies. The grains are combed out and packed onto a flattened segment of their hind legs called the corbicula, or pollen basket. Pollen constitutes a bee's only source of protein. Relatively speaking, enormous amounts of pollen are needed to maintain a hive. Approximately 300 clover flowers, for example, are needed to provide enough pollen for a single bee-load, and ten bee-loads are needed to raise a bee from egg to adulthood.

Bees use their tongue (a lapping-sucking organ consisting of highly modified mouthparts) for taking up sugar-rich nectar from flowers. They swallow the nectar, using some for their own nourishment and shunting the rest into a sacklike "honey stomach." Back at the hive, nectar can be regurgitated directly to bees in residence. Bees on the receiving end can then begin the process of honey manufacturing. Most nectars are in the neighborhood of 90% water and honeys are only 13 to 20% water. In order to concentrate the nectar, worker bees repeatedly regurgitate droplets onto their tongues 100 to 200 times to aerate it and dry it out. To facilitate the curing process, other workers circulate air throughout the hive by fanning their wings (up to 26,400 times a minute); this air movement accelerates the evaporative process. Other changes are chemical in nature. In the bee's stomach are enzymes that break down the complex nectar disaccharides and polysaccharides into simple sugars. In particular, the enzyme invertase converts the disaccharide sucrose into its component monosaccharides, fructose and glucose. The incipient honey is then placed into wax cells, where it continues to lose water until it reaches its final concentration. Once closed with a wax cap, the cell holds the honey until it is required by the bees.

Bees in business

To make a prairie it takes a clover and one bee.
—EMILY DICKINSON, No. 1755

BEES GO TO ENORMOUS trouble to make honey—and for good reason. Honey is the principal source of nourishment for both immature and mature bees. To produce 1 pound of wax, for example, workers must consume about 8.5 pounds of honey. Manufacturing a kilogram (a little over 2 pounds) of honey requires on average 130,000 loads of nectar—about 10 million visits to individual flowers. If the flowers are within a mile of the hive, the total distance flown to visit these flowers will amount to something on the order of 400,000 kilometers—roughly ten times around the earth (Ioyrish 1974).

Small wonder that bees have an image of being busy. In fact, they appear often as heraldic symbols of industry on the coat of arms or corporate seal of many a business enterprise or town. At the same time, however, also appearing on these coats of arms are beehives, artificial structures designed by humans and symbols of the oppression of bees by humans intent on ruthlessly exploiting the enormous effort expended by bees just trying to keep their families happy.

Why do humans exploit bees? It's clear that humans have had a taste for honey for literally millenia. There are cave paintings dating back to about 6000 B.C. that clearly depict humans robbing bee hives of honey. Stealing honey from bee hives is not without risks; bees are notorious for their short temper and for their ability to defend themselves against intruders. Why risk life and limb for what is essentially regurgitated bee spit? In fact, honey is truly a remarkable substance. Among its most appealing attributes is that it is sweet; in nature, such intensely sweet substances are few and far between. One reason that honey is so sweet is that, as a result of the action of the enzyme invertase, honey is richer in the more intensely sweet monosaccharides glucose and fructose than in the more frequently encountered sucrose, better known as table sugar (fructose is about sixteen times sweeter than an equivalent amount of sucrose). Honey also contains other sugars, including maltose, melezitose, and dextrins. All of this carbohydrate not only makes for happier meals in terms of taste but it also provides a tremendous source of caloric energy. One kilogram of honey contains on average 3,150 to 3,350 calories; by contrast, an equivalent weight of apples supplies only 400 calories, of oranges 230 calories, and cucumbers 140 calories.

For those concerned with their general health, honey is also full of vitamins and minerals. Among the vitamins, the water-soluble B_1, B_2, B_6, pantothenic and nicotinic acids, and vitamin C, as well as the fat-soluble vitamins E, K, and A, are present in significant amounts. Minerals present include calcium, phosphorus, potassium, iron, copper, manganese, magnesium, and sulfur. The concentration of some of these minerals roughly parallels the concentration in human blood serum, making honey an easily metabolized source of essential nutrients.

In addition to calories, vitamins, and minerals, honey contains many other components that may promote good health. Its enzyme content is quite high, relative to other forms of food; these enzymes may aid digestion. In addition, the organic acid content is high—malic, citric, tartaric, and oxalic acids can all be found in honey. These, along with the enzymes catalase and peroxidase, may be responsible for the pronounced antibacterial properties of honey. There are also trace amounts of essential oils and other volatile components of floral nectars that add aroma and taste to honey. One final positive feature of honey is that bees generally provide large quantities of it in a single place. A single hive can contain over 300 pounds of honey on a good day.

Not everything about honey, however, is necessarily beneficial. Honey often contains floral substances that are narcotic or even downright toxic. Xenophon recounts the experience of 10,000 mercenary Greek soldiers returning from Persia in 400 B.C. after an unsuccessful attempt by Cyrus the Younger to usurp his brother's throne. Heptakometes, the leader of the opposition, poisoned the soldiers by setting out hives filled with honey made from rhododendron flowers. Rhododendron nectar contains a substance called andromedotoxin, a potent toxin. The soldiers, unable to resist such a treat, consumed a sufficient amount to induce violent vomiting and diarrhea. After resting for a few days, they were sufficiently recovered to continue their march home, much the worse for wear. Presumably, such substances in nectar deter inappropriate flower visitors from stealing nectar without providing proper pollination service. However, humans can get caught in the cross fire between flowers and their pollinators.

Despite the occasional toxic substance, honey is in general certainly worth the effort expended to swipe it from bees, a conclusion reached by humans long before recorded history. As the ancient cave paintings depict, the first stage of the interaction between humans and bees was simply that humans stole honey from bee hives. There are even contemporary cultures that still steal honey from bees. Among the Veddas of the forested areas of Sri Lanka there exist the vestiges of an almost pure "honey civilization." The Veddas are one of a number of southeastern Asian natives whose diet, culture, and economy rely heavily upon bees. Unlike Europeans, who must content themselves with only one species of bee, the Veddas have five from which to choose: *Apis dorsata*, on bare hillsides in high trees; *Apis indica*, in rock clefts and in hollow trees; the dwarf bee, *Apis florea* in rock ledges; and two other species utilized by relatively few Veddas. The collecting process is rather complicated, owing to the remote location of the hives. Long ladders, bamboo canes, and smoking torches are all used accompanied by various sorts of incantations. For half the year, the Veddas live almost exclusively on honey and, since it's a rather dangerous way of making a living, a whole host of spirits are invoked to oversee honey hunts. The Gurungs of the southern Himalayan slopes in west central Nepal are also honey hunters, specializing in obtaining honey, wax, and grubs (to eat) from *Apis laboriosa,* the giant bee of Asia.

Beekeeping—the semidomestication of bees—began very early in human history. There are tomb and temple illustrations from ancient Egypt, dating between 2400 and 600 B.C., depicting beekeepers. In the region of Abusir, there are bas-relief sculptures of beekeeping; the one in the sun temple of Neuserre, built in the fifth dynasty of the Old Kingdom at Abu Ghorab (not too far from Cairo), shows honey handling and the sealing of filled vessels. In tombs near Luxor in the Valley of the Nobles, wall paintings from 1450 B.C. show that horizontal hives made of unbaked clay or mud were stacked one on top of another. In one of these paintings, the use of a smoker to calm the bees is depicted. Hieroglyphic annual records from 3,000 years ago show that the price of honey was set at the equivalent of about five cents a quart. One of the uses ancient Egyptians had for honey was as an embalming fluid (due no doubt to its antibacterial properties). The body of Alexander the Great, who died during a foreign campaign, was transported back to Macedonia immersed in a golden coffin filled with honey. Egyptians also probably invented mead, a wine produced by fermenting honey (contamination with water leaves honey susceptible to microbial fermentation). Mead has the distinction of being the oldest form of alcoholic beverage and antedates grape-based wine by many years.

In Greece, red pottery hives have been excavated in archeological investigations in at least five sites. These sites range in age from 400 B.C. to Roman times. These hives are thimble-shaped and etched with ridges internally to provide bees with a rough surface on which to attach honeycomb. They were laid on their side to allow bees access to the interior. Hippocrates, the "Father of Medicine," recommended honey in combination with ass's meat and milk as an aphrodisiac and Homer mentioned honey as an important ingredient of pottage. Greek athletes consumed honey before Olympic events.

Romans also kept bees; Virgil's *Georgics* contains a complete guide to beekeeping in verse. Romans used at least nine types of hives: log hives, cork bark, wooden boards, woven wicker, fennel (called *ferula*), dung (although these were advised against because they were prone to catch fire), earthenware, brick, and a transparent substance that was possibly some kind of horn, among them. Romans considered honey one of the privileges of the upper classes. Centenarian Pollio Rumilius confided to Julius Caesar that his good health and sobriety was due to *interius melle exterius olio* ("honey on the inside, oil on the outside") and Nero's wife Poppea used honey and ass's milk for a facial mask. Greeks and Romans alike extolled the medicinal value of honey, especially as a vision restorative, general antidote for poisons, and emollient for ulcerated or inflamed wounds. Roman soldiers kept honey in their packs for the treatment of wounds obtained in battle. As did the Egyptians, the Romans fermented honey with water to produce a honey wine they called hydromel.

When Romans invaded ancient Britain, they found the locals busy in the enterprise of beekeeping, usually in hives of woven osier. The ancient Phoenicians in fact referred to Britain as the Isle of Honey, due to the prominence of honey in the diet. Druid bards, too, referred to Britain as the "Honey Isle of Beli," Beli being the British name for Bile, a Celtic hero. Honey hunting, as opposed to beekeeping, was also practiced (by the less enterprising) and is specifically mentioned in the Domesday Book.

By A.D. 400 or 500, a major innovation in beekeeping had taken place—the upright hive. The prototype upright hives were simple hollow logs, used in Russia, Poland, and Germany. Skeps, woven baskets used with the open end down, were introduced in the first century A.D. These are frequently depicted in medieval manuscripts and in early printed books. Coiled straw proved superior to wicker due to its weathertight properties, and examples have been found in England dating back to the twelfth cenutury, although the style is thought to have originated earlier with Germanic tribes living west of the Elbe. Skep hives weren't exactly managed, since in order to extract the honey the bees often had to be destroyed, either by immersing the entire operation in water or by poisoning the bees with sulfur fumes. More talented skep operators could remove honey without killing the bees by a process called driving. The skep was turned upside down, and an empty skep placed above it; bees were driven out by thumping on the sides with the hands. Bees are more reluctant to leave their hive if there are numerous larvae present in the comb, so the success of driving depends on the time of year during which it is attempted. Combs full of honey were then cut from the sides of the skep in chunks. Often skeps were kept in boles, or recessed shelters, to protect them from the weather. Protection was also afforded by wooden benches or stone stands, and by clooming, or covering the whole affair with cow dung.

One of the greatest innovations of beekeeping was the movable frame hive, in which a comb could be removed without destroying the rest of the hive. The first reference to movable frames was in a book published by Sir George Wheeler in London in 1682 titled, "A Journey into Greece"; the hive was mentioned in a description of the beekeeping operation of a monastery in Attica. The advantages of movable frames are numerous—movable frames make easier the tasks of removing honey, cleaning the hives, and feeding bees. Cabinet makers in Europe began to improve on hive design by experimenting with tiered hives equipped with separate sections to house the honey chamber, but the ultimate achievement in movable frame hives actually took place in Philadelphia, Pennsylvania, in 1851.

The Reverend L.L. Langstroth can rightly be called the father of modern beekeeping. Langstroth was among the first to take into consideration the concept of bee-space in designing a hive. Bee-space is the 1/4 to 3/8-inch clearance that bees permit themselves in order to walk around in their hive. Any spaces less than this size are invariably plugged up or sealed with propolis, a resinous material collected by bees from trees; bees build comb in any spaces larger than 3/8 inch. Langstroth discovered that by allowing 3/8-inch clearance around a wooden frame suspended from a top bar in a box, bees would not be tempted to anchor the frame to the walls of the box; unencumbered, the frame, and the comb it contained, could thus be moved in and out with ease. Langstroth's hive therefore had the advantages of movable combs, frames around the comb, separate honey compartments, and tiered boxes that could be moved individually.

Throughout the Middle Ages, honey was virtually the only sweetener available to Europeans, rich or poor. However, in the early sixteenth century, Spanish and Portugese explorers and merchants introduced the cultivation of sugar cane to the New World tropics; with the swift spread of sugar cane cultivation, sugar extraction became a cost-effective commercial enterprise. Prior to that time, due to the expense involved in its production, sugar had been available only in apothecary shops as a prescription medication. For the next four centuries, honey faded from prominence as the principal sweetener of Western civilization. When the sugar supplies of the French were cut off from the colonies during the Napoleonic Wars, the emperor actively encouraged the cultivation of sugar beets, rather than support a return to honey production. Times are changing, however, and honey is enjoying a renaissance along with the natural foods movement. Today, annual U.S. honey production is over 100,000 metric tons. China, Mexico, Canada, Argentina, and Australia each produce over 10,000 metric tons (Fig. 4.4).

Honey, as everyone no doubt realizes, is stored in honeycombs, an assemblage of hexagonal cells that most bees call home. These cells are made of wax, another insect secretion of considerable economic value. Beeswax is the secretion of a series of glands located on the ventral side of the honey bee's abdomen. Liquid wax is forced out into scalelike sheets, removed with the legs, masticated, and shaped into the characteristic hexagonal cells. Workers in the wax brigade gorge on honey and then hang suspended in clusters, called festoons, for the energy-intensive operation (production of 1 pound of wax in a hive requires the consumption of about 5 to 10 pounds of honey).

Like many insect secretions, beeswax is a rather amorphous assemblage, consisting of various fatty acids and hydrocarbons. It melts at a fairly low temperature and is an excellent lubricant, yet,

Figure 4.4
Uncapping and extracting honey in the field in Yucatan, one of the world's most productive honey centers (G. Robinson).

since it will not melt at temperatures less than 150° F, it is relatively resistant to natural weathering processes. Between 6 and 7 million pounds are produced each year in the U.S., mostly for mildew prevention, waterproofing, adhesives, insulation, cosmetics, and candles. The Catholic church alone is said to use 3 million pounds annually for candles (whereas once only beeswax exclusively could be used for church candles, today church candles need contain only 25% beeswax). Although nowadays synthetic waxes such as paraffin have largely supplanted beeswax, it has historically been used in ointments, suppositories, cosmetics, candies, industrial lubricants, dental products (such as in wax impressions), adhesives, paint and varnish removers, shoe and furniture polishes—the list is virtually endless. Waxed thread is still used by shoemakers for its durability and by sailors for its resistance to weathering.

Wax has been important for centuries as an artistic medium. It's easily molded and shaped at room temperature, it mixes well with coloring agents, and, due to its low melting point, lends itself well to casting and molding. Egyptians modeled figures of their deities in wax and enclosed them in tombs; hieroglyphs also indicate that household objects were often made of wax. There is a biblical reference to the use of wax in bronze casting. Lost wax casting, or *cire perdue,* is at least 6,000 years old.

Greeks modeled votive offerings, children's dolls, and wax images with magical properties. These wax figures have come down through the centuries as both love and vengeance charms. Those who considered pin-sticking voodoo rituals strictly Caribbean will be surprised to learn that this sort of activity went on throughout history in European nations as well.

Romans also made extensive use of wax in ceremonies and daily life. At the end of the Saturnalia came the period of festivities called the Sigillaria, the time at which presents of wax figures and statuettes were exchanged among friends. This was probably the progenitor of medieval votive offerings of wax figures. Sigillarii were the fellows who made the statuettes. Wax death masks of past ancestors were Roman household treasures brought out on ceremonial occasions. Romans also made use of beeswax as sealing wax for documents. Wooden boards coated with wax provided them with erasable writing surfaces.

Wax in art hit its heyday during the Italian Renaissance. Great sculptors like Michelangelo and Giovanni da Bologna used wax to model figures to serve as prototypes of sculptures. In Florence, Andrea del Verrocchio (1435–1488) made his name by making lifesize

figures in wax complete with costumes, glass eyes, and wigs.
Madame Toussaud in 1823 capitalized on the technique and gained
fame with her celebrated wax museum. Toussaud began her career
of wax casting as a young girl. Adopted by her maternal uncle Dr.
Matthew Philip William Curtius, young Marie Grosholtz grew up
admiring his collection of wax and ceramic miniature portraits.
The sister of King Louis XVI, Princess Elizabeth, summoned her to
Versailles at the age of 18 to serve as art tutor and companion.
When the French Revolution broke out, Marie saved her own head
by casting wax death masks of guillotine victims. When her uncle
died in 1793, she inherited his collection of wax portraits, which
provided the basis for her now-famous museum.

It isn't only Western cultures that appreciate the aesthetic properties of wax. In Indonesia, textiles are handprinted by a technique
called batik, in which parts of a pattern on fabric not to be dyed are
covered with wax, which is then melted off. Over 30% of commercial batik comes from Indonesia; India, Malaysia, and Thailand
contribute most of the rest. It's still very much a cottage trade, with
people painstakingly applying the wax, melting it off, and applying
the next coat of color in sequence.

Royal jelly, propolis, and bee pollen are products of bee labor
that are becoming increasingly more valuable in today's markets.
All are consumed for putative health benefits in various guises. In
Asia, the economic value of royal jelly (a tonic that is purportedly
invigorating) now exceeds the value of honey, although there is little if any scientific evidence to support its reputation. Propolis, the
resinous material gathered by worker bees from tree exudates (particularly trees in the poplar family), is used in the hive to plug
cracks and holes; small mammals such as mice that get trapped and
die inside a beehive are also encased in propolis, which appears to
have antimicrobial properties. It is sold in health food stores as a
restorative, athough its actual health benefits have never been conclusively demonstrated.

Not only have the products of bees been of great value to humans, the bees themselves have provided useful services. For centuries, people all over the world have taken advantage (every cloud
has a silver lining) of the propensity of bees (and other stinging hymenopterans) to sting ferociously and painfully. There's even a Biblical injunction to do so: "And I will send hornets before thee, which
shall drive out the Hivite, the Canaanite, and the Hittite, from before thee" (Exodus 23:28). In medieval days, beehives were kept behind the walls of fortified cities. When a city was under siege by an

enemy, the beehives were hurled from the walls onto the invading army, which added to the discomfort of those already being fired upon.

One historic occasion on which beehives actually determined the outcome of a battle was the siege against King Henry I of France mounted by the Duke of Lorraine: cascades of beehives tossed in among the duke's troops panicked the horses and led to an ignominious and rapid retreat. In 1528, in Hohenstein, Germany, a local peasant mob assailed the house of Elend, the local clergyman, in order to pillage and steal; when words failed to dissuade the mob, the clergyman compelled his servants to throw beehives into the crowd and succeeded in dispersing it quickly. This technique has been used in the last century to break up picket lines. Beehives were also used aggressively during the Civil War in the United States and also by Germans in East Africa against the British in World War I. Although in World War II there is no record of the use of beehives in warfare, there were efforts made to "draft" bees into service as messengers; tiny capsules were to be strapped to their legs so that, once released, they could unobtrusively cross enemy lines with vital information.

Pollination

> Birds, butterflies, and flowers
> All make one band of paramours.
> —WILLIAM WORDSWORTH, *Green Linnet*

PROBABLY THE MOST important contribution made by insects to human health and well-being is one for which they get little credit: pollination. Plants, from which people get so much food (and, for strict vegetarians, all of their food), are generally firmly rooted in the ground and incapable of all but the most minute movements. Sexual reproduction is just as important for plants as it is for animals when it comes to generating genetic variation, but plants have a singular disadvantage compared to animals when it comes to sex: they can't just get up and find themselves a mate. For plants, sex is a matter of getting male gametes—pollen grains—in contact with female gametes. In plants, the pollen is produced in anthers, the plant equivalent of male genitalia. Anthers are borne on structures called stamens, which collectively make up the androecium. The female gametes are housed in the pistil, consisting of the stigma (the receptive surface), the style (a tube through which the pollen grain must grow to reach the waiting gamete), and the ovule (the structure that will house the embryo resulting from fertilization). Collectively, all of the ovules and their supporting structures make up the ovary.

For some plants, getting the appropriate body parts together is a simple matter in that they are hermaphrodites capable of fertilizing themselves; flowers contain both male and female organs and the pollen and stigma are compatible. For most flowering plants, however, moving pollen to stigma—the process of pollination—is an operational problem requiring outside assistance of one form or another. Some species depend on nonliving, or abiotic, means to move their pollen; grasses, for example, are wind-pollinated. Pollen grains are shed into the air and carried around randomly with the hope that a few get lucky and run into the right kind of stigma. Corn is pollinated by wind, which is one reason in the Midwest "hay" fever usually strikes when the corn sheds pollen and fills the air with hopeful gametes. Wind pollination is generally a risky sort of business with a lot of pollen grains effectively wasted by landing in inappropriate places (like the nose of a hay fever sufferer rather than a stigma of the same plant species).

Many other plants induce animals to carry around their pollen. The process of pollination has proved in most cases to be mutually beneficial to both plant and animal. Most often plants offer an incentive or reward to the pollinator for carrying out the appointed task; on their part, pollinators often possess morphological or behavioral traits that enhance the accuracy and efficiency of pollen delivery.

Flowers offer a variety of enticements to attract potential pollinators. The most direct reward is pollen itself. Pollen is high in protein (up to 30%) and fat (up to 10%). There is a catch, however, to eating pollen—it is encased in a tough indigestible protein coat called the exine. Well that it might be, since a plant that loses all of its pollen to the stomach of an insect visitor is not destined to be successful as a parent. Nectar, on the other hand, was made to be consumed. Nectar, a relatively new plant development in evolutionary terms, is restricted to flowering plants and appears to play no role in the life of the plant other than as a reward for pollinators. Primarily a dilute solution of sugar and water, fructose, glucose and sucrose are the primary constituents. Variable amounts of amino acids, lipids, vitamins (mostly B complex), organic acids, and other trace materials are present. Nectar is often concealed in the base of long tubular projections, accessible only to potential pollinators with appropriately long tongues. Probably of all floral rewards, nectar has the greatest appeal among both vertebrates and inverte-brates; many cheaters take advantage of the system, stealing nectar without moving any pollen around. While some thieves content themselves with simply stealing the nectar, nectar robbers not only steal nectar but destroy the flower in the process so that an appro-priate pollinator can never visit.

Since pollen and nectar, the principal rewards offered by flow-ers to visiting insects, aren't conspicuous to a potential pollinator, plants have developed accessory structures that serve as advertise-ments. Petals, which collectively make up the corolla, are often characteristically pigmented to attract species sensitive to particu-lar colors. Individual flowers are often grouped together into an inflorescence; this complete structure makes a far more conspicu-ous display than does a single tiny flower. The typical daisy "flower," for example, is actually an inflorescence, consisting of white ray flowers arranged around an assemblage of yellow disc flowers. Fra-grances also act to catch the attention of passing pollinators. Floral fragrances can be quite complex, containing hundreds of different odoriferous chemicals. Flowers also can come equipped with vari-ous sorts of anatomical features that act as trip wires, releasing pollen only when a visitor of the appropriate size, shape, and weight happens by.

Some flowers successfully "cheat the system" and trick insects into visiting them without providing a real reward. Orchids in the genus *Ophrys* look and smell like a female wasp or bee. Males of many of these wasp or bee species emerge a few days earlier than fe-males; desperate for action, these males notice the orchid flowers, land on them, and then frantically try to mate with them. In the process, they pick up the pollen package (pollinia) prepared by the orchid. Understandably disappointed, the male wasp or bee de-parts, and, like as not, repeats the whole process with another mimic flower, this time depositing his pollinia right where the or-chid needs it. The whole process has been called pseudocopulation by biologists; nobody really knows just how the male wasps feel about it.

Most insect pollinators are from orders that arose just about the time when flowering plants put in an appearance on earth, and flowering plants have been influencing insect evolution ever since. Many groups are generally associated with flowers of a particular type and structure, reflecting eons of mutual accommodation. Bee-tles, moths, butterflies, and flies are involved in pollinating a wide variety of plants, including many species of economic importance.

However, melittophily, or bee pollination, is in many respects the most specialized form of pollination and is easily, from the human perspective, the most important. Morphologically bees are well adapted for pollen transfer—hairy bumble bees can carry as many as 15,000 pollen grains and the number of ovules in bee-pollinated plants is accordingly high. Honey and bumble bees have a specialized structure on their hind leg called a pollen basket for collecting and transporting pollen. Bee-pollinated flowers tend to have irregular outlines, strong landing surfaces, bright vivid colors (including ultraviolet, a color that bees can see and that we humans cannot), and subtle odors.

Humans have known about the importance of pollination from the point of view of raising crops for a long time, at least 3,500 years. A relief found at Nimrud in Egypt shows two winged creatures holding male flowers above a female date palm, the first documented case of artificial pollination. In ancient times, plant sexes were distinguished and named somewhat haphazardly—individuals with "feminine" characteristics were deemed females, and individuals that were not overtly female were assumed to be male. The concept of plant sexuality (in the sense that seeds are the result of sexual congress and fertilization) wasn't discovered until the seventeenth century. The first suggestions came in 1676, in a paper read to the Royal Society of London by the British botanist Nehemiah Grew. Grew argued that "the attire [stamens] doth serve, as the male, for the generation of the seed."

After extensive meticulous experimentation, Rudolph Jakob Camerarius, professor of physic at Tubigen, Germany, published *Epistola de Sexo Plantarum* in 1694—a thesis in which he demonstrated that two different floral parts, stamen and pistil, were needed to produce ripe seed and that in all probability the two parts represented the plant's sexual organs. Curiously, the phenomenon of hermaphroditism, the presence of functional male and female parts in the same individual, was recognized in animals, in which it is extremely rare, well before it was identified in plants, in which it is commonplace. However, Jan Swammerdam's contemporaneous description of hermaphroditic snails provided Camerarius with an analogy to draw upon in his investigations of plants. Although Camerarius's work was not immediately accepted everywhere, the great systematist Carolus Linnaeus picked up on this idea and constructed a classification of flowering plants in 1752, *Systema Naturae,* based on their sexual organs.

The importance of insects in cross-pollination, or the transfer of pollen from one individual to another, was not recognized until almost a century after plant reproductive organs had been identified. Insect visitation of flowers had not gone unnoticed over the years: Aristotle in his *Historia Animalium* had described in detail the tendency of bees to visit the same type of flowers consecutively. He also described in great detail (*Hist. Anim. V,* 32:26) the process of caprification: the movement of pollen-covered female fig wasps from wild caprifigs in which they developed, to other incipient fig fruits.

In 1751, Grew rightly recognized that pollen and other floral products served as insect food:

> The Attire . . . [is] not only Ornament and Distinction to us, but also Food for a vast number of little Animals, who have their peculiar provisions stored up in these Attires of Flowers; each Flower becoming their Lodging and Dining-Room, both in one.

Joseph Gottlieb Kolreuter, professor of natural history at the University of Karlsruhe, was perhaps the first to demonstrate incontrovertibly that insect visitation was a prerequisite to pollination and fruit production in a number of fruits, vegetables, and ornamental flowers. "In flowers in which pollination is not by immediate contact, *insects* are the usual agents, and consequently they alone bring about fertilization. It is probable that they render this important service if not to the majority of plants, at least to many, for all the flowers which we discuss have in them something agreeable to insects." His careful and precise observations were published in a book titled *Vorläufige Nachricht von einigen das Geschlecht der Pflanzen betreffenden Versuchen und Beobachten, Preliminary Information on Several Experiments and Observations that Pertain to the Sex of Plants* (1761).

Not long after the publication of Kolreuter's landmark opus, Christian Konrad Sprengel began his scientific observations of the interactions between flowers and insects. Sprengel was rector of a Lutheran school at Spandau and was a botanist by avocation only. In 1793, the results of his years of research were published in the book *Das entdeckte Geheimniss der Natur im Bau und in der Befruchtung der Blumen* (*The Secret of Nature Revealed in the Structure and Fertilisation of Flowers*). In this work, he postulated functions for characteristic floral structures in the context of insect pollination. The son of a clergyman, his religious background was in one sense the source of his botanical insights: "Convinced that the wise Creator of nature has brought forth not even a single hair without some particular design, I considered what purpose these hairs might serve." Thus inspired, he correctly identified the function of nectar guides (pigment patterns directing pollinators to nectar stores), among other plant features. So fascinated was he with the diversity of plant structure and function that he neglected his duties at the school and was forced to resign. He eked out a living conducting field trips and giving lectures and language lessons; at the going rates of two to three groschen per hour, he remained poor for the rest of his life and his remarkable insight went unrecognized for half a century.

Charles Darwin did more for publicizing the importance of insects as pollinators of plants than anyone else in the nineteenth century. Beginning in 1858, with the publication of his paper "On the agency of bees in the fertilization of papilionaceous flowers" in *Gardeners' Chronicle,* Darwin published literally hundreds of pages filled with insightful observations on pollination. Among his publications were three successful books, *The Various Contrivances by which Orchids are Fertilized* (1862), *The Effects of Cross and Self Fertilization in the Vegetable Kingdom* (1876) and *The Different Forms of Flowers on Plants of the Same Species* (1877). His notoriety as the author of the famed *Origin of Species* and his highly readable style ensured that these later works found a wide and receptive audience.

Thus, less than two hundred years ago, the importance of insects as pollinators of plants was virtually unknown to the world. This basic biological fact, obvious today, was discovered long after such seemingly more mysterious natural phenomena as the circulation of blood. Yet insect pollination of crop plants has ramifications that affect not just human nutrition but even the geopolitical climate. Darwin, for example, pointed out that the ascendancy of the British Empire was determined by its spinsters: spinsters kept cats, which kept down populations of mice, which were therefore unable to destroy the nests of bumble bees, which could then pollinate red clover, which could be used to feed cattle, which provided beef to the armed forces which defended and maintained the Empire. Almost ten billion dollars' worth of crops are pollinated by bees alone every year in the United States. These include plants the fruits of which result directly from the action of pollinators—among these are almonds, apples, avocadoes, blueberries, cherries, cranberries, cucumbers, citrus fruits, kiwifruit, melons, nectarines, peaches, pumpkins, squash, and strawberries. Many crop plants are grown from seed that is the result of bee pollination; these include asparagus, broccoli, carrots, cauliflower, celery, and onions. Some plants, while not requiring bee pollination, nonetheless benefit from visits by bees, which increase yields, uniformity of a crop, and fruit quality. Such plants include grapes, peanuts, beets, olives, and soybeans. Finally, bees pollinate plants that are used as forage or sources of hay for cattle and other livestock; alfalfa and clover, bee-pollinated plants, thus contribute to the production of dairy and meat products.

Of all the species used for commercial pollination, the western honey bee *Apis mellifera* is unquestionably the most valuable. One of its most important attributes is its high flower constancy—its tendency to visit flowers of the same species on a foraging trip (thereby insuring cross-fertilization). On average, the pollen load of a honey bee is over 90% pure (that is, 90% of the pollen grains are from a single species of plant). With 20,000 to 30,000 workers, honey bee colonies contain large numbers of foragers and can move large quantities of pollen (20 kilograms of pollen are required to maintain a colony through a single year). Amply endowed with branched hairs, honey bees can carry from 8 to 29 milligrams of pollen on a single trip. Relatively large with respect to the flower-visiting insect fauna, they have long tongues that can reach inaccessible nectar and are big enough to come into contact with pollen in a large variety of differently configured flowers. They range far and wide—in hard times, foragers will fly over 11 kilometers away from the hive in search of food—and they willingly visit an enormous diversity of species (they are referred to by some bee biologists as "pollen pigs"). Perhaps their most desirable attribute is their manageability; semidomesticated, they can be transported to where they are needed and in this way be manipulated to visit preferentially crops in need of pollination services.

The migratory beekeeping industry is an important contributor to the production of many crops, particularly in the western United States. Migratory beekeepers in California move their bees from orange groves in spring to sage fields and then to bean fields as flowers become available. Operators in states as diverse as Texas, Idaho, and Montana make a practice of moving bees to California when local agricultural production stops due to winter weather and in this way can keep their bees foraging year-round.

Bumble bees, *Bombus* species, are also economically important pollinators. Unlike honey bee colonies, bumble bee colonies are annual rather than perennial; only the fertilized queens successfully live through the winter. In spring, the queen begins to forage for nectar and pollen; when her ovaries mature, she constructs a nest, usually taking over a bird or rodent nest or burrow. She forms a pollen mass in the nest and tops it with a wax cup; eggs and pollen are then sealed in. When larvae hatch, they consume the provisions and then are progressively fed by the queen as they grow. The queen also continues to construct wax cells and deposit eggs. After about a week to ten days, the larvae spin cocoons and pupate. Within three or four days of eclosing (or emerging) as adults, they are out foraging for nectar and pollen to provide to sibling grubs in the nest.

A large bumble bee colony consists of only 150 to 200 members, so there are far fewer foragers to work a field than there are from a honey bee colony. But what they lack in number, they more than make up for in size. The large size of bumble bee foragers means they can get to hidden pollen and nectar stores and they can set off trip mechanisms in such important forage crops as red clover and alfalfa (some bumble bees, with short tongues, are "robbers," however, who prefer to cut the flowers at the base of the corolla and steal nectar without tripping the pollination mechanism). Although they are less than ideal as general pollinators due to their low flower constancy (pollen loads are often less than 50% pure), their small numbers, and their unpredictability from season to season (due to their annual habit), their contributions as legume pollinators are inestimable. For example, the livestock industry in New Zealand owes its existence in part to the fact that in 1885–1886 three species of bumble bees were introduced from Britain to help establish pastures of red clover, *Trifolium pratense.*

To a lesser extent, solitary bees are also important in pollinating forage plants. The alfalfa leafcutter bee *Megachile rotundata* was imported from eastern Europe or Asia into North America in 1930. Leafcutter bees live in burrows, hollow stems, nail holes, or plastic tubes. Adults emerge in late May; mated females make cells in tube-like burrows, lining and then capping off each cell with leaf clippings held together with saliva. An egg in each cell is provisioned with a honey-pollen mass. Since a single female can produce only a few dozen offspring and rarely forages farther than 100 meters from her nest to find pollen and nectar, approximately 10,000 bees would be needed to pollinate 2 hectares of alfalfa. Nonetheless, these bees have proved useful and have been introduced around the world, as far south as Chile, for pollination of alfalfa. Other solitary bees that have commercial potential include *Nomia melanderi*, the alkali bee, and *Osmia seclusa,* an Asian species, introduced for pollination of cole crops and alfalfa, respectively, in the U.S. *Osmia cornifrons,* the horn-faced bee, has been imported from Japan to improve U.S. apple production. This bee can pollinate twenty-five apple flowers for every flower pollinated by the honey bee—and its sting is far less painful than that of the honey bee.

In the interest of equal time, it must be stated that nonsocial insects can also be pollinators of economic importance. Social species dominate the pollination business for good reason: because of their population size and communication skills, they can mobilize large numbers of individuals to gather pollen quickly and efficiently, desirable attributes in a pollinator-for-hire from the human point of view. Yet, due to anatomical or structural constraints, some economically important plants have their own unique and irreplaceable pollinators. Midges, for example, normally thought of simply as annoying pests, happen to play a vital role in the production of chocolate.

Chocolate comes from the shiny red seeds or "beans" of the cacao tree, *Theobroma cacao.* The tree is native to tropical America, where it has been cultivated for 4,000 years. The beans are produced inside large green elliptical pods that form on the main trunk and limbs of the tree; the pods are collected and split open and the beans, covered with sticky pulp, are removed. After the pulp is extracted, the beans are sun-dried and fermented to allow the characteristic flavor and color to develop. The dried beans are then roasted, ground, and further processed into chocolate products.

Cacao trees produce flowers after the tropical rainy season comes to an end. The flowers emerge directly off the main trunk; each flower has five petals and each petal surrounds a cluster of anthers. At the base of each pocketlike petal is a series of tiny glands that secrete nectar. At the center of the flower is a rod-shaped stigma, in its turn surrounded by a barrier of five long filaments called staminodes. Few insects can negotiate this floral maze in order to pollinate the flower, and indeed most cacao flowers are never pollinated; according to one estimate, only 30 of every 10,000 flowers go on to produce seed pods. It wasn't until 1941 that the identity of the major pollinators of cacao were described. The principal pollinators of *T. cacao,* the individuals to whom the world owes chocolate, are tiny flies in the family Ceratopogonidae. None of the species measure over 1/6 inch in length. Their small size allows them to crawl down the staminodes to reach the nectar glands hidden in the petal pockets. In the process, they get dusted by pollen; when they travel down the staminodes of a neighboring cacao plant, enough pollen from their bodies brushes against the stigma to fertilize the flowers.

Killer bees

ONE OF THE MOST familiar insects in the world, *Apis mellifera* currently has a cosmopolitan distribution; its ubiquity, however, is the result of its semidomestication by humans, who have transported bees with them all over the globe. The species is thought to be native to Europe, western Asia, and Africa. Although bee biologists are still debating the finer points of bee taxonomy, most agree that there are four other species in the genus—*A. cerana, A. dorsata, A. laboriosa,* and *A. florea*—all more or less Asian in distribution, which suggests an Asian origin for the genus as a whole. Throughout its natural range, *A. mellifera* has differentiated over time into geographic races, each uniquely adapted to a particular set of climatological and ecological conditions characteristic of its distribution. Although opinion varies, there are about two dozen clearly identifiable, genetically distinct races currently recognized.

North America (in fact, the entire New World) was virgin territory as far as *Apis* species go—no honey bees existed on this continent before Europeans settled it. They didn't wait long, however, before bringing them over—the first honey bees to cross the Atlantic and join the settlement of the New World arrived in 1622 from England. These bees were *Apis mellifera mellifera,* the so-called "dark bee" of northern Europe. They quickly acclimated in North America, adapted as they were to the relatively cold winters of Scandinavia. They were carried along wherever pioneers settled; Thomas Jefferson remarked that they were known to the native American Indians as "white man's flies." The introduction was so successful, and honey bees became so widely established, that 150 years later a spirited debate was carried on in scientific and apicultural publications over whether the species was native to the New World in the first place.

The dark bee was, however, in many ways less than ideal from the point of view of beekeeping. It was, and still is, aggressive, and very sensitive to all sorts of diseases. Thus, there was considerable desire among American beekeepers to introduce other races of honey bees into the mix. Transport across the Atlantic proved a formidable barrier until the availability of regular steamship service across the Atlantic between Europe and the United States. Importation of colonies then became feasible and a flurry of activity took

Figure 4.5
Advertisements from the *American Bee Journal,* during the heyday of importantion of bee races from around the world.

place in the second half of the nineteenth century. There is much debate over who should get the credit for the most spectacularly successful introduction—of *Apis mellifera ligustica,* the Italian bee, the world's most popular bee. Captain von Baldenstein of Switzerland was first to tout its virtues in the literature after traveling to Italy and introducing the bees to his native land.

Jan Dzierzon in 1853 established a thriving business in Germany raising Italian bees and importing Italian queens to happy customers. The first attempt to bring Italian bees to America, in 1855, was unsuccessful; the bees arrived dead, apparently because a ship's officer swiped honey from the colony during the voyage. On September 22, 1859, however, fourteen living colonies arrived in New York from Germany; although many of these colonies did not survive the winter, six belonging to Phineas Mahan of Philadelphia did and he began to advertise Italian queens the following spring. The purity of his stocks, however, was questioned (by one S.B. Parsons, who had himself imported Italian bees from Italy that same spring). Nonetheless, the Italian bee quickly rose to prominence as the race of choice for North American beekeepers.

This didn't prevent breeders from importing other strains, however (Fig. 4.5). Most of the races proved unpopular, including the Egyptian bee *A. m. lamarckii,* the Cyprian bee *A. m. cypria,* and the Syrian bee *A. m. syriaca.* Importation of *A. m. carnica,* the Carniolan bee, by Charles Dadant, was more successful; carniolans are today the second-most popular race for beekeeping in the world, although they never really caught on in the United States (possibly due to their resemblance to the unpopular dark bee *Apis mellifera mellifera*).

Importation and experimentation with different races came to an abrupt end in 1922 when the U.S. government, prompted by reports of massive deaths of honey bees in Europe due to Isle of Wight disease resulting from infestation by the parasitic mite *Acaropsis woodi,* passed the Honey Bee Act, "A Bill to Regulate Foreign Commerce in the Importation into the United States of the Adult Honeybee (*Apis mellifera*)." This bill prohibited the importation of adult honey bees from other countries unless arranged through the U.S. Department of Agriculture. The act not only remains on the books, it was strengthened in 1976 by a clause prohibiting the importation of eggs and semen as well as adult bees.

The passage of a bill prohibiting the importation of bees, however, did not stop bees from entering the United States on their own power. The United States is presently facing such an introduction of *A. m. scutellata* with less than great enthusiasm. *A. m. scutellata* is popularly known as the "African bee," although there are at least ten other races found in Africa that could just as easily lay claim to that title. *A. m. scutellata* is in fact native to Africa; morphologically it is virtually indistinguishable from European bees but behaviorally it is quite different (Fig. 4.6). In Africa, bees were not fully domesticated to the extent that they were in Europe and honey is frequently stolen not only by humans but by a vast array of animal life. Millennia of unremitting attack have meant that the African bee today tends to be far more easily disturbed and as a result is more likely to abscond (or abandon a nest) or to sting than most European strains. This trait makes it unattractive to beekeepers. However, due to the relative frequency with which flowers become unavailable in their native habitat, these bees are also energetic nectar gatherers and phenomenal honey producers, foraging for longer periods during the day and gathering nectar even during inclement weather. Due to these habits, African bees are capable of producing more than twice the amount of honey than an Italian colony, especially when nectar sources are few and far between. This trait makes them very attractive to beekeepers.

In 1956, a noted Brazilian bee geneticist, Warwick Kerr, traveled to Africa with the idea of importing African bees into Brazil to incorporate their genetically based favorable attributes into a more tractable strain. Dr. Kerr, winner of the Dreyfus Prize for his work in bee genetics, was commissioned by the Brazilian Ministry of Agriculture to import bees. His first attempt was unsuccessful, foiled by an overzealous Portuguese customs inspector who sprayed his cargo with DDT. He returned in 1957 and collected seventy queens, of which forty-seven survived the trip. He set them up in his laboratory equipped with queen excluders to ensure that they wouldn't escape. However, at some point in 1957, according to Dr. Kerr, visitors to the apiary, concerned that the excluders were reducing the efficiency of pollen collection, removed the excluders from twenty-six hives, the residents of which proceeded to swarm and establish themselves in the wild.

The bees made remarkable progress, possibly aided by intentional releases by beekeepers, reaching Rio de Janeiro in 1963 and Recife on the northeast coast by 1965; by 1971, they crossed the Amazon. Virtually all Brazilian honey bees were "Africanized" by the end of 1971. By 1980, they reached Venezuela and Colombia. They continue to extend their range, both north and south, by 200 miles every year, and they established populations in the United States (in the vicinity of Brownsville, Texas) in 1990. The first death attributed to killer bees in the United States (Lino Lopez, an 82-year-old rancher with emphysema who was attacked after poking a burning burlap-wrapped stick at a colony that had nested inside the wall of an abandoned building), was reported on July 20, 1993, in Harlingen, Texas. At the time he died, approximately forty stingers remained in his body.

Figure 4.6
The African bee, *Apis mellifera scutellata* (right), is, despite its fearsome reputation, actually smaller than its more placid European cousin *Apis mellifera ligustica* (left).

African bees are a matter of genuine concern because of their aggressive nature. These bees are known in the popular press not as African bees but as "killer bees." The venom of *Apis mellifera scutellata* is no more toxic than is the venom of European honey bees, but *A. m. scutellata* is far more likely to sting. They react three times faster to the presence of an intruder in the hive and are far more sensitive to jarring or vibration. Once roused, workers are more likely to respond to alarm pheromone (at lower concentrations) by stinging. In one quantitative test of aggressiveness, in which a black flag is waved outside a hive, over ten times as many stings accumulated on the black flag outside an African colony than outside a European colony. African bees are also more likely to pursue nest invaders for greater distances once disturbed (reportedly for over half a mile).

The name "killer bee" was coined in Brazil, where by 1965 reports of unprecedented attacks not only in Saõ Paulo but throughout Brazil became numerous. The *abelhas assassinas* had purportedly caused as many as 150 deaths in Saõ Paulo by 1965. How many of these deaths were directly attributable to African bees, as opposed to the longtime resident Italian bees, is impossible to discern. The term "killer bee" may have actually been promoted by the Brazilian government in an attempt to discredit Kerr, an outspoken critic of the military regime. In 1972, the National Academy of Sciences published the results of a study on the situation in South America, bringing the situation into public awareness in the United States. *A. mellifera scutellata* more effectively attracted public notice by making a personal appearance in the United States that same year. On May 2, 1972, the SS *Argomester* docked in Richmond, California, with a load of Japanese cars. En route to Richmond, it had docked briefly in Los Angeles next to a molasses boat from South America. When the boat was unloaded, a nest of very aggressive bees was discovered in one of the holds. Identification of African bees is extremely difficult, since they are physically so similar to European bees; nonetheless, these bees were considered to be among the first uninvited arrivals of African bees on American soil.

The discovery of stowaway African bees on container ships prompted a series of studies designed to distinguish Africans from other races. One of the more successful early methods developed involved morphometrics, the use of dozens of measurements of body parts to characterize size and shape. Morphometric analysis confirmed that a bee nest found on June 6, 1985, in Lost Hills, California, in a kit fox den by a bulldozer operator was indeed occupied by *A. m. scutellata.* Government officials commenced an intensive search-and-destroy mission throughout the area, inspecting over 16,000 colonies in a 1,200 square kilometer area. In total, six African colonies were discovered and destroyed in the dragnet.

American concern over the introduction of African bees is heightened due to their proximity; already established throughout Mexico, the bees are expected to continue to colonize American soil throughout the 1990s. Of most immediate concern is the human health problems posed by the introduction of extremely aggressive honey bees. At least 40 people already die every year in the United States from bee and wasp stings (mostly due to allergic reaction); this number may rise to as many as 100 a year. The predisposition of African bees to nest in urban sites—including such cavities as termite nests, cracks and crevices in office buildings, and in hollow cement posts or walls—means that they are likely to come into contact with humans with far greater frequency than do feral, or wild, populations of European bees.

An additional, less-well-publicized, source of concern is the impact that African bees will have on the agricultural community. *A. m. scutellata* is extremely aggressive genetically; African colonies increase in size much faster than do their European equivalents and Italian drones can't seem to compete with African drones during nuptial flights and may even be less likely to inseminate their own queens. The tendency of managed African colonies to abscond (leave a nest site after disturbance), to swarm, and to rob honey supplies from neighboring colonies means that honey production may suffer at least initially from the Africanization of American bees. The introduction of African bees into Venezuela almost eliminated the beekeeping industry there; annual production of honey fell from over 580 metric tons per year to less than 100 metric tons per year. In addition, difficulties with managing hives may interfere with migratory beekeeping, which is essential to the pollination and production of many agricultural crops of great importance.

The economic impact of African bees on American agriculture is difficult to estimate. In Brazil, management practices have been modified to accommodate the aggressive nature of the bees and honey production has actually increased in recent years. Beekeepers wear more protective clothing, keep fewer hives in each apiary, and keep hives separate from each other and away from human dwellings. Potential rewards for changing management practices are great; whereas in the United States, European colonies average approximately 24 kilograms of honey per hive each season, African colonies can produce 55 to 90 kilograms of honey per hive in a season. In addition, African bees are extraordinarily resistant to certain bee diseases and parasites that are threatening U.S. beekeeping. *Varroa jacobsoni* is a mite that parasitizes bee pupae in their brood cells. This introduced pest is widely established in the United States and routinely weakens and kills colonies of European bees despite efforts to control it; however, infested African colonies conduct business as usual, so the Africanization of American bees may prove to be a boon in at least one way.

Control of African bees is a difficult proposition since they are so physiologically and behaviorally similar to European honey bees, the activities of which are responsible for some 80% of agricultural production in this country (over and above honey and wax production). Considerable debate is under way as to what strategy should be taken with respect to encroaching swarms. Optimists believe that cold weather conditions may either cause substantial mortality or else select for greater docility; pessimists advocate controlled breeding experiments and release of desirable European genetic stocks.

References

The joys of group living/Social structures

Anderson, M., 1984. The evolution of eusociality. *Ann. Rev. Ecol. Syst.* 15: 165–189.

Atkins, M.D., 1980. *Introduction to Insect Behavior.* New York: Macmillan.

Ito, Y., 1989. The evolutionary behavior of sterile soldiers in aphids. *Trends Ecol. Evol.* 4: 69–73.

Wilson, E.O., 1971. *The Insect Societies.* Cambridge: Harvard University Press.

Bees in business

Anonymous, 1976. Honey bees used in warfare. *Am. Bee J.* 116: 28–30.

Butler, Charles, 1623. *The Feminine Monarchie: or, The History of Bees.* London: John Haviland for Roger Jackson.

Crane, E., 1983. *The Archaeology of Beekeeping.* Ithaca: Cornell University Press.

Dutta, T.R., R. Ahmed, S.R. Abbas, and M.K. Vasudeva Rao, 1985. Plants used by Andaman aborigines in gathering rock-bee honey. *Econ. Bot.* 39: 130–138.

Hultgren, K., and R. Hultgren, 1985. Madame Toussaud's London Wax Museum. *Gleanings in Bee Culture* 113: 365–366.

Ioyrish, N., 1974. *Bees and People.* G.A. Kozlova, trans. Moscow: Mir.

Koster, J., 1980. The classical bee. *Am. Bee J.* 120: 645–647, 660.

Mellor, I., 1981. *Honey.* New York: Congdon and Lattes.

Naile, F., 1976. *America's Master of Bee Culture: The Life of L.L. Langstroth.* Ithaca: Cornell University Press.

Ransome, H.M., 1986. *The Sacred Bee in Ancient Times and Folklore.* Burrowbridge: Bee Books Old and New.

Reese, K.M., 1988. Honey intoxication zaps Turks and many others. *Chem. Eng. News* 66 (April 25): 48.

Reese, K.M., 1988. Honey story fouled up. *Chem. Eng. News* 66 (June 13): 76.

Root, A.I., 1972. *The ABC and XYZ of Bee Culture.* Medina, OH: Root.

Sanford, M.T., 1985. Bee pollen: boon or bust in the 80s. *Gleanings in Bee Culture* 113: 64–65.

Stadler, H. and D. Metzler, 1944. Let's draft our bees. *Travel* 82 (April): 28–34.

Valli, E. and D. Summers, 1987. *Honey Hunters of Nepal.* New York: Harry N. Abrams, Inc.

Pollination

Bodenheimer, F.R.S., 1958. *The History of Biology: An Introduction.* London, Dawson.

Faegri, K. and L. van der Pijl, 1978. *The Principles of Pollination Ecology.* Oxford: Pergamon Press.

Free, J.B., 1970. *Insect Pollination of Crops.* New York: Academic Press.

Kevan, P.G. and H.G. Baker, 1983. Insects as flower visitors and pollinators. *Ann. Rev. Entomol.* 28: 407–433.

Meeuse, B.J.D., 1961. *The Story of Pollination.* New York: Ronald Press.

Proctor, M. and P. Yeo, 1972. *The Pollination of Flowers.* New York: Taplinger.

Robinson, W.S., R. Nowodgrodski, and R.A. Morse, 1989. The value of honey bees as pollinators of U.S. crops. *Am. Bee J.* 477–487.

Singer, C., 1959. *A History of Biology to About the Year 1900.* London: Abelard-Schuman.

Killer bees

Camazine, S. and R. Morse, 1988. The Africanized honeybee. *Am. Scientist* 76: 465–471.

Landsburg, A., 1978. *The Insects Are Coming.* New York: Warner Books.

Potter, A., 1977. *Killer Bees.* New York: Grossett and Dunlap.

Rinderer, T., 1986. Africanized bees: the Africanization process and potential range in the United States. *Bull. Entomol. Soc. Am.* 32: 222–227.

Schumacher, M.J., J.O. Schmidt, and N.E. Egen, 1989. Lethality of 'killer' bee stings. *Nature* 337: 413.

Sheppard, W.S., 1989. A history of the introduction of honey bee races into the United States. *Am. Bee J.* 129: 664–667.

Smith, D.R., O.R. Taylor, and W.M. Brown, 1989. Neotropical and Africanised honey bees have African mitochondrial DNA. *Nature* 339: 213–215.

Taylor, O., 1985. African bees; potential impact in the United States. *Bull. Ent. Soc. Am.* 31: 14–24.

Winston, M.L., 1992. The biology and management of Africanised honey bees. *Ann. Rev. Entomol.* 37: 173–193.

Winston, M.L. 1992. *Killer Bees: The Africanized Bee in the Americas.* Cambridge: Harvard University Press.

Chapter 5

VEGETARIAN LIFESTYLES

Green meals

IF NUMBER IS any way to measure success, then the most successful insects on the planet are the ones that eat plants. Of about 800,000 known species of insects, over 350,000 are herbivores, or consumers of plant tissues. In fact, the number of insect species consuming green plants is slightly larger even than the number of species of green plants (at last count around 308,000). Insects aren't the only arthropods that have adopted a vegetarian lifestyle in a very big way; among the Arachnida, many species of mites are herbivorous as well.

One thing that has contributed to the diversification of herbivorous insects is that there are many ways to eat any particular plant. Leaf-chewers, like most grasshoppers and walkingsticks, simply tear off plant leaf tissue and swallow. Other insects with chewing mouthparts feed inside plant tissues. Leaf miners eat the layers of plant cells between the inner and outer layers of leaf cuticle. Some caterpillars, beetle grubs, and even maggots can make linear mines, which look like pale wavy lines on a leaf, or blotch mines, in which large leaf areas look white, since all the good green stuff has been eaten away. Many morphological features of leaf miners reflect their space limitations inside a leaf. Generally, they are very small and compressed or flattened. In these species, the mouthparts generally point forward instead of down, to give them a lower profile inside the leaf. Other herbivorous insects that lead a concealed existence bore into the stems of weeds or the trunks of trees or tunnel their way underground through the roots of plants. Concealed feeding means that miners and borers are relieved to some degree from predation by visually orienting enemies, such as birds.

Sucking insects generally feed on the vascular sap, or phloem, of plants, that moves from the leaves where photosynthesis takes place and which is therefore rich in carbohydrates, and on the xylem, which moves up from the roots carrying water and mineral nutrients from the soil. Sucking herbivorous insects are concentrated in the orders Hemiptera and Homoptera and include in their ranks (among many, many others) leaf bugs, stink bugs, chinch bugs, aphids, whiteflies, jumping plant lice, scale insects, treehoppers, leafhoppers, and cicadas. One notable accomplishment of these sucking insects (although a dubious one by human standards) is that they can manufacture two different types of saliva. One sort is injected into the plant and hardens to form a sheath in which the thin probing mouthparts can move around freely. Another kind of saliva, like more typical salivary secretions, breaks down starches and cell walls to make sucking easier.

Some insects can eat plant tissue by commandeering the plant's hormonal system in such a way that the plant is induced to produce bizarre and unusual growths, which provide the insect with a place to live and with nice nutritious tissue on which to feed. These feeding structures are called galls. Gallmaking insects can be found among the thrips, the aphids, beetles (notably weevils), moths, wasps, and flies; as well, there are some gallmaking noninsects, like mites (for that matter, galls can be induced by a whole host of other organisms, including viruses, bacteria, and nematodes). Although some plant species appear to tolerate galling insects with few ill effects, some gallmakers, like the Hessian fly, a serious pest of wheat, can greatly reduce yields or even outright kill the host plant. The scientific name of the Hessian fly—*Mayetiola destructor*—reflects its impact.

Some insects concentrate their plant feeding on the fruits or seeds of the plants. Seed feeders in particular can have a tremendous impact on plant populations because, by destroying a seed, a seed predator is actually destroying a whole potential plant (which would be quite a mouthful under other circumstances). Seeds are generally very nutritious, high in both fats and protein, because they contain just about everything a growing seedling will need. Seed feeders include in their ranks seed-sucking bugs as well as masticating caterpillars, beetle grubs, and even maggots and wasp grubs. Many seed predators can feed on seeds not only when they are growing on the plant but also when they fall to the ground, or when human beings harvest them and store them (or grind them up, make flour, bake cakes and cookies, and store *them*).

All plant-feeding insects require some basic equipment for adopting a plant diet. First, insects need sensory equipment for finding edible plant food. In some cases, this means finding anything green; polyphagous insects are capable of feeding on many different families of plants. Despite the fact that some of the world's most conspicuous plant pests are polyphagous (locusts and armyworms come to mind), about 90% of all herbivorous insects feed on three or fewer plant families; such species are said to be oligophagous. There are even some insects that can feed only on a single species of plant. The level of sophistication or complexity of the sensory systems employed for food-finding varies with the breadth of the diet. In general, herbivorous insects use olfactory receptors to detect volatile substances given off by the plants (such as essential oils). These receptors are often located on the antennae. Once on a plant surface, an insect can obtain additional information about its suitability as food by touching or tasting it. Insects may touch the plant with contact chemoreceptors located on the tarsi (legs) or even, in the case of females looking for a place to lay eggs, on the ovipositor. Finally, taste is the ultimate arbiter of acceptability. To taste plants, insects have mouthparts liberally equipped with gustatory chemoreceptors.

The next job of an insect herbivore is to consume the plant tissue, and, again, mouthparts are indispensible for this purpose. Insect mouthparts have undergone a remarkable number of changes to suit the different sorts of diets insects have adopted. For chewing plant tissue, the basic mouthpart plan has been modified only a little. These basic mouthparts can be clearly seen in the strictly vegetarian grasshoppers. The mandibles, the most anterior pair of segmented appendages associated with the head, do the bulk of the chewing. In most species, they are equipped with teeth or ridges to help with cutting, tearing, or shredding plant tissue.

For sucking plant fluids, the mouthparts are modified variously into devices that operate on the same principle as the drinking straw. In both Hemiptera and Homoptera, the mouthparts constitute a beak. The labium serves as a sheath for the remaining mouthparts, and the mandibles and maxillae are modified into piercing stylets. The two maxillae are grooved and fit together to form two channels, one for pumping out saliva and one for pumping in plant sap and the contents of broken plant cells. Lepidopterans consume nectar with sucking mouthparts built on a different plan. The proboscis of a butterfly is a long coiled tube made up of modified parts of the maxillae (the galeae). These are fused together to form a narrow channel. Like the Homoptera and Hemiptera, butterflies and moths have a set of muscles in their heads, the cibarial pump, which gives them the ability to suck up a fluid. Some plant-feeding thrips neither chew nor suck—rather, they rasp away at plant tissue. To do so, they have a cone-shaped set of mouthparts with three stylets: the left mandible (the right one reduced to only a vestige), and modified parts of the two maxillae (the laciniae).

Once food is ingested, it has to be processed, and insects have a very adaptable alimentary system that has been variously modified for different kinds of plant diets. The most generalized system is that found in the grasshoppers (Orthoptera). The alimentary canal is partitioned into three general regions: the foregut, the midgut, and the hindgut. Valves or sphincters can close off each region to control the rate of passage of food. The foregut usually consists of a pharynx (right inside the mouth), an esophagus, a crop, and the proventriculus. In the crop, in some species, food is mingled with secretions from the salivary glands, which lie underneath the front part of the alimentary canal and empty into the mouth. The proventriculus in some species (e.g., in the Orthoptera) is armed with teeth and serves two functions: regulating the passage of food into the midgut and breaking it up into smaller pieces.

The midgut is the site where most food processing takes place. Unlike the hindgut and foregut, it is not lined with cuticle, so nutrient absorption can take place. Midguts are sites of active enzyme secretion. Digestive enzymes that break down fats, carbohydrates, and proteins are secreted into the midgut, and detoxification enzymes, which break down the poisons that are often found in plant food, are located in the cells lining the midgut. The midgut is lined with a peritrophic membrane, which allows dissolved nutrients to pass through but protects midgut digestive cells from damage by food particles and prevents the influx of microbes that may cause infection. It also serves as sort of a wrapper for waste products and is excreted with feces.

In the hindgut, water is absorbed and waste products are concentrated and packaged for excretion. At the front end of the hindgut are the Malpighian tubules. These take up water, salt, and waste products from the blood (hemolymph) and move them on through to the hindgut, where the water is resorbed and feces are formed. The number of Malpighian tubules varies with the species; while scale insects have only two and aphids have none, some grasshoppers have about 250. Wastes containing nitrogen are generally excreted in the form of uric acid, which is relatively nontoxic and which can be excreted dry without polluting the insect's immediate vicinity. Using uric acid as a waste product allows terrestrial insects to conserve water since other forms of nitrogenous wastes, such as ammonia, must be flushed out of the body in large volumes of water in order to avoid toxicity. The feces of most leaf-chewing herbivorous insects are very distinctive; they generally reflect the shape of the gut as they are extruded, like the way cookies are formed in a cookie press. There are actually keys to insect feces (called "frass" by those in the know), which can be used to identify the species that made the deposit.

The alimentary canal of sap-sucking species has been considerably modified in comparison to the basic body plan. Sap feeders consume a very dilute source of nutrients, particularly nitrogen-containing protein. Although phloem is rich in carbohydrates, it contains only about 1% nitrogen; xylem, only 0.05% nitrogen, is even more dilute. As a consequence, sap-feeders have to pass enormous volumes of fluid through their bodies to extract enough protein to survive. They are therefore specialized to eliminate large volumes of water very rapidly, so they don't end up diluting their internal organs. In the Homoptera, a section of the midgut is enlarged into a large bladder-like organ that lies right next to the front end of the hindgut and the Malpighian tubules; the whole arrangement is called a filter chamber. With the midgut lying along the full length of the hindgut, water can pass directly from one section of the gut to the other, for faster elimination. The excess water, along with its load of excess carbohydrate, passes out the anus onto the plant surface, to form a sticky sugary residue called honeydew.

In terms of absolute amounts, plants are the most abundant sort of food on earth. However, much plant material (the vast bulk, it seems) goes uneaten. Now, insects aren't keen on giving up on food for the taking, so it stands to reason that there must be something wrong with plants as a sole source of nutrition. Actually, for any animal, there are many basic problems associated with living on an exclusively plant food diet—these problems are routinely faced by human vegetarians. One of the main problems with plant food is that it is relatively low in protein. At 0.05% nitrogen, xylem sap is barely nutritious enough to sustain life; heartwood and bark, at 0.3% nitrogen, aren't much better. Even relatively protein-rich foliage is rarely greater than 20% protein. Moreover, the protein in plant food is not of the right composition to suit the nutritional needs of most animals (that's why strict vegetarians have to eat a variety of different plants, so that the deficient proteins in each type of food balance or complement each other).

Water can also be in short supply for insect herbivores. Although many weedy species have relatively high water content, the leaves and seeds of woody species—and wood itself—can be critically low in water. Seeds and grains as well are not exactly juicy. On average, herbivorous insect bodies contain about 75% water, substantially higher than most of the plant food they consume, which means that many plant-feeding insects must take a drink of water every now and then (as many grasshoppers do) or they must carefully conserve the water that they take in with their food.

Plants are also loaded with indigestible substances. The glucose molecules that make up the carbohydrate cellulose, for example, are linked in such a way that the enzymes of most insects (and most other animals) can't break them apart (silverfish and a few of the higher termites are excepted). Also found in plants like grasses is the substance silica. Not only is this hard rocklike material (of which sand is made) indigestible, it also wears down even the relatively tough cuticular surfaces of insect mandibles that come in contact with it.

Plants tend to be relatively deficient in the vitamins and minerals that animals need to survive. Sodium, for example, is present at very low concentrations in plant tissue. All animals with nervous systems require significant amounts of dietary sodium to ensure proper nerve function. The low sodium content of plants is why conscientious farmers put up salt blocks or salt licks in their cow pastures for their cows. Some insects have "licked" the low sodium problem in a remarkable way. Butterflies of many species can often be seen gathering in so-called "puddle clubs" around mud puddles, bird droppings, and animal urine. It's believed that what attracts them to the puddles in the first place is the high level of sodium in these materials. Puddling in most species of Lepidoptera seems to be exclusively male, for reasons that are not totally understood; it may be that, when male butterflies mate and pass on a spermatophore, or sperm package, to females, they pack it full of sodium and thus require more sodium than do females to make it through their adult life. Since they can't get enough from nectar, the usual diet for a butterfly, they must go to alternative food sources, unattractive as they might appear at first glance.

One of the major obstacles that limits the activities of herbivorous insects is that all plants produce a number of chemical substances that appear to have no function other than to discourage insects and other would-be herbivores from consuming them. These secondary compounds (so named because they play no role in primary life processes of plants, such as respiration or photosynthesis) are extremely diverse; at least 100,000 different chemical structures have already been characterized from flowering plants. They can act in a variety of ways. Some are feeding deterrents that either prevent an insect from recognizing plant tissue as edible or render the plant tissue unpalatable. Other secondary compounds interfere with an insect's ability to extract nutrients from plant tissue and hinder growth and development. A tremendous number of plant chemicals are out-and-out poisons that can kill insects if consumed in sufficient quantity. The insecticidal nature of some plant chemicals has actually been exploited economically; the commercial insecticides nicotine and pyrethrin are both of botanical origin.

Despite all the formidable barriers to herbivory, though, just about every species of plant has its share of insect herbivores. Insects have a phenomenal capacity for overcoming the nutritional and chemical barriers that plants have erected to protect themselves. And once they overcome those barriers, they can eat a lot of plants. Even though individually insect herbivores are much smaller than, say, mammalian herbivores (like herds of buffalo), they still are capable of consuming amazing amounts of plant tissue. More plant material cycles through a population of grasshoppers, for example, than cycles through a population of meadow mice; the total energy flow through insect herbivores can be three to four times greater than the energy flow through mammalian herbivores. Insects can eat up to 75% (or more) of a given plant species in a particular spot during outbreak situations.

Not only do insects eat a lot of food, they also excrete a lot of waste, and they play an important role (out of proportion to their small size) in recycling nutrients in ecosystems. Under outbreak conditions, by defecating insect herbivores in forests account for 2 to 5% of the carbon, potassium, nitrogen, sodium, and calcium and over 8% of the phosphorus that eventually returns to the forest soil. Insect frass may thus represent a new source of plant fertilizer for the truly industrious entrepreneur.

Down on the farm

FOSSIL EVIDENCE SUGGESTS that insects first appeared on earth some 400 million years ago and it is likely that many of the early colonists of dry land were herbivorous. The angiosperms, or flowering plants, the dominant form of plant life on earth today, first appear in the fossil record about 100 million years ago, and humans, or *Homo sapiens,* only about 250,000 years ago. Humans, angiosperms, and herbivorous insects more or less peacefully cohabited the planet until about 10,000 years ago; at that time, the human perspective on the plant-eating habits of insects changed irrevocably. Concurrently on at least four continents, humans hit upon the idea that growing plants was a far more reliable way to obtain food than gathering plants. Thus, humans invented agriculture. Probably the first plants to be domesticated were the cereals—corn in the Americas, rice in southeast Asia, wheat and barley in the Near East, and sorghum in Africa. Subsequently, additional plants were domesticated—legumes such as beans, peanuts, and soybeans; fruits such as coconut and banana; and roots and tubers, including beets, potatoes, sugarcane, yams, and cassava.

Nobody knows the precise details that led to the invention of agriculture, but it's generally thought that it was the result of an anonymous act of carelessness. During the Paleolithic period, or Old Stone Age, humans made their living by collecting seeds or grains from wild plants whenever they were available—a sort of catch-as-catch-can existence necessarily nomadic because there was no single locality that produced enough food to support a human population year-round. It's likely that the whole world changed when somebody tossed some of the collected grain accidentally into a garbage pile. In the high-nutrient environment, the seeds germinated and some nameless genius saw the solution to the age-old solution of finding enough food to eat—don't go out and search for it but grow your own. The invention of stone tools around 3000 B.C., which greatly facilitated working the soil as well as gathering the harvest, marked the beginning of the New Stone Age, or Neolithic period, and agriculture became firmly established and a viable enterprise.

One of the consequences of cultivating plants is that, for the first time in human existence, people were able to generate a surplus of food—more than they could eat in a short period of time. This surplus could be stored in anticipation of the hard times ahead. Storage of food allowed people to abandon a nomadic lifestyle and to settle, in larger numbers, in one place. Surpluses of grain meant that domestic animals could be maintained year-round and a regular source of meat and dairy products improved people's diet (cholesterol notwithstanding). Soon towns and cities arose and civilization really began to take off. Art, science, culture, and eventually even university courses in entomology would develop, all as a result of someone tossing some grain in the garbage a long time ago.

In addition to affecting human lifestyles, the domestication and cultivation of plants greatly affected insect lifestyles. Selective breeding of plants, either deliberate or accidental, led to the establishment of agronomic traits that were highly desirable from a farmer's point of view. Dispersal mechanisms, for example, were bred out of plants, so seeds remained on the stem instead of falling to the ground and scattering. Growth and development became more synchronous, so harvesting activities could be concentrated into a relatively short time period. Edible seeds got larger and more numerous on each stem. Such changes were well-suited to insect herbivores as well, who, when feeding on cultivated plants, were faced (like farmers) with an abundant, predictable, highly localized food supply. In addition to altering the biology of the plants themselves, early agriculturalists also altered the ecology of the plants. Crop plants were irrigated and fertilized, so, from the insect perspective, water and nitrogen limitations on the nutritional suitability of plants were greatly reduced. Insect populations could increase to levels totally without precedent in the natural world—much to the chagrin of hungry people, who for the first time in about 250,000 years on the planet had cause to resent the herbivory of insects.

This resentment grew all the more acute during and after the second Agricultural Revolution, which began around the fifteenth century and is continuing today. With the beginning of the Renaissance, world exploration introduced peoples of all countries to new and different crop plants. Plants were exchanged and established in places far distant from their places of origin. Sugarcane, for example, was indigenous to southeast Asia and was introduced extensively to North, Central, and South America; potatoes, native to South America, became staples in northern Europe. Thus, as cultures acquired new foods and fiber sources, plants acquired new insect enemies. Developments in science also created new markets for hitherto economically unimportant plants—as sources of medicine and as sources of recreational drugs (including coffee, tea, and chocolate, as well as the more insidious tobacco and opium). Today, hundreds of plant species are producing products of economic importance in virtually every nation and as a consequence insect herbivory and its yield-reducing consequences have far greater economic importance than ever before.

Population growth (locust plagues and insect exorcisms)

> The locusts have no king, yet they go forth all of them by bands.
> —PROVERBS 30:27

IN ORDER TO UNDERSTAND the impact insects can have on plants, it's important to know something about the staggering ability that insects have to increase their numbers. Insects of the same species living in the same place at the same time constitute a population. Populations characteristically change in size over time; birth and immigration increase the size of a population while death and emigration decrease its size. The resulting average population size varies widely with species and locality. While there are some social species, like mound-building termites, that live in populations containing literally millions of individuals, there are, on the other hand, some populations of butterflies that contain only a few dozen individuals.

There are several aspects of insect reproduction that determine how fast populations can increase in size and how large populations can grow. One factor that can vary from population to population is how many offspring are produced each time an insect reproduces. A female German cockroach (*Blattella germanica*), for example, packs anywhere from thirty to forty-eight eggs into each ootheca, or egg capsule, that she produces. Cockroaches in an apartment with a careless tenant who leaves faucets dripping and crumbs all over the kitchen floor will be healthier and hardier and consequently can pack more eggs into an ootheca than the less fortunate cockroaches next door in an apartment occupied by a compulsively tidy tenant. Another important factor is how many times offspring are produced over the lifetime of an insect. German cockroaches produce on average about five oothecae during their reproductive life but more prolific individuals may produce twice that number. Finally (and not really obviously), the age when insects reach sexual maturity and begin to reproduce can have a big impact on the rate at which a population grows.

The age at first reproduction (or generation time, the time it takes to run through the generational cycle of egg, larva, pupa, adult, egg) doesn't seem at first to be so important in determining rates of population growth, but an example with more familiar vertebrates may make the concept more graspable. Mice reproduce when they're 18 days old, dogs and cats at 7 months or so, and humans not until they're at least 14 *years* old. Think how many litters a cat can have by the time it takes a person to even figure out what's going on. The actual record for individual cat reproduction is on the order of 420 kittens (produced by one Dusty of Texas between 1935 and 1952); in contrast, the record for humans was set in the eighteenth century by the first of the two wives of Theodore Vassilyev, who had sixty-nine children (including sixteen sets of twins, seven sets of triplets, and four sets of quadruplets). This is one record that not too many people are vying to break.

These mammalian records pale in comparison with insect potential for reproduction. Some fly species have a generation time of less than a week. Take *Drosophila melanogaster,* the humble fruit fly, for instance. Under ideal conditions, a generation can take a little less than two weeks, so up to twenty-five generations a year are possible. Every female can lay at least 100 eggs. So, even on the conservative side, starting with a single pair of fruit flies, assuming all offspring survive and all female offspring produce 100 eggs, in only one year there will be 10^{41} fruit flies. Packed tightly together at 1,000 per cubic inch, these flies would form a ball 96 million miles in diameter, a distance approximating that from the earth to the sun (Borror et al. 1976).

Entomologists love to make these sorts of calculations and in fact often get into disputes about them. Estimates on the number of offspring a single pair of house flies (*Musca domestica*) can produce in a year led to one such controversy. In 1906, L.O. Howard took a shot at estimating the potential number of offspring from a single house fly in one season in Washington, D.C. According to his calculations, if the fly began laying eggs in the beginning of the year, there would be 5,598,720,000 by September 15—enough, according to Howard, to cover the earth to a depth of 47 feet. These figures went unchallenged for over half a century until 1964, when Harold Oldroyd took exception and recalculated that "a layer of such thickness would cover only an area the size of Germany, but," as he observed, "that is still a lot of flies."

That we're not ankle-deep in house flies suggests that something out in nature is keeping populations in check. That something is the environment. Any given environment has only enough resources to support a limited number of individuals of one species indefinitely. This number is the carrying capacity of the environment. Population growth tends to increase until the carrying capacity of the environment is reached and then it either drops precipitously, levels off, or fluctuates cyclically around the carrying capacity.

There are two kinds of environmental factors that can cut down on population growth. One group of forces opposing unlimited growth acts independently of the size of the populations. Bad weather, for example, can wipe out everybody. A hurricane, tornado, or earthquake can bring about a huge toll in insect populations that is never reported on the nightly newscasts. Even less dramatic climatic vagaries can limit insect population growth. The temperature tolerance of insects varies with species. Generally there is an optimal range over which insects function most efficiently. Since insects are cold-blooded (heterothermic); their body temperatures are often dictated by the environment and sudden changes in temperature can prove deadly. The majority of insects cannot function in temperatures below freezing (32°F, 0°C). There are remarkable exceptions, of course. Grylloblattids (or rock crawlers), which live high atop mountains in the western U.S. and in eastern Asia, are most active at temperatures just around freezing. They are, however, not very tolerant of warm temperatures—the heat of a human hand can kill them. In contrast, many species adapted to desert or thermal spring environments are extraordinarily tolerant of high temperatures; there are brine flies that live in hot springs in water temperatures exceeding 140°F.

Moisture is also a determinant of environmental suitability. As with temperature, most insects function most efficiently over a relatively narrow range of humidities. Again, there are exceptions—stored-product insects, for example, that can live at effectively zero percent humidity by extracting water from the foods they eat, and, at the other extreme, insects adapted to aquatic environments. Other determinants of climatic conditions—wind, light, and soil and air chemistry, for example—are also determinants of the range and population sizes of individual insect species.

Some forces, usually biological in origin, become more severe as population density increases. These so-called density-dependent factors include such sources of mortality as disease; the more crowded individuals are (even insects), the greater the chances for a sick individual to come into contact with and infect others. Like mammals, insects suffer from diseases caused by protozoa, bacteria, viruses, fungi, and nematodes. Other density-dependent mortality factors include competition for food. Plant-feeding insects compete for foliage or fruit to eat, and for places to lay their eggs; predaceous insects compete for prey. The greater the number of competitors, the more intense the competition. It's important to remember, however, that insects don't necessarily have to die in response to deteriorating conditions. One effect of higher population densities is that, when the carrying capacity of the environment is exceeded, the tough get going—individuals can disperse to find greener, less crowded pastures, leaving behind them lower, more manageable population densities.

Possibly the most conspicuous example of the potential for insect population growth to get out of hand involves *Schistocerca gregaria,* the migratory locust. This species of grasshopper is native to the Middle East. Periodically, populations build up to enormous levels and swarms of individuals take off, a fantastic phenomenon called mass streamaway, in search of food to sustain themselves. Locust swarms can be up to 1,000 square kilometers (400 square miles) in extent and can contain up to 10 billion individuals. These swarms are maintained as cohesive units for days or weeks over thousands of miles (Fig. 5.1).

Two factors are instrumental in leading to mass streamaway: crowding, which increases the rate of physical contact among growing nymphs, and inadequate nutrition during development. The complex transformation that leads to swarming in locusts was not well understood until the early part of the twentieth century, when detailed studies by Boris Uvarov revealed the mechanism.

Figure 5.1
Locusts blocking a train on the Athens-Salonika line (Department of Entomology, University of Illinois archives).

Normally, the migratory locust takes the form called *Schistocerca gregaria* var. *solitaria*. These individuals lead solitary lives, getting together only for mating and egg-laying. However, when numbers increase in an area, nymphs are crowded and food becomes scarce. Under these conditions, a series of physiological changes takes place during the process of development. Nymphs begin to aggregate and form small groups that venture short distances from their place of birth. These treks increase in frequency with age. After five nymphal instars, when the nymph finally molts into the adult stage, it becomes clearly recognizable as the form called *Schistocerca gregaria* var. *gregaria*. This form differs in both color and morphology—the adults are pinkish, sexually undeveloped, and have relatively long wings for the length of their body. These individuals bask in the sun all day and begin to make short flights in apparently random directions as air temperatures rise. One day, when atmospheric conditions are just right, mass streamaway takes place; as if on signal, about two hours after sunrise one day, millions of individuals head off in the same direction within fifteen to twenty minutes of each other. As swarms pass overhead, other locusts join them, and eventually millions of locusts take to the air en masse.

Most swarms appear to move in a downwind direction. Locusts are not, however, just passively blown along for the ride—even when wind speeds are less than 20 kilometers per hour (9 to 12 miles per hour), the typical air speed for a locust, locusts actively fly downwind. They can sustain these speeds for twenty hours or more, and individual swarms can move over 3,000 miles from their place of origin. Flying downwind, particularly in Africa and southwest Asia, is really a good move on the part of the locusts. Here's why: desert locusts invade some 11 million square miles of territory in that part of the world but breed in only about 5 million square miles. The area is seasonally dry and breeding is restricted to areas of seasonally high rainfall. Weather patterns are tied closely to the Intertropical Convergence Zone, a narrow zone 25 to 300 miles wide where northern- and southern-hemisphere air masses (the trade winds) converge. At convergence zones, air flows upward and, as it moves upward, it expands and cools. Because cool air can't hold as much moisture as warm air, this so-called adiabatic cooling results in heavy precipitation. Locust breeding grounds are concentrated in areas of convergent air flow. Swarms move downwind in spring from north to south; traveling downwind virtually guarantees deposition in the Intertropical Convergence Zone, where locusts can successfully concentrate for summer breeding. One reason for heading for this area of heavy rainfall is that locust egg cases are particularly prone to desiccation, so dry soil is totally unsuitable for producing children. Over her 6-month lifespan, a female can lay from 300 to 400 eggs in several batches.

Locusts are unwelcome visitors even when they travel in small numbers because of their voracious and undiscriminating appetites. To "eat like a horse" is an understatement—a single locust can eat its own weight in food (0.10 ounce) every day, a feat few horses (at a thousand pounds or more) are capable of. Pound for pound, that's 60 to 100 times more than even a gluttonous human can eat in a day. Each square kilometer of locusts, amounting to 40 to 80 million individuals, can eat up to 250 metric tons of food daily—enough to support a city of 80,000 people for the same period. In nonmetric terms, one square mile's worth of locusts—100 to 200 million individuals—can consume 220 to 270 tons of food, or enough to feed 200,000 people.

Locusts have been literally plaguing humanity for centuries. Perhaps the most well-known literary allusion to locusts is in Exodus 10:3–6, 12–17, 19, the description of the eighth of the ten plagues of Egypt, inflicted upon the Egyptians when they would not grant freedom to their Israelite slaves:

> And Moses and Aaron came in unto Pharaoh, and said unto him, thus saith the Lord God of the Hebrews . . . tomorrow will I bring the locusts into thy coast: and they shall cover the face of the earth, that one cannot be able to see the earth: and they shall eat the residue of that . . . which remaineth unto you from the hail, and shall eat every tree which groweth for you out of the field: and they shall fill thy houses and the houses of all thy servants, and the houses of all the Egyptians; which neither thy fathers, nor thy fathers' fathers have seen, since the day that they were upon the earth unto this day.

The ninth plague, darkness, following on the heels of the eighth plague, is thought by some to result from the darkening of the sky by the enormous locusts swarms.

Throughout the Old Testament, there are frequent references to locusts, particularly when the idea of multitudes is to be conveyed (e.g., Jeremiah 46:23, "They shall cut down her forest, saith the Lord, though it cannot be searched; because they are more than the grasshoppers and are innumerable" or Judges 6:5, "For they came up with their cattle and their tents and they came as grasshoppers for multitude; for both they and their camels were without number.") So the ancient Hebrews were quite familiar with the ability of insect populations to increase to enormous sizes. They also nad some awareness of locust lifecycles and biology. The Hebrew language has a number of different names for locusts, which may represent different developmental stages, sizes, or color phases.

The typical response to sudden visitation by locusts, before the advent of insecticides, was to seek help from divine powers. King Solomon, for example, issued a prayer of protection from locusts at the opening ceremony for the first temple (I Kings 8:37). It goes without saying that the Hebrews weren't the only ancient peoples with locust problems. The ancient Assyrians prayed to Ashur, a god always accompanied by an oversize locust companion, and wore locust talismans and amulets. The Greeks also had to deal with outbreaks of locusts and similarly resorted to divine intervention for assistance. The brass statue of Apollo Parnopios erected on the Acropolis at Athens was in recognition of his promise to drive away locusts (*parnopes* in Greek), which he evidently kept.

One safe thing about a promise to drive away locusts (or indeed any insect that suddenly undergoes a population explosion) is that in most cases it is easy to keep; eventually, density-dependent population regulation factors kick in and bring down the numbers in a given locality. So, no matter what people did to drive away locusts, it eventually appeared to be successful. It's not surprising, then, that prayers, spells, incantations, and denunciations continued to be regarded as effective anti-insect devices even down to contemporary times. Historically, Christians have subscribed to this belief. In the early days, court proceedings involving insects were strictly ecclesiastical because it was reasoned that insects were not answerable to civil authorities, rather only to divine authority. Thus, early on, it became the responsibility of the church to fathom the wishes of God or to foil the schemes of Satan (depending on who was responsible for the visitation in question). Excommunication was the ultimate weapon against insects, although there was a bit of confusion on this point; excommunication is effective only for church members and, since insects didn't pay taxes and weren't baptized, they technically couldn't claim church membership. Nonetheless, excommunications against insects could be obtained for a fee from Rome with great ease, although there was some controversy as to the advisability of such actions—because they are God's creatures, to anathematize them was tantamount to criticizing His good judgment. Clever lawyers found their way around these formidable obstacles, however.

In one early case, at the close of the ninth century (A.D. 886), Rome and its surroundings were hit by a plague of locusts. Although peasants gathered and destroyed millions of locusts (a bounty was placed on their heads), the numbers continued to grow. In desperation, Pope Steven VI sprinkled the entire area with holy water, whereupon the locusts (according to contemporary accounts) disappeared.

Ecclesiastical explanations of outbreaks differ a bit from ecological explanations (Evans, 1987). According to church doctrine, as explained by a Jesuit priest, Father Bougeant, in 1739 in the form of a letter, entitled *Amusement Philosophique sur le Langage des Bestes*, to a female acquaintance, the death of any nonhuman organism sets free its devil, which is constrained to enter the egg or embryo of another animal to continue its "penal bondage to the flesh," deprived for eternity from a spiritual existence. This doctrine explains first of all why "all species of animals produce many more eggs or embryos than are necessary to propagate their kind and to provide for a normal increase." Thus, an outbreak of any particular species is the result of an excess of devils and a corresponding housing shortage.

> This accounts for the prodigious clouds of locusts and countless hosts of caterpillars, which suddenly desolate our fields and gardens. The cause of these astonishing multiplications has been sought in cold, heat, rain and wind, but the real reason is that, at the time of their appearance, extraordinary quantities of animals have died or their embryos been destroyed, so that the devils that animated them were compelled to avail themselves at once of whatever species they found most ready to receive them, which would naturally be the superabundant eggs of insects.

Whether or not Father Bougeant was serious, many of his contemporaries were. An elaborate legal system was developed for the criminal prosecution and punishment of animals, including insects, by the Catholic Church. In the late fifteenth century, when a plague of locusts in Northern Italy was poised to devastate crops, the locusts were excommunicated. Recording this account, the Savoyan jurist Gaspard Bailly wrote:

> The deadliest sword, wielded by our holy mother, the Church, to wit, the power of excommunication, which cutteth the dry wood and the green, sparing neither the quick nor the dead, and smiting not only rational beings, but turning its edge also against irrational creatures; since it hath been shown at sundry times and in divers places, that worms and insects, which were devouring the fruits of the earth, have been excommunicated and in obedience to

the commands of the Church, have withdrawn from the cultivated fields to the places prescribed by the bishop who had been appointed to adjudge and to adjure them.

Ingenious explanations could be devised for the seeming lack of success of such excommunication. In 1338, locusts appeared in Botzen in the Tyrol, causing massive destruction to crops and fields. At the ecclesiastical court of Kaltern, about 10 miles south of the swarm, the parish priest was told to initiate proceedings against these locusts. The tribunal settled on a sentence of excommunication, which the parish priest executed. The failure of the locusts to disappear immediately was explained by the sins of the local people and their failure to keep up with paying their tithes.

Lest anyone think that the rights of locusts were trampled in these proceedings, there was an elaborate legal procedure to take them into account. First came the petition of the townspeople seeking redress of grievances, followed by the official declaration of the townspeople. Then the plea for the defense of the insects was entered, followed by the response of the townsfolk, and the reply of the defendant. The decision of the bishop's proctor was followed by the sentence of the ecclesiastical judge. Locusts and other insects were duly appointed legal representation and many of the arguments raised on their behalf included the fact that they were owed some alternate place to make their living (being God's creatures); in some cases, the insects were ordered by the court to relocate to a specific piece of land, usually outside of town, designated for their exclusive use (although generally the townspeople retained water and mineral rights to the property). Other legal arguments in defense of the insects maintained that their presence in an area was not regarded as undesirable by all creatures, and birds on several occasions were called in as friendly witnesses by lawyers representing the insects.

Resorting to divine intervention to deal with a locust outbreak is hardly a vestige of the distant past. As recently as 1875, when the American Midwest was hard-hit by an outbreak of the Rocky Mountain locust *Melanoplus spretus* (Fig. 5.2), the Governor of Missouri proclaimed,

> Wherefore, be it known that the 3rd day of June proximo is hereby appointed and set apart as a day of fasting and prayer, that the Almighty God may be invoked to remove from our midst those impending calamities, and to grant instead the blessings of abundance and plenty; and the people and all the officers of the States are hereby requested to desist, during that day, from their usual

employments, and to assemble at their places of worship for humble and devout prayer, and to otherwise observe the day as one of fasting and prayer (Evans, 1976).

Granted, by the nineteenth century other, more scientific, measures were taken to combat the locust menace. Bounties were offered for collected eggs and nymphs (in Missouri, five dollars per bushel of eggs), and American ingenuity was put to new tests in designing devices to collect and destroy the adults (such devices included the King suction machine, a horse-drawn tank equipped with a fan, and the Adams locust-pan, a modified harvester that tossed locusts up into a pan full of oil).

By the start of the twentieth century, the development of insecticidal chemicals, along with a greater understanding of locust life history and reproduction, greatly improved the ability of humans to head off locust plagues—for the first time in millennia. Poison baits, consisting of bran or other attractive food laced initially with arsenicals and subsequently with more potent insecticides, killed larger numbers of immature grasshoppers than was feasible by hand collecting (even with the assistance of a King suction

Figure 5.2
Life cycle of the Rocky Mountain locust *Melanoplus spretus*
(Webster, 1908):
  a. females ovipositing
  b. egg case, opened up to display eggs
  c. eggs
  d. cross-section to show deposition of egg case
  e. egg case in place

machine). The use of aircraft to dispense insecticides also increased the mortality that humans could inflict on the locust multitudes. One aircraft, for example, was responsible for 400 tons of dead locusts in only fourteen minutes in Somalia in the early stages of swarm formation.

One of the hardest lessons to learn in combating locust swarms is that, because locusts pay no attention to political boundaries, effective control requires international cooperation. A series of outbreaks in Africa south of the Sahara in 1928 led to the First International Locust Conference in Rome in 1931. One outcome of that meeting was the establishment in London of the Anti-Locust Research Center, a facility dedicated to investigating the life (and death) of migratory locusts. In an example of the spirit of international cooperation, the British, deeply embroiled in World War II, sent several hundred troops into neutral Saudi Arabia in 1943 at the invitation of King Abd al'Aziz al Sa'ud to assist in an international locust eradication effort. Although the focus of the war had by then shifted away from the Middle East, the British invested troops, time, and money into heading off the locust plague to protect the civilian populations there and to free Allied ships for transport of military troops and supplies to the Pacific, rather than for transport of humanitarian aid to the Middle East.

Today, locust surveillance is carried out not only by European and American satellites but also by African nomads; spray programs are coordinated by the United Nations Food and Agriculture Organization's Emergency Center for Locust Operations in Rome. By preemptive strikes and preventative maintenance, the world has been free of major plagues of red locust and African migratory locusts for almost forty years, although desert locusts, with a much broader range, continue to wreak havoc with crop production throughout the sixty-five countries in which they breed.

Useful homopterans (working for scale)

A Route of Evanescence
With a revolving Wheel—
A Resonance of Emerald
A Rush of Cochineal.
—EMILY DICKINSON, No. 1463

BY AND LARGE, members of the insect order Homoptera, as strict plant-feeders, are economically undesirable associates of humans.

Taxonomists divide the order into two suborders. Members of the Auchenorrhyncha are generally active and mobile and include the cicadas, treehoppers, froghoppers, and leafhoppers. In contrast, species of Sternorrhyncha are inactive or sedentary. Aphids, whiteflies, and scale insects all spend much of their lives within a few feet of where they first began to feed as crawlers, or newly hatched nymphs.

There are species in both suborders that are serious economic pests of cultivated plants, reducing yields by removing large quantities of sap or causing mortality by mechanically blocking the vascular system and interfering with nutrient transport. Some of these species compound the damage they do by acting as vectors of plant diseases. In fact, over 90% of plant viruses are transmitted by homopterans. Leafhoppers can serve as vectors of a variety of viral pathogens of crop plants and are involved in the transmission of such diseases as alfalfa dwarf, alfalfa witch's broom, aster yellows, clover phyllody, clover stunt, corn stunt, elm phloem necrosis, Pierce's disease of grapes, peach X disease, phony peach disease, potato yellow dwarf, sugarbeet Argentine curly top, sugarbeet yellow wilt, and tomato-Brazilian curly top. Aphids, on the other hand, are vectors of sugarcane mosaic, parsnip yellow fleck, carrot redleaf, cauliflower mosaic, cabbage black ringspot, and a number of other plant viruses.

Plant viruses can be transmitted by homopterans in two ways. Stylet-borne viruses are carried mechanically from one plant to another on the long piercing mouthparts of the insect. Circulative viruses, in contrast, remain with the insect for a considerable length of time—often through the entire life cycle. Some circulative viruses (especially those transmitted by leafhoppers) actually reproduce inside the body of the insect vector.

Although in general homopterans are considered economic pests, several species have proved to be economic boons. These species are almost exclusively members of the Sternorrhyncha. As sedentary as they are, they would be vulnerable to predators but for the fact that they can produce various forms of body armor: waxes, fluffs, powders, filaments, or hard plates. Mealybugs owe their name to the farinaceous white "meal" that covers the soft vulnerable bodies of all life stages except the first instar nymph and the winged male. Armored scales are protected by a hard resinous coat secreted by glands and permanently attached to the body. One proof that human ingenuity knows no bounds is the fact that over the centuries people have devised uses for the various fluffs, powders, and waxes that homopterans developed to keep themselves safe from harm—to the point that some of the economically more profitable species have actually been "domesticated."

One of the more spectacular success stories involves lac and its derivatives. Lac is the waxy platelike secretion of a small scale insect called, appropriately enough, *Laccifer lacca,* in the family Lacciferidae (Fig. 5.3). The secretion is the only animal resin of any commercial value, and it is chemically a complex mixture of various polyhydroxy acids. The word "lac" derives from the Hindi word meaning "hundred thousand" and refers to the huge aggregations that these insects form on their host trees (an etymological aside: the word "lox," referring to smoked salmon, is derived from the same root and describes the population densities of spawning salmon). Host species include fig, banyan, and a number of other species in Indochina, the Philippines, and Sri Lanka. As far as their biology goes, nymphs of both sexes crawl out from under the lac-covered shell of their mother and quickly plug into whatever food source they can find. In densities of about 150 per square inch, they sit and suck, secreting lac from almost all over their bodies. As they suck, they go through three molts, progressively losing seemingly vital parts of their bodies, including among other things, legs, eyes, and antennae. For the females things go on pretty much as they did before, but males un-dergo a pseudopupation stage in which they develop new legs, eyes, antennae, and two wings. They then emerge and go out looking for sex. They really don't have too far to look—the ratio of males to females in

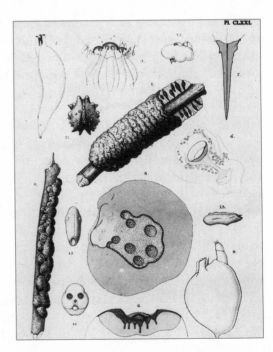

Figure 5.3
Life cycle of *Laccifer lacca,* the lac scale (Department of Entomology, University of Illinois archives). 1. Brood-lac, the coalesced secretions of many females.

most lac bug populations is somewhere around 1:5,000, and the females, which produce sex pheromones to guide their prospective mates' search, are absolutely immobilized inside their little resinous cases, so the challenge to the male is really minimal. In any case, after mating, the males die and cease to be of interest to anybody.

The females, on the other hand, are of great economic interest and importance. They produce an orange-colored resinous secretion, or stick-lac, which is simply the coalesced bodies of breeding females. Huge populations can coat tree limbs with resin up to one-half inch in thickness. This stick-lac is ground, washed, and filtered through muslin bags; the filtrate is passed over water-cooled drums to form flat flakes, the "shell-lac" of commerce. Alternatively, seed or grain lac can be ground by hand in stone mills or stuffed into long thin bags, heated over a fire, scraped, basted with water, spread onto a porcelain cylinder, and pressed flat and stretched wide with hands, feet, and teeth. Shellac color ranges from "ruby" to "blonde"; orange-brown is the purest form and white shellac is obtained by bleaching with chlorine.

Most of the world's lac is produced in India, which makes about 40 million pounds annually. It's an important source of subsidiary income to many of the poorer districts in Bihar, Orissa, Assam, and several other states. Lac's uses, even in today's synthetic society, are many and varied due in part to its physical properties, which include thermoplasticity, film-forming capacity, high resistance to organic solvents, excellent adhesion to a variety of surfaces, and good electrical properties. Moreover, unlike synthetic resins, it's completely biodegradable, nonpoisonous, and odorless. Lac has been used in insulators, electrical apparatus, and sealants; since it dries dust-free in ten minutes, it is invaluable in shoe and floor polishes and printing inks. Chippendale, Hepplewhite, and Sheraton, all famous furniture makers, used nothing else; championship bowling alleys are usually shellacked before an official match. Shellac can be found in playing card gloss, chocolate glazes, jewelry settings, billiard balls, mirror backing, and ironically enough, insecticide sprays (where it seals in the active ingredients).

Historically, the most important use of lac was in the record industry. The first shellac-based record was invented in 1907 by T.H. MacDonald and was a rather primitive affair consisting of two shellac-coated pieces of paper sandwiched together. World War I made shellac rather difficult to obtain, so there were a few abortive attempts to produce a synthetic replacement. None were very successful, so after the war shellac became very important for the growing record industry. In 1927–1928, Great Britain, Germany, and France collectively produced 260 million records, representing 18,000 tons of shellac. World War II, though, fixed it for the shellac industry. The unavailability of the product encouraged American scientists to come up with synthetic substitutes (such as vinyl), which proved to be far superior to the real thing, since shellac records are bulky, breakable and subject to extraneous surface noises (although, ironically, today vinyl records barely hold their own against compact discs for the very same reasons). About the only place in the world today where shellac is used at all in record-making is, not surprisingly, India. Lac remains an important commodity in Thailand as well, and in China, where lac production was reinvigorated by government investment in 1962. Not only is lac resin important in the production of varnish, but the fire-resistant host trees of the lac scale are important in maintaining the integrity of swidden (slash and burn) agroecosystems.

Scale insects have commercial value in art and artifacts not only as as a source of varnish but also as a source of dye. *Dactylopius coccus,* the cochineal scale, is not much more impressive at first sight than is the lac scale, but at least it has legs and eyes. The cochineal scale confines its feeding to cactus, specifically to species of prickly pear (*Opuntia,* particularly *O. cochinillifera*) which grow in Mexico and parts of the southwestern United States. Aztecs in Mexico and Pima Indians in Arizona cultivated them and used them in the manufacture of dyestuffs. The product was so valuable to them that cochineal was demanded as a medium of tribute from conquered peoples. Hernando Cortez was the first Westerner to report its use, in 1518, on his visit to Mexico. In 1523, Cortez received orders from the Spanish ministry to "multiply the commodity" and within a few years the dye was a major article of European commerce.

It wasn't for another 150 years, however, that people realized that cochineal was an insect. Anton von Leeuwenhoek and his newly invented microscope were needed to establish the fact. A man's entire fortune was lost in 1725 when he wagered that cochineal was not an insect; the story was published by the jubilant winner of the bet (in Dutch, French, and Spanish, no less) in a book titled *The History of cochineal proved by authentic Documents.* Spain attempted to monopolize the supply and forbade exportation of live insects. Several attempts by French and Dutch adventurers to smuggle the stuff out met with failure. It wasn't until 1833 that the French successfully introduced cochineal cultivation into French

Algeria (Fig. 5.4). Spain, faced with continental competition, then established a cochineal industry in the Canary Islands.

In Europe, cochineal was used extensively in cosmetics, specifically rouge and fingernail polish. It found further use in medicine and fabric dyeing until, at the turn of the twentieth century, it was largely replaced by synthetic aniline dyes. Cochineal had just about disappeared from the economic forefront when interest in natural dyes was rekindled as a result of the discovery that many of the more popular synthetic dyes have carcinogenic properties (a finding that quickly led to a ban on their use in many foodstuffs). Cochineal is used occasionally (particularly in Europe) in medicine and food; Peru and the Canary Islands are the principal producers. Its continued use in cosmetics has caused some interesting cultural problems; lipsticks made with cochineal are not considered kosher by Orthodox Jews, who have strict rules about the use of animal products, even those that are not eaten.

The life cycle of the cochineal scale is fairly typical, as the lives of scale insects go. Females form small groups (from four to ten) on the nopal cactus and secrete a white cottony wax material (which

Figure 5.4
Cochineal (Blanchard, 1883). Cultivation of cochineal in nineteenth-century Algeria.

protects against desiccation in the dry desert climate). After a few weeks of sucking cactus fluids, the female becomes sexually mature, although in the process she does not undergo any conspicuous changes in morphology other than enlarging to the size of a small pea (this sort of metamorphosis, in which immature morphology persists into adulthood, is known as neoteny). The male, in contrast, at its final molt acquires wings, flies to find females, mates with them, and dies (all in about a week, since he has no mouthparts with which to feed). The fertilized eggs hatch quickly into nymphs. The females are especially prized for dye production—they come equipped with a maroon pigment throughout their bodies, identified in 1910 as carminic acid. It amounts to about 10% of the total body weight of the female scale insect. This pigment may serve not only as a repellent to would-be predators such as ants—it tastes repulsively bitter—but also as an advertisement of the unsuitability of the cochineal as a menu item.

To produce the dye for commercial use, the cochineal are scraped off their cactus leaves, plunged into boiling water to kill them and to remove their wax coating, dried in the sun, and ground into a fine powder. About 25,000 insects are needed to produce about a pound of cochineal, and about three pounds of powdered insect make a pound of dye. Material to be dyed is boiled in a water extract of the powder along with leaves of a plant called *tejute,* which contains oxalic acid, a substance that is an intensifier and mordant (a substance that can combine with a dye to make an insoluble compound fixed in the fiber of fabric) for carminic acid. In pre-Columbian days, one of the more commonly used mordants was human urine; the number of chamber pots unearthed in the Aztec capitol was puzzling to archaeologists until the procedures for producing cochineal dyes were discovered.

Yet another scale insect has brought color into the drab lives of humans for centuries. Use of kermes dye, from *Kermococcus vermilius,* antedates Moses; in fact it probably was the source for one of the three colors specified for use in the curtains of the Tabernacle and for the robes of Aaron. Evidence exists for the use of kermes by the ancient Sumerians. The name "kermes" is mysterious in derivation (it seems to come from the Sanskrit meaning "worm") but it is fairly certain that the insects were native to the Mediterranean and Near East, where they feed on species of oak. The Arabs, by the way, are the ones who called the insect *kermes.* To Dioscorides, the Greek, the insect was known as *coccus.* In the Middle Ages, the name *vermiculus,* or little worm, came into vogue. To some extent, the dye has left its mark on our own culture through its linguistic reincarnations. From *kermes* comes the words "carmine" and "crimson," and from *vermiculus* comes "vermilion." The word "scarlet" in fact is derived from yet another name for this species—*sikarlat,* a Persian term meaning "red" and latinized to *scarlatum.* At one time, kermes was the reddest red color known and was used in many of the medieval Flemish tapestries. In Venice, dye production was strictly regulated. Adulteration of kermes with madder by dyers in the city of Lucca was punishable by a fine of 100 lira or loss of the offender's right hand. The Red Coats of British soldiers and the red fez characteristic of the Turks (and later Shriners) owed their brilliant red color to kermes.

The females, globular in shape and about 7 millimeters in diameter (less than 1/3 inch) at maturity, can produce up to 3,000 eggs. Since there's only one generation annually, production is necessarily limited. Moreover, harvesting kermes is rather labor-intensive—each insect is individually scraped off its twig by fingernail. On a good day, a person can harvest a little over two pounds of scales, which will lose 70% of their body weight in the drying process to yield only a little over half a pound of dry scale for a hard day's work. The principal color component of kermes dye is an anthraquinone pigment, kermesic acid, similar in structure to carminic acid, the principal color component of cochineal.

Scale insects have historically been important as a source of food as well as a source of dyes or pigments. A species familiar to desert travelers in the Middle East was *Trabutina mannipara,* a species that feeds on the twigs of tamarisk bushes growing in the central Sinai mountains. This insect was treasured not for its color but for its excrement—the sweet honeydew produced as a result of the rather inefficient method homopterans use for extracting nutrition from plant sap. Several lines of evidence suggest that the manna that rained down from heaven to sustain the Israelites during their ordeal in the desert was actually the honeydew of *T. mannipara.*

Location is one thing—beginning in Elim and ending at Rephidim, the biblical appearance of manna is consistent with the northern limit of distribution of the manna scale today. Timing is another suggestive factor—manna was discovered on the fifteenth day of the second month after the Exodus (late May or early June), exactly when honeydew production commences in contemporary times. The Bible also states that manna fell from heaven at night, a time when honeydew is likely to accumulate, since ants and other insects fond of sweetness are not actively harvesting it. Manna was said to resemble cumin seeds in appearance and bdellion resin in its sticky texture, an accurate description of the honeydew on both counts. Finally, the taste of manna is consistent with its identity as honeydew, since there are very, very few other sweet things that one is likely to run into when wandering in the desert.

Dyes, varnishes, and desert snacks notwithstanding, the greatest economic contribution made by scale insects is, in all probability, weed control. Biological control is the term for the use of natural enemies to eliminate an undesirable species. If the undesirable species targeted for elimination is a weed, then insects are generally the control agents of choice. Employing biological control is simply exploiting the fact that many herbivorous insects have a devastating impact on the plants on which they feed. Many species of scale insects have been deliberately introduced into areas where weeds have run rampant. They're ideal for this purpose—they are often highly oligophagous and, passionately devoted to their host plants, do not pose a threat to any other possibly valuable plant species in areas of introduction. Species of cochineal have been successfully introduced into India, Sri Lanka, South Africa, Australia, Mauritius, New Caledonia, Celebes, Java, Hawaii, East Africa, the Leeward Islands, and even the United States in order to control prickly pear cactus (*Opuntia vulgaris* and other species). The nymphs and adult females inflict mortality on their prickly pear hosts due to the fact that, while they are withdrawing sap, they are at the same time injecting a toxic substance that can cause premature fruit drop, loss of pads, chlorosis (loss of chlorophyll), necrosis (tissue death), and eventual collapse of the plant.

The successes of weed biocontrol by homopterans and other herbivorous insects have often been spectacular and any of these success stories could serve to illustrate the fact that insects really can regulate plant populations. Probably Klamath weed control provides the best example. *Hypericum perforatum* was called Klamath weed because it was first discovered in the United States along the Klamath River in California in 1900 (it's otherwise known as St. Johnswort). By 1944 there were 2 million acres of Klamath weed choking off quality rangeland. Cattle and sheep died after eating it, due to the presence of a light-activated toxin, hypericin, in the foliage, and land prices dropped to less than one-third of uninfested neighboring areas. In Australia, the same plant was being devastated by a small leaf beetle called *Chrysolina quadrigemina*.

Shipments of the beetle from Australia commenced in 1944 (the beetle is actually native to Europe, but World War II prevented direct transport across the Atlantic). The first problem with settling the beetles in their new home was that the beetles were seasonally reversed and had to be acclimated to the Northern hemisphere summers from June to September. By 1946, after numerous tests on crop plants to ensure that, as reported, *C. quadrigemina* accepts only Klamath weed as a host plant, five thousand beetles were released. After three generations, there were thousands of beetles and by 1950 there were millions. Beetle grubs feed on overwintering plants and prevent them from flowering and fruiting in subsequent years. After a few years, there are no more root reserves and the plants die. By 1956, Klamath weed was a dim memory throughout much of California. The plant persists in shady areas only (beetles won't fly into shade to oviposit). From 1953 to 1959, increases in land values and savings in herbicides amounted to 9 million dollars and cattle grazing on the reclaimed land were worth $2 million per year, an impressive return on an investment amounting to no more than a couple of thousand dollars.

Gallmakers ("gall enough in thy ink")

> Let there be gall enough in thy ink.
> —WILLIAM SHAKESPEARE, *Twelfth Night*

A GALL IS AN AREA of abnormal plant growth produced in response to a "foreign agent," including, among other things, plant-feeding insects. Galls can occur on just about any part of a plant—from root tips to flowers and seeds. On leaves alone there are galls on stipules, petioles, veins, and leaf blades. They are often recognized (and classified) by their appearance; on leaf blades alone there are Filzgalls, fleshy galls, kammergalls, fold galls, roll galls, roll and fold galls, pit galls, pocket galls, oak-apple galls, and lenticular galls, to name but a few.

For thousands of years, the origin of galls was quite a mystery; Theophrastus, the "father of botany," remarked over two thousand years ago that most of the structures seemed to contain insects inside them but he thought that the appearance of these cecidozoans, or gall inhabitants, was strictly coincidental. Even the great Francesco Redi, who in the seventeenth century disproved that maggots appear spontaneously in rotting meat, was sufficiently baffled by the appearance of insects inside galls as to maintain a strict silence on the subject in his publications.

In point of fact, the reason that there is so often an insect inside a gall is that the insect is at least partly responsible for gall formation in the first place. This fact was experimentally confirmed in the seventeenth century by both Martin Lister in England (nephew of the physician to Queen Anne) and Marcello Malpighi in Italy (and described in his book *De Gallis*). Gall formation begins when a female lays an egg inside the plant; in response, the plant tissues begin to grow abnormally. Plant cells enlarge and multiply wildly. Insect galls are generally built on a similar plan. Lining the central chamber, where the insect makes its home, is a layer of protein-rich nutritive tissue; the cells in this layer often have multiple nuclei and are physiologically active, synthesizing carbohydrates and proteins. Immediately outside the central chamber is a layer of hard, thick-walled cells. These cells are rich in indigestible substances, such as lignin and tannin. This layer has been called the "protective layer," although its function is still not really certain. Outside this layer are layers of normal plant parenchyma and epidermal cells.

For a long time, no one could agree on whether plants produced galls to protect themselves from further damage by an insect enemy—to seal it off and prevent its access to other plant tissues— or whether insects stimulate a plant to produce a gall in order to secure a safe haven and easy meals. As is so often the case in scientific controversies, it turns out that gall development depends on both the insect and the plant. When a female inserts an egg into plant tissue, she also injects a few other things along with it; some of these substances may have hormonal effects on plants and alter growth patterns. The suggestion has even been made that the insect injects a mutualistic virus into the plant, which can insinuate itself into the plant's genetic material and take over regulating the genes that control growth. While the disruption of growth patterns is the most conspicuous consequence of gall insect infestation, plant physiology is substantially altered as well. Generally, the production of nutritive substances such as carbohydrates is greatly increased. In fact, the galls of certain species of mites, aphids, wasps, and flies actively secrete a sugary secretion or honeydew. Such a secretion may attract ants and other predators that could protect the gall inhabitant from potential enemies.

The interaction between a gall-forming species and its host is unique; even if many different species of gall-forming insects infest the same host plant, each insect species produces a unique, characteristic gall morphology. Not all insects that lay eggs in plants can induce gall formation. Among the talented few are aphids, caterpillars, beetle grubs, thrips, wasps, and maggots; in addition there are a couple of families of mites that can cause galls to form on plants. The lion's share of gall-formers is found in only two families: the Cecidomyiidae, a family of gall flies, and the Cynipidae, a family of tiny gall wasps. In North America, these two families make up 70% of about 1,700 species of gallmaking insects. The different families of gallmaking insects are not at all alike in their plant preferences; while the cynipid wasps, for example, concentrate their efforts on oaks, the gallmaking aphids attack a tremendous variety of plants. In addition, species that induce gall formation vary in their ability to infest their hosts—while there are rarely more than one or two goldenrod ball gall flies on a single plant, each housing only one maggot, there can be as many as half a million cynipid wasp galls on a single oak tree.

By the same token, not every plant responds to the presence of an insect egg in its vital parts by forming a gall. Gall formation is relatively rare among the ferns, mosses, conifers, and even the grasses; the vast majority (over 90%) of species that do form galls are flowering plants. Although galls have been found in at least half the families of flowering plants, the families hardest-hit by gallmaking insects are the Rosaceae, or rose family; the Compositae, or daisy family; and the Fagaceae, or oak family. In North America, the species in the daisy family make up about one-sixth of all plants susceptible to gallmaking insects.

Gallmaking is a specialized lifestyle to which not all insects are suited. Insects that develop inside galls have had to make a number of adjustments. A problem faced by gallmakers is relatively unusual in the insect world. Since gallmakers are confined in a small space, from which they cannot escape until development is complete, there is a premium on hygiene and sanitation—they can't afford to contaminate their living quarters. As a result, many species of gallmakers have no functional anus. All of the waste products accumulated over the course of development are stored in the body and this collected waste material, or meconium, is voided only when the gallmaker metamorphoses into an adult.

On the other hand, unlike most plant-feeding insects, the gall-inhabiting stages of gallmakers have no need to seek out food—they're literally surrounded by it. As a result, gall inhabitants often lack legs or have reduced numbers of legs; gallmaking mites, in the family Eriophyiidae, have only two pairs of legs instead of the typical arachnid four pairs. Also frequently absent are antennae, and with the antennae the sensory equipment they would use for locating host plants at a distance. The more mobile adult stages of these species are generally fully equipped with wings, legs, and sensory apparatus, since it is their responsibility not only to find host plants for their offspring but to find potential mates as well.

From the perspective of an insect gallmaker, living in a gall is a mixed blessing. The lignified outer wall prevents entry by many types of enemies. Gallmaking sawflies, for example, are free from attack by parasitic flies, who lack ovipositors long or strong enough to penetrate the protective layer. Unlike their close relatives that feed externally on leaf tissue and pupate in cocoons in the soil, gallmaking sawflies are also free from parasitism in the cocoon stage by wasps. However, the protective layer that keeps predators out can also prevent the gallmaker from escaping in case a predator does manage to gain access to the central chamber. Mortality of gallmakers can be very high as a result. While most gall midges are proper gallmakers, a few rogue species eat fellow gall midges that do make galls. Not only are there interloping predators, there is also an assortment of inquilines, or "uninvited guests"—other plant-feeding species attracted to the rich layer of nutritive tissue lining the central chamber. Aside from the fact that these inquilines rob gallmakers of their food, many of these species are not averse to consuming the gall occupant along with the plant tissue.

One final problem faced by gallmakers is escaping from the gall after the final molt to adulthood. Whereas mouthparts of immature stages of gallmakers are ideally suited for tunneling through plant tissue, the mouthparts of the adult stages of many gallmakers (particularly moths and flies) are woefully inadequate for this task. As a result, mature larvae often prepare escape tunnels immediately before they pupate, excavating a path to the outside and scraping away at plant epidermis until only a weak "window" remains behind. Upon completion of development, the adult can simply push its way through the window to the outside world.

Plants differ in the extent to which they suffer from attack by gallmakers. Despite the fact that at least six species of jumping plant lice can form hundreds of galls on hackberry trees, the trees never seem to suffer any ill effects. In contrast, a single goldenrod ball gall fly can reduce the production of seeds by a goldenrod by almost half. Gallmakers that attack crop plants, like the alfalfa gall midge *Asphondylia websteri,* can significantly reduce yields and cause economic damage.

In one case, the ability of a gallmaking insect to kill an economically important plant had world-shaking consequences. The wine industry of France was dealt a devastating blow in 1860, when *Phylloxera (Daktulosphaira) vitifoliae,* the grape phylloxera, was accidentally introduced from North America—an event regarded as a viticultural disaster of unparalleled magnitude. The grape phylloxera is a tiny, wingless aphid, less than 1/20 inch long, that forms galls on both the leaves and roots of grapevines. While the leaf galls rarely cause much damage, the root galls can stunt or kill vines in three to ten years. Within only twenty-five years, over 2 1/2 million acres of grapevines (amounting to fully one-third of France's wine-producing grapes) had been ruined.

That France has maintained its reputation for producing fine wines is the result of careful entomological observation and great ingenuity on the part of French enologists. Noticing that the native North American fox grape *Vitis labrusca* did not readily succumb to grape phylloxera, the French imported rootstocks of this resistant species; in France, stems of the French wine grape *Vitis vinifera* were grafted onto the hardy American rootstocks. The rescue mission cost the French nation some 10 billion francs but, by restoring the French wine industry to prominence, it saved that nation untold billions and allowed the world to continue, with but a brief interruption, to enjoy fine French vintages throughout the twentieth century.

American wine, however, has not fared as well of late. In 1986, a new and virulent form of phylloxera made an appearance in California vineyards. This particular phylloxera reproduces only asexually and feeds exclusively on roots. By 1993, over 10,000 acres in Sonoma and Napa counties (premier wine country) were infested. Failure to control phylloxera either by pesticide applications or development of resistant rootstocks has led at least one enterprising vinter to market a rosé made out of infested Cabernet Sauvignon vines called "Bug Creek" wine, with 10% of the proceeds from sales of the rosé earmarked for research sponsored by the American Vineyard Foundation to stamp out phylloxera.

Gall insects aren't entirely bad news. Their ability to kill certain plants has been put to use in a number of weed biocontrol programs. Among the notable successes was the introduction of *Procecidochares utilis,* the pamakani gall fly, into Hawaii to control Mexican devil weed, *Eupatorium adenophorum,* an invasive weed of pastureland that was accidentally introduced into Maui in the mid-nineteenth century. The products of gallmakers, as well as their plant-consuming services, have been of economic importance for centuries. Galls, particularly those on certain species of trees, are very rich in tannic acid, a mixture of compounds, including gallic and ellagic acids, esterified to glucose. When the structure of tannic acid was determined and its component parts identified, they were named in honor of their source—gallic acid and ellagic acid (from the French word *galle* spelled backwards). The Turkish oak gall-nut is about 50 to 60% tannic acid and the Chinese gallnut amost 70%.

Tannic acid has several rather distinctive properties. Among other things, it is highly toxic to microbes. This may be the reason that gallmakers so often induce tannin production by their hosts—high tannin levels may protect gall insects from fungal or bacterial invasion. Yet another distinctive property is its tendency, when combined with iron salts, to form a blue-black precipitate. This is the substance that has, since about 2000 B.C., been used as ink. Ink made from galls over the centuries has been considered highly desirable. Dioscorides published a recipe for a gall-based ink in A.D. 50. Ink made from "oak apples" was used extensively by medieval monks in the ninth and tenth centuries for manuscript illumination. In some cases, certain types of documents could by law be recorded only with a gall-based ink. The Aleppo gall, produced on oaks in eastern Europe and Western Asia by the cynipid wasp *Cynips gallae-tinctoriae,* was particularly prized for its quality and was "specified in formulas for inks used by the United States Treasury, the Bank of England, the German Chancellery, and the Danish government" (Frost 1942). As recently as 1891, a report from the Massachusetts Record Commission on Record Inks and Paper concluded that ink from gallnuts was permanent and, if faded, even restorable by application of a solution of tannic acid.

Galls have also been useful in the dyeing of leather and of other materials. Again, the Aleppo gall has over the centuries been recommended for producing permanent black dye. As recently as 1914, over $17,000 worth of gallnuts were imported from Baghdad for the purpose of dyeing sealskin furs. Another gall, produced on oaks (such as *Quercus infectoria*), has been used in the production of a dye known as Turkey red. The gall, formed by the wasp *Cynips insana,* is variously known as the "mad apple of Sodom, Dead Sea Fruit, or Mecca or Bussorah gall" (Fagan 1916). It is used locally in Asia Minor (in Basra). Because of their ability to produce a permanent pigment or dye, galls have a long historical association with cosmetic use. Over 2,000 years ago, Pliny the Elder wrote that the

Aleppo gall was widely used not only for dyeing furs or leather but for dyeing hair black.

Due to their tannic acid content, galls also have been used in the process of tanning leather. The tanning process is one in which proteins in animal hides are complexed and rendered impervious to rot—it's the fundamental process of making leather. Tannic acid is an excellent binder of proteins; historical records of the use of galls for tanning purposes date back to the first century. More recently, the binding properties of tannic acid have been exploited to the detriment of plant-feeding insects in general; nicotine and other organic insecticides have historically been complexed with tannic acid to stabilize them for longer shelf life.

While gall products such as tannic acid have been important throughout the world, galls themselves have been of value to people as well. The most direct use of galls for ornamentation is certainly the use to which the Aguaruna of Peru put galls found on a tree they call duship (*Licania cecidiophora*). The Aguaruna, who live in rain forests and farm manioc, collect the gall-infested leaves of the duship, which fall from the tree twice a year, and remove the galls in order to use them to string, like beads, in necklaces. These necklaces can contain as many as forty looped strands and contain as many as 41,600 galls. While not as durable as gemstones, galls are remarkably sturdy as far as plant products are concerned; duship capes collected half a century ago are still in good shape, possibly because the high tannin content of the galls protects them from decay.

Silk and silkworms

> The silkworm in the dark green mulberry leaves
> His winding sheet and cradle ever weaves.
> —PERCY BYSSHE SHELLEY, Letter to Marie Gisborne

SILK, IN ITS many guises (satin, shantung, peau de soie, brocade, organdy, taffeta, or chiffon) a fabric of elegance and expense, is basically nothing more than insect saliva. More precisely, silk is the threadlike viscous secretion of the salivary glands of a caterpillar called *Bombyx mori*, a member of the moth family Bombycidae (Fig. 5.5). Now all caterpillars make silk, but few in as grand a style as the silkworm caterpillar. The salivary glands, stretched to their full length, are almost ten times the length of the mature caterpillar itself and about half its total weight. *Bombyx mori* uses the silk to spin a cocoon at pupation time; it can spin at a rate of about

Figure 5.5
Life cycle of *Bombyx mori* (Department of Entomology, University of Illinois archives).
a. adult male (top) and female (bottom)
b. eggs
c. silkworm caterpillars
d. pupa inside silken cocoon

a.

b.

c.

d.

6 inches a minute in continuous strands of up to 1/2 mile in length. The silk itself consists of two materials: fibroin, a horny proteinaceous substance that makes up the inner layer, and sericin, a gummy outer covering that is boiled away to give silk its characteristic luster. What appears to be a single thread is in reality a double one; the paired salivary glands each secrete a filament through the minute openings of appendages called spinnerets, beneath the

mandibles of the caterpillar. It's spun as a dry fiber, a feat that cannot yet be accomplished by modern technology.

Silk has a number of properties that make it desirable for textile use. Compared to wool and cotton, its density is very low, so it's light and comfortable, not bulky. Its tenacity and elasticity are relatively high and it is therefore highly crease-resistant—more so than almost any natural fiber. Silk insulates very well; in its natural state, it probably serves to protect the metamorphosing pupa from environmental unpleasantness. Its insulative properties thus make silk garments warm in winter and cool in summer. It's lustrous and shiny and has high affinity for dyes. It's also highly absorbent, almost three times more so than nylon. Like wool (and unlike synthetic fibers), silk is noncombustible. In proportion to its weight it's the strongest natural fiber known; a silk cable can sustain heavier weights than a metal cable of comparable dimensions. It's elegant, comfortable, and beautiful and has been the most sought-after, the most jealously guarded, and most furiously defended fiber ever.

Silk is such an ancient fiber that its discovery is lost in legend. Ancient literature credits the discovery of silk to Lady Si-Ling, wife of the emperor Huang-ti in the year 2640 B.C. According to one version of the story, Si-Ling was having tea in her garden when the cocoon of a pupating caterpillar dropped into her teacup. In attempting to remove the cocoon from the hot water, Si-Ling removed instead a mass of shimmering fibers. In any event, she was largely responsible for putting the silkworm into production and, in her honor, the fiber was call *si,* from which eventually comes the English word "silk" and virtually every other name for silk: *sirket* in Mongolian, *seri* in Japanese, *serik* in Arabic, *sericum* in Latin, *seta* in Italian, *soie* in French, *Seide* in German, and *silke* in Swedish.

The secret of silk production, or sericulture, was fanatically protected for centuries. Indeed, silk was the exclusive privilege of the imperial family until 1200 B.C., at which time silk privileges were extended to wealthy nobles and merchants. A class of "silk serfs" arose to keep them well-stocked. When finally, around A.D. 300, the Chinese allowed the release of four Chinese women to Japan to instruct the royal court in the art of sericulture, the Japanese were so grateful that they immediately built a temple in Settsu province in honor of the instructors. In the two hundred years that followed, silk weavers from China moved to Japan in force to found what was until recently the world's largest silk industry.

The western world, meanwhile, was in frustrated ignorance of the secrets of silk. Silk, though not the silkworm (penalty for export

Figure 5.6
Silk production at the Souzhou Silk factory, Souzhou, China (R. J. Leskosky): (right) a factory worker combines fibers from several cocoons to form a single thread, which is wound onto a reel; residual sericin causes the individual threads to adhere to one another; (above) caterpillars are fed mulberry leaves on round, stacked woven trays.

of these from China was death), was introduced to the Greeks in the aftermath of Alexander the Great's conquest of Persia and India in the fourth century B.C. With the capture of Babylon, Alexander acquired silken robes of a value of 5,000 talents, roughly equivalent to $7 million. Determined to be more than a ruthless conquerer, Alexander ordered the dispatch of a thousand educated Greeks on his Asian campaign to gather information that would enhance Greek culture. From this thousand came the first account in Western civilization of the silkworm. Although far from accurate, particularly in the details of the silkworm's life cycle, it was considerably better than ideas prevalent at the time—ideas that persisted into the Roman era. Even repectable Roman scholars such as Virgil believed that the Chinese "spin their fleecy forests into a slender twine." Dionysius Perigetes, circa A.D. 200, claimed that silk was combed from "the various colored flowers of the land."

The fact that they didn't know what it was didn't stop any Romans from lusting after it. Silk, perhaps because of its exorbitant cost, was the fabric of choice for the Roman aristocracy. At first, its use was restricted by law to the imperial family and meritorious military men but its use soon permeated all levels of society. To put a halt to such business, the Emperor Augustus ruled in A.D. 14 that ordinary citizens were no longer permitted to wear silk, especially "those of lower situations who copied the costumes of their betters." Two years later, an edict was issued forbidding men to wear silk (except as trim) as it was deemed effeminate and decadant. Women of the aristocracy reveled in the luxury of wearing silk, especially the diaphanous gowns from the island of Cos. The passage of the Oppian laws, an attempt to reestablish a moral lifestyle in Rome, restricted the use of silk for ladies' gowns. One account records the reaction:

> Not unlike their sisters of every age, the Roman women resented this interference with their freedom and especially in connection with their dress, so they gathered outside the Senate and successfully lobbied for repeal of the obnoxious law, a feminine political action which up to that time was without parallel in history. Thereafter Roman ladies were permitted to dress as they pleased, without restrictions as to silk, color of dye, gold threads, or transparency (Cowan 1865).

In the reign of Aurelian, A.D. 270 to 275, silk was worth its weight in gold; its price reflected not only its popularity but the difficulty in transporting it from India and the Orient. Most of the silk in Europe was brought from China along the great Silk Road, a trade route extending from China through Turkestan to Iran and eventually terminating at the northeast coast of the Mediterranean Sea. The Silk Road, opened in 126 B.C., was one of the longest roads in the world, stretching over 6,000 miles. Although closed in the ninth century, it was again reopened in the thirteenth century, when Marco Polo used the ancient route to travel to China.

It wasn't until the sixth century that Westerners were able to break the Easterners' secret. In A.D. 555, the emperor Justinian sent two Nestorian monks from the order of St. Basil, longtime residents of China thus privy to the secrets of sericulture, back to China to smuggle out some *Bombyx* eggs. Smuggling was an absolute necessity even for an emperor in view of the death penalty for trading or transporting eggs or caterpillars. The monks managed it by concealing the eggs and mulberry seed inside the bamboo staffs that they as pilgrims were expected to carry around with them. The ruse was successful and as a result all the silkworms in Europe and the United States (until 1865) were descendants of the eggs carried back to Rome by the monks.

Sericulture spread rapidly through Europe. In A.D. 946, King Roger of Sicily conquered Greece and forced a number of silk weavers to settle in Palermo and ply their trade there. Knowledge spread quickly and soon Italy dominated the silk industry in the western world. Weaving and then silkworm raising were introduced to France in the reign of Henry IV (although kings before him attempted to do this—Henry II at his 1520 coronation was the first king of France to wear silk stockings). Religious discrimination was in part responsible for the further dissemination of sericulture throughout Europe. In 1685, the revocation of the Edict of Nantes by Louis XIV sent thousands of French Protestants fleeing to Germany, Switzerland, and England. Since most of the silk weavers of France were Protestant, the revocation almost led to the total annihilation of the silk industry there.

England briefly picked up the lead in silk production in Europe. James I did much to encourage its growth—he ordered mulberry trees from France and offered free seeds to anyone who would grow them. Outside Buckingham Palace, the king had a mulberry plantation laid out for public display. In 1718, England began manufacturing silk fabrics when John Lombe of Derby, disguised as a worker in an Italian throwing mill, drew pictures of the machinery and returned to England on government subsidy. Lombe subsequently died in great agony, poisoned by an Italian woman sent to England for the purpose.

It was also James I who attempted to introduce sericulture to the New World. In 1609 seven boats set sail for Virginia with mulberry trees and seeds; the eggs and seeds were lost in a terrific storm at sea, a storm of such magnitude it is said to have inspired Shakespeare's play *The Tempest*. Eventually eggs and seeds arrived, along with a few French people to instruct the colonists, and sericulture was more or less forced on the settlers. The colonial governor ordered that ten mulberry trees be planted on every hundred acres at the risk of a penalty of 20 pounds of tobacco. Ministers extolled its attributes in sermons, politicians enacted huge monetary incentives and poets waxed lyrical:

"Where worms and wood doe naturally abound
A gallant silken trade must there be found
Virginia excells the world at both
Envy nor malice can gaine say this troth."

Nonetheless, the Virginia planters stuck stubbornly to tobacco, the colonial medium of exchange.

The establishment of reeling mills in Pennsylvania stimulated interest in sericulture in the middle Atlantic states and sericulture was conducted on a modest scale throughout Pennsylvania, New Jersey, and Connecticut. At Yale's 1789 commencement, President Ezra Stiles wore a gown made of silk produced from his own silkworms. By the beginning of the nineteenth century, Connecticut was producing between thirty and forty thousand dollars' worth of silk annually.

Silk production took a downturn in the early nineteenth century, at least in part due to a speculative frenzy that gripped would-be silk growers. Traditionally, the silkworm, which feeds only on mulberry trees, was raised on white mulberry *Morus alba,* imported from Asia. Neither the species of mulberry native to eastern North America, *M. rubra,* nor the imported and widely grown European species *M. nigra* were especially suitable for raising silkworms. A new type of mulberry, however, was imported into the United States in 1824 by G.S. Perrottet, called *M. multicaulis.* Under the mistaken belief that this tree was used in China for raising silkworms, people became desperate to buy and plant large stands of *M. multicaulis.* In the heat of what has become known as the multicaulis craze, trees were bought and resold days later for many times their original purchase price. It soon became glaringly apparent, however, that the tree was not hardy enough to withstand northern winters and the bottom fell out of the market; literally millions of trees were simply abandoned by cultivators and would-be silk czars.

In lieu of the production of raw silk, particularly in the aftermath of the multicaulis disaster, many American communities turned to silk manufacture. The government gave this fledgling industry a great assist in 1857 when it increased import duties on silk fabric almost threefold (from 24% to 60%); raw silk, however, could be imported duty-free. The metropolitan Northeast, already firmly established in the textile industry, became the center of silk manufacturing. In Paterson, New Jersey, known as "Silk City," there were at least thirty companies employing 600 people in all aspects of silk manufacture by the mid-nineteenth century. By the beginning of World War I, the numbers had grown to over 24,000 people working in over 300 mills. Paterson was particularly well-suited to the growth and development of the silk industry—located near New York, a major distribution center for raw silk, accessible by railroad, and blessed with a water supply peculiarly conducive for dyeing. Many of the mill workers were recent immigrants from Europe who had previously been employed in the silk trades in their countries of origin. Every aspect of silk processing was carried out in Paterson; enterprises engaged not only in production of fabric but also in peripheral businesses such as loom manufacture, dyeing, and even jacquard carding, the creation of punch cards that program looms to weave particular designs. These cards are generally acknowledged to be the conceptual predecessors of computer punch cards. The Jacquard loom was invented in nineteenth century France. Holes were punched in a roll of paper, corresponding to a decorative pattern. This roll pressed up against a series of wire hooks; wherever a wire hook lined up with a hole, it picked up and pulled through a thread. When American Herman Hollerith was faced with the prospect of collecting data for the 1890 census, he adapted the loom design, using electrified wires in place of hooks; when a wire encountered an opening, it transmitted an electrical impulse rather than a thread. This device, a primitive digital computer, was the basis for the formation of IBM.

Labor trouble in Paterson in the early part of the twentieth century led to the breakup of the major mills; manufacturing magnates could not make what they considered a reasonable profit when constrained by union rules and wage scales. Coupled with the invention of synthetic fiber such as nylon in the 1930s, the silk industry faded from prominence not only in Paterson but from the metropolitan Northeast in general. World War II firmly established synthetic fibers in American markets when action in the Pacific theater all but eliminated imports of raw silk from the Far East. Today, however, silk is still very much a valuable commodity, primarily as a luxury fabric (much as it was in its earliest days).

Even without the geopolitical disruptions of World War II, nylon may well have driven silk out of the market; one of the reasons that nylon and other synthetics so quickly replaced silk is that silk is basically a pain in the neck to produce. *Bombyx mori* will eat only the leaves of the mulberry tree—preferably *Morus alba,* but it can be coerced into eating a few other *Morus* species. And it eats a lot of mulberry leaves—up to 40,000 times its own weight from hatching to pupation. One pound of caterpillars can consume 12 tons of mulberry leaves over the course of a lifetime. *B. mori* begins life as a tiny black caterpillar—in Japanese, the first instars are called *kego,* or, hairy babies—but within four weeks they grow 25 times larger and 12,000 times heavier. While they're caterpillars, they're susceptible to a number of diseases (one of which, pebrine, proved to have some beneficial consequences—it was at least in part due to his investigations of this disease that Louis Pasteur was inspired to formulate his famous theory on the microbial origin of diseases). The mulberry silkworm is effectively a domestic animal—no wild specimens of *B. mori* are to be found anywhere in the world—and they exhibit the same sorts of physiological frailties that other domesticated animals exhibit.

When the caterpillars are mature, they are transferred to racks where they can spin without interfering with one another. Once the cocoons are collected, they are taken from the farm to a filature, a factory where they are processed. The cocoons are boiled in hot water, which acts both to kill the pupa inside and to remove the outer sericin layer, to give the silk its characteristic gloss. The next step is reeling, or finding the end of the cocoon filament and unraveling and winding it onto a spool. The filaments of several cocoons are reeled together to make threads of particular sizes (e.g., seven cocoons for 20/22 denier silk, Fig. 5.6). Throwing, or twisting threads onto a spool, is followed by soaking and dyeing; silk threads are then reeled into skeins, skeins bundled into books, and books into bales. These bales are then shipped to mills for weaving.

About 350 cocoons are required to make a pair of silk stockings; for one dress, about a pound of silk, or 1,700 cocoons. To obtain 1,700 cocoons, one needs about 125 pounds of mulberry leaves to feed to hungry caterpillars. The 1,700 cocoons require about 10 hours to reel and more time to weave. Sericulture is thus definitely not for the lazy or unenthusiastic.

China leads the world in silk production, its annual output of 30,000 tons of raw silk accounting for 80% of the world's supply. Although for decades Japan led the world in silk production, today it is a distant second to China, producing only about 7,000 tons of raw silk annually. Other world-class producers include India (which produces large amounts of silk but exports very little), South Korea, Indonesia, Brazil, Thailand, and Uzbekistan. Among European nations, Italy takes the lead with an annual output of about 20 tons of raw silk. At the other end of the commercial exchange, the United States is the world's leading importer of silk fabrics and finished goods.

Although the vast majority of silk in the world comes from the cocoons of *Bombyx mori*, other species of insects contribute in a lesser way to the world silk market. Most of the other silk sources are close relatives of *B. mori*. One of the chief nonmulberry silkworms in Asia is *Antheraea yamamai,* the Japanese oak (eri) silkworm. A member of the family Saturniidae, the Japanese oak silkworm spins a silken cocoon comparable in many ways to the silken cocoon of *B. mori*. Unlike the cocoon of the mulberry silk-worm, however, the cocoon of the Japanese oak silkworm frequently contains colored pigments (depending upon the light intensity experienced by the larva); pigment differences, as well as some structural differences in the silk itself, mean that dyeing conditions are more critical. Moreover, with current rearing practices, only one generation of the Japanese oak silkworm per year can be produced. Other saturniids have been examined periodically as potential commercial silk sources; in Turkey, the moth *Saturnia pyri* has been proposed as a silk source preferable to the mulberry silkworm because of its polyphagous habits and its larger cocoon.

The search for a better method of silk production is not without its own pitfalls, however. In the mid-nineteenth century, an amateur lepidopterist named Leopold Trouvelot living in Cambridge, Massachusetts, imported eggs of several moth species in an attempt to develop by hybridization a silk-producing moth resistant to the many diseases that afflict the mulberry silkworm in culture. Through an oversight, caterpillars of one species, *Lymantria dispar,* accidentally became established in his neighborhood. As caterpillars, *L. dispar* disperse by ballooning, or spinning a thin thread of silk, which provides the tiny first instars with lift and buoyancy. Thus, these so-called gypsy moth caterpillars can cover a great deal of territory. From Trouvelot's backyard, the gypsy moth went on to colonize all of New England by the end of the century, reaching epidemic proportions in some forested areas. In less than a hundred years, it had expanded its range from coast to coast, aided in its American conquest by the tendency of the flightless females to lay masses of several hundred eggs in the wheel wells of campers and other vehicles, which transported the species across the country. As a major defoliator of dozens of forest tree species, ranging from white pine to black oak, the gyspy moth has made an economic impact far different from that envisioned by the well-intentioned but misguided Trouvelot.

Another group of arthropods of some importance in commercial silk production are spiders. Unlike caterpillars, spiders spin silk not from a pair of salivary glands but rather from a series of silk glands of various descriptions in their abdomen. Silk is pulled out through openings in the spinnerets, which vary in number depending on the family, either by the legs or by the entire body, after the spider anchors down one end to the ground. Virtually all spiders spin silk, which plays a vital role in many aspects of spider biology. Different types of silk are spun for different purposes. For example, even within an orb web, sticky strands serve to snag prey while smooth strands are used by the spider to negotiate its own web and so avoid entanglement.

Depending on species, spider silk varies in thickness from 1/50,000 inch to about 1/5,000 inch; in comparison, a blonde human hair, at 1/250 inch, is a veritable tree trunk. Elastic and resilient, some spider silks can be stretched over 20% of their full length. Historically, spider silk was used in constructing fishnets, bird snares, or woven into small pouches; cobwebs in particular have found use as bandage material (to stanch blood flow) and even, stretched and doubled over, as artists' canvas. The most commercially important use of spider silk today is as material for constructing crosshairs in optical instruments, including surveying equipment, range finders, microscopes, bombsights, and telescopic gunsights. For gunsight crosshairs, silk from the immature stages of *Latrodectus mactans,* the black widow spider, is desirable; despite the potentially dangerous effects of handling this spider, which as an adult is capable of inflicting a fatal bite, some brave souls harvest silk from semidomesticated specimens.

At least partly because of the difficulties attendant upon collecting sufficient quantities of silk from potentially lethal spiders, the U.S. Army has funded research aimed at producing silk in the laboratory by genetic means. Genes that encode different silk proteins have been cloned into bacteria, which then manufacture these proteins in mass quantities. Future military applications of silk fibers, once they can be manufactured cheaply and rapidly, include incorporation into bulletproof vests, lightweight helmets, and parachute cords. Spider silk is particularly well-suited for this last application due to its extraordinarily low "glass transition temperature," or temperature at which the fiber becomes brittle. Because it is still elastic and resilient at temperatures as low as $-40°C$, silk fiber strands have a lower risk of breaking in freezing weather than do most other fibers.

Stored-product pests and home invaders

ALONG WITH ALL the good things that resulted from the domestication of plants for food came a few brand-new problems. By storing food, people were unintentionally revolutionizing the life of insects that had heretofore made their living feeding on foodstuffs that were typically few and far between. Many arthropods inhabited the nests of birds and mammals, particularly rodents, and fed on the grain caches assembled by their unwitting hosts. Others were associated with dead animals and dried animal carcasses, never a resource in tremendous supply, and still others lived on whatever they could find in leaf litter or soil. For these arthropods, the human practice of stockpiling food created a bonanza beyond their wildest imagination. From the human point of view, stockpiling food and other materials created a monster in the form of stored-product pests. For virtually every kind of product stored in any quantity by humans, even manufactured goods, there are insects that eat it, almost to the exclusion of anything else.

Probably the most numerous and economically destructive stored-product species are those that feed on grain and grain products. That this mooching has been going on for some time is evidenced by the fact that remains of stored-product species, such as the beetle *Tribolium castaneum,* have been found in alabaster vases in Tutankhamen's tomb dating back to around 1350 B.C. *Sitophilus granarius,* another grain pest, has been found in grain dating back to the ninth to seventh centuries B.C. in Israel. In fact, one minor grain pest, a colydiid beetle called *Thaumasphrastus karanisensis,* was first discovered in grain from an Egyptian tomb; it was subsequently found breeding in south Texas in a rice mill. Many pests of stored grain actually do feed on grain in the field in places where the weather is nice enough to allow it. There are dozens of species, however, that make their living in mills, elevators, and kitchen pantries today.

Stored-product insects differ in their approach to raiding larders. Some, like the weevils, prefer to feed on whole grains; these so-called primary pests include the grain weevils, the grain borers, and the bruchids, as well as some species of Lepidoptera (like the Angoumois grain moth). These are mostly pests in granaries, elevators, corn cribs, barns, and in any other place where intact kernels are stored. Others prefer to feed on processed or damaged grains; these secondary pests include the flour beetles and mealworms. These grubs are often called "bran bugs" and are pests in flour mills and in processed food. Many species of Lepidoptera are also secondary pests, particularly those in the Pyralidae, or snout moth family. These include the Indian meal moth *Plodia interpunctella* and the Mediterranean flour moth *Anagasta kuehniella.*

One characteristic that most grain-feeding stored-product pests share is an extraordinary tolerance to dry, hot conditions. The Mediterranean flour moth *Anagasta kuhniella* can develop on foods with less than 1% water content without ever taking a drink of water. Many stored-product species meet their dietary requirement for moisture simply by extracting water from the carbohydrates in their food. Mealworms, many of whose relatives in the family Tenebrionidae live in the desert, can actually absorb moisture from the air, even when the relative humidity is below 50%, to supplement their water intake. Adapting to feeding on a constant and predictable food supply has led in some species to the loss of a means of dispersal, or getting around; there's no need to. The granary weevil, *Sitophilus granarius,* for example, has only rudimentary vestiges of what once were functional wings.

In some ways, when humans sieve, winnow, chop, blend, and otherwise process plant material, they render it far more to the liking of insects, who don't then have to contend with a lot of indigestible chaff. Processing can also remove a lot of the nasty poisons that plants tend to accumulate to discourage herbivory. But not always, and many stored-product species have a remarkable tolerance for what would otherwise seem to be totally unsuitable food. The cigarette beetle *Lasioderma serricorne* (Anobiidae) has an extraordinary predilection for cigars, cigarettes, and chewing tobacco, even though the tobacco plant, in the field as well as chopped and rolled, is loaded with nicotine, a substance that is lethal to most insects. Drugstore beetles (*Stegobium paniceum*) have been found happily breeding in jars of cayenne pepper, a concentrated diet of which is certainly beyond the capabilities of even the most diehard pepper fans. And confused flour beetles (*Tribolium confusum*) have been found contentedly infesting bales of marijuana confiscated and stored by the Federal Drug Enforcement Administration (although the name "confused" refers to the fact that the species is often confused with *T. castaneum,* the red flour beetle, and doesn't refer to any diet-related mental state in this instance).

Paper is one plant-derived product particularly vulnerable to insect depredation. Over the centuries humans have used a variety of substances in the manufacture of books and paper and associated with this material are a number of arthropods collectively known as bookworms. One of the earliest literary references to the bookworm appears in the Apocrypha of the Old Testament; Moses instructs Joshua to take care of the Torah by anointing it with cedar oil, a highly efficacious insect repellent to this day. Aristotle remarked in Book V of his *Historia Animalium* on finding "a creature resembling a scorpion . . . [and] also other animalcules resembling the grubs found in garments." The scorpionlike creature may well have been a pseudoscorpion or a book mite

(*Cheyletus eruditus*); if so, it was more likely feeding on book-worms than on books per se.

Principal members of the literary guild, as it were, are species of Lepidoptera, particularly in the families Oecophoridae and Tineidae; species in the latter family are generally associated with dried carrion and so would likely have little problem accommodating to a diet of vellum or parchment, prepared from dried animal skins. Paper, plant fibers felted together into sheets, was not widely used in book production until the twelfth century. Paper, as well as the glues used to bind sheets of paper together, is vulnerable to the depredations of *Lepisma saccharina,* the silverfish. As one of the few animals that can manufacture its own cellulases, it is capable of extracting nutrients from the cellulose fibers in paper products. Other book associates include *Ptinus fur,* a tiny spider beetle, and many species of anobiid beetles.

Control measures for bookworms today are largely preventative; simply replacing wooden shelves with metal ones in libraries is a good preventative measure since eggs and larvae have no place to hide. Modern control techniques, particularly for preserving rare books, include fumigation with insecticides, sealing off infested books in carbon-dioxide-saturated environments, and even flash-freezing infested books.

References

Green meals

Jones, D.P., 1973. Agricultural entomology. In: *History of Entomology.* Smith et al., eds. Palo Alto: Annual Review Inc., 307–332.

Ohmart, C.P., L.G. Stewart, and J.R. Thomas, 1983. Leaf consumption by insects in three *Eucalyptus* forest types in southeastern Australia and their role in short-term nutrient cycling. *Oecologia* 59: 322–330.

Schowalter, T.D., 1981. Insect herbivore relationship to the state of the host plant: biotic regulation of ecosystem nutrient cycling through ecological succession. *Oikos* 37: 126–130.

Slansky, F. and J.G. Rodriguez, 1987. *Nutritional Ecology of Insects, Mites, Spiders, and Related Invertebrates.* New York: Wiley.

Weis, A. and M. Berenbaum, 1989. Herbivorous insects and green plants. In: *Plant-Animal Interactions* (W. Abrahamson, ed.). New York: McGraw-Hill.

Down on the farm

Heiser, C.B., 1973. *Seed to Civilization.* San Francisco: Freeman.

Klein, R.M., 1979. *The Green World.* New York: Harper and Row.

Population growth (locust plagues and insect exorcisms)

Bodenheimer, 1960. *Animals and Man in Bible Lands.* Leiden: Brill.

Borror, D., D. Delong, and C. Triplehorn, 1976. *Introduction to the Study of Insects.* New York: Holt, Rinehart and Winston.

Draper, J., 1980. The direction of desert locust migration. *J. Animal Ecol.* 49: 959–974.

Ellis, P.E., D.B. Carlisle, and D.J. Osborne, 1965. Desert locusts: sexual maturity delayed by feeding on senescent vegetation. *Science* 149: 546–547.

Evans, E.P., 1906. *The Criminal Prosecution and Capital Punishment of Animals.* Reprint, London: Faber and Faber, 1968.

Evans, H.E., 1976. *Life on a Little-Known Planet.* New York: Dutton.

Hobson, R. and J. Lawton, 1987. New battle in an ancient war. *Aramco World Magazine.* 38: 6-13.

1954. Insects in the Bible. In: *Insect Facts.* Commemoration Committee for 100 years of Professional Entomology. SB 931, C74.

Johnson, C.G., 1969. *Migration and Dispersal of Insects.* London: Methuen.

Kennedy, J.S., 1951. The migration of the Desert Locust (*Schistocerca gregaria* Forsk.). I. The behaviour of swarms. II. A theory of long-range migrations. *Phil. Trans. R. Soc. Lond. [B]* 235: 481–489.

Newman, L.H., 1966. *Man and Insects.* Garden City: Natural History Press.

Oldroyd, H., 1964. *The Natural History of Flies.* New York: Norton.

Riley, C.V., 1880. *Second Report of the United States Entomological Commission for the Years 1878 and 1879, Relating to the Rocky Mountain Locust and the Western Cricket.* Washington, D.C.: Government Printing Office.

Waloff, Z., 1976. Some temporal characteristics of Desert Locust plagues. *Anti-Locust Memoir* 13: 1–36.

Waloff, Z., 1966. The upsurges and recessions of the Desert Locust plague: an historical survey. *Anti-Locust Memoir* 8: 1–111.

Webster, F.M., 1908. Some things that the grower of cereal and forage crops should know about insects. *Yearbook of the USDA.* Washington: Government Printing Office, pp 367–388.

Useful homopterans (working for scale)

Baranyovits, F.L.C., 1978. Cochineal carmine: an ancient dye with a modern role. *Endeavor* 2: 85–92.

Ben-Dov, Y., 1988. Manna scale, *Trabutina mannipara* (Hemprich and Ehrenberg) (Homoptera: Coccoidea: Pseudococcidae). *Syst. Entomol.* 13: 387–392.

Blanchard, R., 1883. *Les Coccidés Utiles.* Paris: Librairie J.-B. Balliere et Fils.

Bodenheim, F.S., 1951. *Insects as Human Food.* The Hague: Junk.

Clausen, C.P., ed., 1978. *Introduced Parasites and Predators of Arthropod Pests and Weeds: a World Review.* USDA Agricultural Research Service Agricultural Handbook 480.

Cowan, F., 1865. *Curious Facts in the History of Insects.* Philadelphia: Lippincott.

Eisner, T.S. Nowicki, M. Goetz, and J. Meinwald, 1980. Red cochineal dye (carminic acid): its role in nature. *Science* 208: 1039-1042.

Graham, F., 1984. *The Dragon Hunters.* New York: Dutton.

Kearney, P.W., 1958. Shellac. *Illustrated Library of the Natural Sciences.* E.M. Weyer, ed. New York: Simon and Schuster, pp. 1543–1551.

Kosztarab, M., 1987. Everything unique or unusual about scale insects (Homoptera: Coccidae). *Bull. Entomol. Soc. Amer.* 33: 215–220.

Lloyd, A.G., 1980. Extraction and chemistry of cochineal. *Food Chem.* 5: 91–107.

Madsen, H. L., H. Stapelfeldt, G. Bertelsen, and L. H. Skibsted, 1990. Cochineal as a colorant in processed pork meat. Colour matching and oxidative stability. *Food Chem.* 40: 265–269.

Muckhopandhyay, B. and M. Muthana, eds., 1962. *A Monograph on Lac.* Bihar, India: Indian Lac Research Institute.

Roberts, D. A. and C. W. Boothroyd, 1975. *Fundamentals of Plant Pathology.* San Francisco: Freeman.

Ross, G.N., 1986. The bug in the rug. *Nat. Hist.* March: 67–73.

Saint-Pierre, C. and O. Bingrong, 1994. Lac host-trees and the balance of agroecosystems in South Yunnan, China. *Econ. Bot.* 48: 21–28.

Watt, G., 1908. *The Commercial Products of India.* London: John Murray, Albermarle Street.

Gallmakers ("gall enough in thy ink")

1993. A loaf of bread, a jug of wine, and phylloxera. *Entomol. Soc. Newsletter* 16: 1–4.

Berlin, B. and G. Prance, 1978. Insect galls and human ornamentation: the ethnobotanical significance of a new species of *Licania* from Amazonas, Peru. *Biotropica* 10: 81–86.

Cornell, H.V., 1983. The secondary chemistry and complex morphology of galls formed by the Cynipidae (Hymenoptera): why and how? *Am. Midl. Nat.* 110: 225–234.

Fagan, M., 1916. The uses of insect galls. *Am. Nat.* 52: 155-176.

Felt, E.P., 1940. *Plant Galls and Gall Makers.* Ithaca, New York: Comstock.

Frost, S.W., 1942. *General Entomology.* New York: McGraw-Hill.

Hartnett, D.C. and W.G. Abrahamson, 1979. The effects of stem gall insects on life history patterns in *Solidago canadensis. Ecology* 60: 910–917.

Mani, M.S., 1964. *Ecology of Plant Galls.* The Hague: Junk.

Metcalf, R.L. and R.A. Metcalf, 1993. *Destructive and Useful Insects.* New York: McGraw-Hill.

Phipson, T.L., 1864. *The Utilization of Minute Life.* London: Groombridge and Sons.

Riley, W.A., 1919. A use of galls by the Chippewa Indians. *J. Econ. Entomol.* 12: 217–218.

Taper, M.C. and T.J. Case, 1987. Interactions between oak tannins and parasite community structure: unexpected benefits of tannins to cynipid gall-wasps. *Oecologia* 71: 275–280.

Weis, A. and M. Berenbaum, 1989. Herbivorous insects and green plants. In: *Plant-Animal Interactions.* W. Abrahamson, ed. New York: McGraw-Hill, pp. 123–62.

Silk and silkworms

Arbousett, M., 1905. *On Silk and the Silkworm.* Leeds, England: Moorland.

Beard, J., 1992. Warding off bullets by a spider's thread. *New Scientist* 136 (14 November): 18.

Calvert, P., 1985. The spinning of silk. *Nature* 315: 17.

Capsadell, L., 1883. *Complete Guide to Silk Culture.* New York: Smith.

Cherry, R., 1987. History of sericulture. *Bull. Entomol. Soc. Am.* 33: 83–84.

Chou, I., 1980. A History of Chinese Entomology. *Entomotaxonomia,* c/o N. W. College of Agriculture, Wugong, Shaanxi, People's Republic of China.

Corticelli Silk Mills, 1902. *Silk: Its Origin, Culture and Manufacture.* Florence, MA: The Nonotuck Silk Co.

Cowan, F., 1865. *Curious Facts in the History of Insects.* Philadelphia: Lippincott.

Feltwell, J., 1990. *The Story of Silk.* New York: St. Martin's.

Hook, N.S., 1958. Spiders for profit. *Illustrated Library of the Natural Sciences.* E.M. Weyer, ed. New York: Simon and Schuster, pp. 2731-2742.

Hyvarinen, A., 1989. *Silk: A Survey of International Trends in Production and Trade.* Amsterdam: ITC Report.

Johnson, S.A., 1982. *Silkworms.* Minneapolis: Lerner.

Kelly, H. A., 1903. *Silkworm culture.* United States Department of Agriculture Farmers' Bulletin 165. Washington, D.C.: Government Printing Office.

Leggett, W. L., 1949. *The Story of Silk.* New York: Lifetime Editions.

Minns, S., 1929. *Book of the Silkworm.* New York: Nat. Americana Society.

Peigler, R.S., 1993. Wild silks of the world. *Amer. Entomol.* 39: 151–161.

Riley, C.V., 1886. *Mulberry Silkworm.* Washington, D.C.: Government Printing Office. USDA Bull. No. 9: 1–63.

Scranton, P.B., ed. 1985. *Silk City.* Newark: NJ Historical Society.

Stored-product pests and home invaders

Berenbaum, M., 1989. *Ninety-nine Gnats, Nits, and Nibblers.* Urbana: University of Illinois Press.

Buckland, P., 1981. The early dispersal of insect pests of stored products as indicated by archaeological records. *J. Stored Prod. Res.* 17: 1–12.

Cornwell, P.B., 1976. *The Cockroach.* London: Hutchinson.

Cotton, R.T., 1943. *Insect Pests of Stored Grain and Grain Products.* Minneapolis: Burgess.

Hickin, N., 1985. *Bookworms: The Insect Pests of Books.* London: Sheppard.

O'Connor, B., 1981. Evolutionary origins of astigmatid mites inhabiting stored products. *Rec. Adv. Acarol.* 1: 273–278.

O'Conor, J., 1898. *Facts about Bookworms: Their History in Literature and Work in Libraries.* New York: Harper.

Smith, R. and C. Olson, 1982. Confused flour beetle and other Coleoptera in stored marijuana. *Pan Pacific Entomol.* 58: 79–80.

Chapter 6

EATING INSECTS

Meat-eating for fun and profit

CARNIVORY—EATING OTHER animals for a living—is one of the most widespread of arthropod lifestyles. With the exception of a few mites, all arachnids do it, as do, among the insects, all dragonflies and damselflies, some stoneflies, a number of crickets, the occasional earwig, quite a few thrips, many bugs, most neuropterans, numerous flies, countless beetles, most wasps, and even a rogue caterpillar or two. One reason that carnivory may be so ubiquitous is that it poses fewer nutritional challenges than do many other types of diet. While predators have a few problems of their own, what they eat is usually very similar in nutritional composition to their own bodies—there are far fewer differences between consumer and consumed, and nutrient acquisition is not as much of a problem as it is with, say, plants. Insect predators in general live by the slogan "You are what you eat"—most (but not all) insect predators are predators of insects (that is to say, they are entomophagous). There are the odd exceptions: belostomatid bugs (toe-biters) eat small fish, large mantids can take on a lizard, mouse, or small bird, dragonfly nymphs have been reported capturing salamanders and tadpoles, and horse fly larvae can kill and consume small frogs, but most insect carnivores feed on other arthropods.

Predators other than insects also make a diet out of insect prey; insects represent an abundant source of high-quality protein that is relatively easily harvested relative to larger, stronger forms of animal protein on the hoof. Among the most fervent predators of insects are their relatives the arachnids, but insects provide an important source of nutrition to all classes of vertebrates, including fish, amphibians, reptiles, birds, and mammals.

Entomophagous (or insect-eating) predators in general, be they arthropod or nonarthropod, share certain anatomical features that facilitate the capture and consumption of insects. For one thing, they generally tend to be larger than their prey. (One exception to this pattern are the ants, which, by foraging collectively in organized groups, can manage to tackle prey much larger than they are individually. Army ants, for example, forage from bivouacs containing up to 20 million individuals weighing a collective 20 kilograms or more). In addition, predators of insects generally possess well-developed sense organs for detecting the presence of their prey. Visually orienting dragonflies, for example, possess enormous compound eyes with over 20,000 facets, which provide them with virtually 360° vision. Most predators are well-equipped for traveling long distances, since insect prey are rarely abundant enough in any one locality to support a large predator population; they tend to be excellent fliers or runners. One conspicuous exception to this pattern are predators that concentrate on consuming sedentary prey such as aphids, which, with their asexual reproduction, can build up locally to enormous population sizes. Finally, most predators of insects are equipped with appendages to assist them in seizing and securing their prey. Modified mouthparts as well as legs play a role in this aspect of predation. Some species greatly reduce the need for physically grappling with prey by injecting paralytic venoms, thus allowing them to tackle prey considerably larger, stronger, and heavier than they are themselves.

Predators (as opposed to other types of carnivores) typically kill their prey, usually consume it themselves (as opposed to killing it and giving it to their offspring), and kill more than one prey individual over the course of their lifetimes. Arthropod predators employ a variety of prey-securing strategies that involve different amounts of energy input. Some predators eliminate the search-and-pursuit aspect of predation by simply concealing themselves and waiting for prey to come to them. In other words, they set a trap and wait. Spiders are typical sit-and-wait predators. Orb weavers (in the family Araneidae) construct large circular webs that act as aerial filters that intercept and snare prey. The sheet webs of agelenid spiders operate on the principle of flypaper and capture almost exclusively insects that crawl and walk carelessly. Trapdoor spiders construct an underground trap with a silken hatch; when a passing insect wanders by and stumbles in, the trap door is pulled shut by a silken thread and the hapless insect is confined with no avenue for escape.

Praying mantids are also ambush predators, which move in the general direction of a potential prey and wait, poised and ready to strike, until it blunders into the vicinity. Mantids (like many ambush predators) are morphologically well-equipped for a sudden strike (Fig. 6.1). They have raptorial, or grasping, forelegs. These legs are so specialized for snagging prey that they are essentially useless for walking. The segment of the leg nearest to the body (the coxa) is greatly elongated and the next segment is serrated and grooved along its lower surface. The spines of the next segment in line fit snugly into the groove when the leg snaps shut, presumably around the body of a hapless meal. The elongation of the coxa causes the mantis to hold its legs folded underneath its body, as if in prayer—hence the name "praying mantis" (and not, as it's often incorrectly but not inappropriately written, "preying mantis"). The praying mantis is unusual among insects in the degree to which it can swivel its head around; if insects can be said to have shoulders, mantids are unique in their ability to look over them.

Both of these traits—the beatific attitude and the disconcerting stare—have attracted considerable attention to mantids worldwide. The sudden and forceful strike of the praying mantis—which takes only fifty to seventy milliseconds to complete—has actually inspired an entire style of kung fu martial arts in which many of the postures and movements mimic the hunting behavior of the praying mantis. The style was founded by a man named Wang Lang almost 400 years ago in the province of Shantung. Inspired while observing a praying mantis dispatch a cicada, Wang Lang, who was

Figure 6.1
Chinese praying mantis *Tenodera aridifolia sinensis* (J. Sternburg).

already a renowned martial artist, refined a series of movements, hand and arm positions, strikes, and grabs based on subsequent studies of mantis predation. The success of his style, demonstrated in several bouts with the world-famous Shaolin monks (who were devoted students of martial arts), led to the widespread adoption and subsequent modification of the praying mantis style.

One of the drawbacks of a sit-and-wait approach is the fact that there may be long periods of waiting with no success. Many sit-and-wait predators, accordingly, have a tremendous tolerance for starving. The spider *Linyphia triangularis* can develop, depending on the amount of food it receives, from egg to adult in anywhere from 70 to 200 days and can weigh anywhere between 2 and 25 milligrams when full grown, depending on the availability of food.

Some sit-and-wait predators enhance their chances of success by not only setting a trap but baiting it as well. *Mastophora* spiders are orb weavers that do not spin an orb web. Rather, they spin only a single thin silken strand equipped with a sticky ball suspended at the end. The spider dangles the ball with one front leg and tosses it at passing insects when they approach. This ball is imbued with spider-synthesized analogues of the female sex pheromones of neighborhood moths; thus, these spiders feed almost exclusively on male moths. The habit of these spiders of tossing their sticky ball at passing insects has earned them the common name of bolas spiders, in honor of the lariat-like bolas used by Argentine gauchos to lasso their prey. In a more scientific vein, the pitching skills of one species earned it the specific epithet *M. dizzydeani*, in honor of the legendary baseball player Dizzy Dean.

Insinuation is the term given to the ability of certain predators to lull their prey into a false sense of security. Insinuation is behind the bait-and-capture technique of *Salyavata variegata* (Reduviidae), the termite-eating assassin bug. The bug captures a termite, sucks it dry, and approaches the colony while carrying the carcass. When a worker approaches to recycle the body, the reduviid backs off until the worker is far enough away from the colony that the reduviid can attack without risk of provoking soldier termites to rescue their nestmate.

Cleptoparasitism is a mode of predation in which search-and-pursuit are carried out by one species and consumption by another—in other words, some predators simply steal prey from other predators. Many species of spiders are cleptoparasites that set up housekeeping within the webs of larger spiders and steal prey before the resident can secure it. *Nephila clavipes* builds a large orb-web with barrier webbing in front of the capture surface. Smaller spiders live in the barrier webbing and steal unwrapped prey. Although there may be as many as three or four cleptoparasites living in a web, the resident *N. clavipes* seems unaware of their presence. *Pholcus phalangioides,* the cellar spider, while a cleptoparasite that frequently raids the nests of other species to steal prey, also lures resident spiders out of their lairs by plucking web strands in such a way as to suggest that it is itself a struggling victim; when the resident spider rushes out to snare the purported prey, the cellar spider pounces and consumes the would-be predator.

Only a little more energy-intensive than sitting and waiting for prey is searching for and attacking relatively sedentary prey. Soft-bodied homopterans such as scale insects or aphids lead a remarkably sedentary existence; moreover, they are small and relatively defenseless. Accordingly, a number of predaceous insects are specialized for feeding exclusively on them. Such predators include ladybugs, the larvae of hover flies, the brown and green lacewings, and even a few caterpillars. Unlike predators of active prey, predators of soft-bodied slow-moving prey generally do not use specific signals—visual, acoustical, or chemical—for locating their prey. Rather, they rely on random searching movements to bring them into the vicinity of a large population, on which they may feed at their leisure.

A tremendous variety of insect predators actively search out and pursue their prey. In the air, dragonflies are a prime example, as are robber flies (true flies in the family Asilidae); on the ground, ground beetles and tiger beetles are typical active predators. Many of the ground-dwelling arachnids obtain prey in this fashion; these include the wolf spiders, scorpions, the sun-scorpions, and the wind-scorpions. In aquatic habitats, diving beetles, whirligig beetles, and belostomatids (giant water bugs) are fast swimmers that chase down their prey. Once large active prey are captured, predators must subdue and secure them. Although physical strength in many cases can suffice (as in the case of mantids), many predators of active prey use paralytic venoms to immobilize their prey and reduce the energy expenditure involved in securing and subduing them. While they are themselves large as wasps go, cicada killers (wasps in the family Sphecidae) would have considerable difficulties transporting their prey, which weigh as much as or more than they do, back to their underground burrows if they were kicking and struggling the whole way; these wasps inject a paralytic venom that instantly immobilizes the cicada not only for the duration of the trip back to the burrow but also throughout the entire developmental period of the grub, which feeds on the still-living prey.

Parasitoids

> The wing'd Ichneumon for the embryon young
> Gores with sharpt horn the caterpillar throng,
> The cruel larva mines its silky course
> And tears the vitals of its fostering nurse.
> —ERASMUS DARWIN, *Origin of Society*

"PARASITOID" IS A TERM coined over eighty years ago to describe insects that are parasitic only during their immature stages; adults are free-living. Those attacking invertebrates almost always kill their hosts and thus should more properly be considered predators, since parasites don't usually kill their prey whereas predators invariably do. Hence the term "parasitoid"—a word to describe an insect lifestyle that has no close analogy in any other phylum. Unlike predators, adults of these species locate prey for their offspring, not for themselves; they like as not content themselves with sipping nectar from flowers.

The vast majority of parasitoids are found in only two orders: Diptera and Hymenoptera. Parasitoids vary in their degree of intimacy with their hosts. Some parasitoids lay their eggs on the body of the host, where the eggs attach and develop externally. Others lay their eggs on leaves; the eggs hatch when they are swallowed by the host and the larvae develop inside the body of the host, consuming bits and pieces of internal organs at their leisure. Still others, the endoparasitoids, lay their eggs directly into the body of the host (Fig. 6.2), so their offspring never have to venture outside the safety of a nice, cozy insect body cavity until they are ready to pupate—at which point they usu-

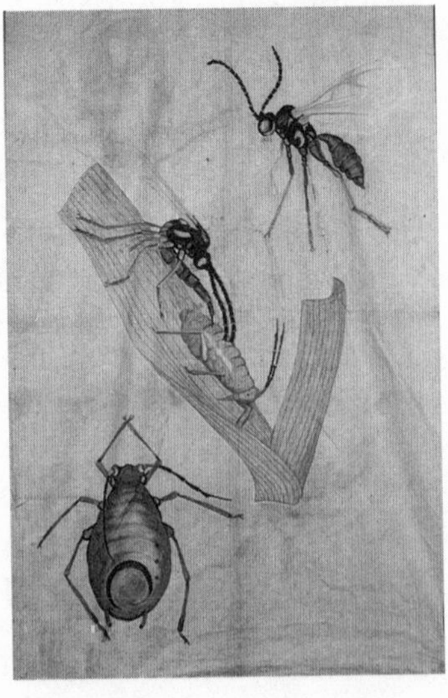

Figure 6.2
Parasitoid wasp attacking an aphid (Department of Entomology, University of Illinois archives).

ally burst through the body wall and kill their still-living but much-depleted host.

Parasitoids share certain ecological similarities with predators. While as larvae they are sedentary, as adults they must search for widely dispersed hosts in which to lay eggs; thus, parasitoid females are well-equipped to travel considerable distances and to locate prey by acoustical, visual, or chemical means. A parasitoid of a gall fly maggot, for example, forages along the stem of the fly's host plant and waits for vibrations resulting from the feeding movements of the maggot mouthhooks. Parasitoids of the cotton bollworm orient to aromatic chemicals released when the bollworms chew on their host plants. Wasp species that parasitize leaf-mining caterpillars can detect leaf mines visually and orient to them specifically in order to locate hosts. Also like predators, parasitoids frequently use paralytic venoms to immobilize their hosts (as in the case of species that deposit eggs on the outside of their hosts). In the case of endoparasitoids, other substances (possibly viruses) are injected that arrest the development of the host so that the parasitoid can complete its development before its host undergoes metamorphosis.

On the other hand, parasitoids also share ecological characteristics with parasites. While predators frequently are highly polyphagous, parasitoids, like many parasites, tend to be highly host-specific. One possible reason for such specificity is that, like parasites, parasitoids must synchronize their development relatively precisely with that of their hosts. They also tend to be smaller than their hosts, whereas predators are often considerably larger than their prey. At least partly due to their small size, parasitoids also tend to have far greater reproductive capacity than do predators, a characteristic shared with parasites.

Biological control of insects

BIOLOGICAL CONTROL IS the use of the natural enemies of an economic pest species for the purpose of reducing the population size of that pest—ideally to the point where the pest no longer has a negative economic impact. In most cases, the pest species is a pest because it's an exotic (introduced from another locality) and in its new habitat, far from its normal contingent of enemies, its populations can grow large and destructive rather quickly. The natural enemy is usually obtained from the pest's native habitat. Traditionally, insect "natural enemies" include parasitoids, predators, and pathogens. Use of these natural control agents falls under the rubric of "classical," or conventional, biological control.

Classical biological control is by no means a new concept. In the thirteenth century and earlier, nests of *Oecophylla smaragdina* (weaver ants) were placed in citrus and litchi trees by farmers in China to control plant-feeding stink bugs (*Tessarotoma papillosa*) and other pests; nests of pharaoh ants (*Monomorium pharaonis*) were brought into barns in China during the same time period for controlling stored products insects. The first long-distance transfer of an exotic predator took place in 1762, when mynah birds from India were brought to Mauritius to control grasshoppers. In more contemporary times, the pest-control crisis that launched biological control into scientific circles was an infestation in California of cottony-cushiony scale on citrus. *Icerya purchasi* was first noticed in Menlo Park around 1868; by 1885, it was a threat to the new citrus industry. C.V. Riley, then State entomologist for Missouri, suggested in 1872 that the pest was Australian in origin; by 1878, as chief entomologist for the USDA, he sent Albert Koebele to Australia to look for natural enemies in the home territory of the cottony-cushion scale. Koebele was sent officially, however, as State Department representative to the International Exposition in Melbourne. Since no foreign travel was allowed for USDA employees Koebele's visit was tacked on as a rider to an appropriations bill just to accommodate Riley.

In Australia, Koebele found a predaceous ladybug, *Rodolia cardinalis,* known as the vedalia beetle. Stocks were sent to Los Angeles to D.W. Coquillet, who reared them and released them onto scale-infested trees enclosed in tents on J.W. Wolfskill's Los Angeles farm. From there they spread to orchard trees nearby. People came from all over the state with little boxes to collect the beetles and take them home. With the assistance of citrus growers and the Los Angeles County Board of Horticultural Commissioners, the beetles were soon established throughout the county. Within one year of their introduction, shipments of citrus went from 700 to 2,000 box cars, the citrus industry was saved, and, best of all, $500 remained of the $2,000 originally allocated for Koebele's trip to Australia. Koebele received a gold watch and his wife a pair of diamond earrings from the grateful growers.

The spectacular success of the vedalia venture catapulted the approach, dubbed "biological control" by H.S. Smith in 1919, into prominence. Examples of successful biological control programs are now legion. The majority involve accidentally introduced pests that form large populations. Many of these targets for biocontrol are relatively sedentary pests (e.g., aphids and scale insects). Most of the successful control agents are parasitoids, at least in part because more parasitoids have been tried in biocontrol programs; there are currently over 1,700 intentionally introduced and established parasitoids, compared with fewer than 500 predators, in the U.S. Host specificity may be the key to a successful biocontrol effort; many of the most successful predators, for example, are exceptionally specialized in terms of range of feeding preference. Ladybugs, for example, feed almost exclusively on aphids or scale insects and include some of the more spectacular biocontrol success stories. Biocontrol agents are now widely available to home gardeners. Both predators and parasitoids can be obtained through mail-order catalogs and arrive complete with instructions for rearing and release.

Insects as prey (studies in bad taste)

THE MAIN PROBLEM faced by any would-be consumer of insects is the interest displayed by virtually all species in self-preservation (i.e., not being eaten). Almost every species of insect has some sort of defense against predators. There are some apparent exceptions: periodical cicadas, for example, which emerge en masse every thirteen or seventeen years seem to lack any kind of morphological or behavioral defense against predators. They may be able to get away with this apparent lack of means for protecting themselves because at least a few individuals escape notice in the mass confusion of a mass emergence; predators quickly eat their fill of cicadas and leave the remaining individuals alone, defenseless though they may be. This strategy is often referred to as predator satiation.

There are several basic defense strategies among insects. Some defenses are in operation all the time, irrespective of whether a predator is nearby. Usually, such defenses are built in as part of the body plan and generally act to reduce the probability that an insect will be discovered by an enemy. Built-in concealment can take several forms. Crypsis, or camouflage, is quite common. In fact, insects are quite remarkable in this regard. There are insects that rest on tree bark that are indistinguishable from tree bark, and insects that rest on rocks that look just like rocks. Crypsis can even be a dynamic process. *Misumena* crab spiders can change color, from yellow to white, to match the color of the flowers on which they sit. There's even a species of grasshoppper that lives in the desert and is basically sand-colored—but when a fire burns through the area, the grasshopper develops black "scorch marks" on its body to enhance its resemblance to the fire-strafed environment.

Background matching can, however, be a tricky business if the environment is subject to change and there is no corresponding flexibility in appearance. The peppered moth is a case in point. Butterfly collectors, at times regarded as eccentric, are if nothing else attentive to variation; they go to great lengths to find rare and unusual forms of species. In the nineteenth century, it was considered a great prize in England to collect a peculiar form of the peppered moth, *Biston betularia carbonaria,* a darker version than the usual form *B. betularia insularia.* But after 1845 or so, the dark forms kept getting more and more common and, within 100 years, in Birmingham, England, 90% of all the forms collected were dark. H.B. Kettlewell decided to investigate the situation. The light form of the peppered moth rests during the day on lichen-covered bark; since it is cryptic and resembles lichen-covered bark, it is almost invisible and is rarely eaten by birds. But when Kettlewell released both dark and light moths around Birmingham, the only ones that weren't eaten were the dark ones. Birmingham is in the middle of the industrial region of England. Soot, smoke, and general pollution covered the trees, not only darkening the trunks but also killing the lichen. Light moths on a sooty trunk were literally dead giveaways and got picked off by birds, while the dark ones, formerly rare because they were conspicuous on light-colored trunks, survived far better in industrialized areas. In Dorset, a rural area with no pollution to speak of, the light forms were more common and, when Kettlewell repeated his experiment there, survived better in the presence of birds and other predators than did the dark ones. Pollution control measures implemented in the 1970s were accompanied throughout the Birmingham area by a resurgence in the frequency of the light-colored peppered moths.

Aside from crypsis, there is another form of concealment known as homotypism, or resembling something inedible. There are insects that resemble bird droppings, which presumably no self-respecting bird would deign to sample; there are even chrysomelid beetles that resemble caterpillar droppings (if possible, an even less appetizing prospect). Three groups of insects are particularly renowned for their reliance on resembling anything but an insect. Many of the treehoppers (Membracidae) look for all the world like thorns and tend to rest on the stem in places where thorns are likely to be found. Most phasmids, or walking sticks, as the name suggests, bear a strong resemblance to sticks or twigs. Many species of caterpillars in the family Geometridae, called inchworms, also resemble sticks or twigs. Both phasmids and inchworms have behavior consistent with their appearance—they can remain motionless, even if prodded, for hours on end. Any twig mimic that begins crawling away after it's touched is likely to lose its credibility (as well as its life). Homotypic insects, like cryptic insects, can have some measure of flexibility in terms of changing their appearance to conform with changes in the environment. In Arizona, an inchworm caterpillar (*Nemoria arizonaria*) feeds on oak catkins (flowers) in the spring and on oak foliage in the summer. In the spring the caterpillars look like catkins and in the summer they look like oak twigs. The change in appearance is apparently due to a physiological response to chemicals in the host plants. Catkins are low in chemicals called tannins; leaves and twigs, in contrast, have high levels of tannin. Caterpillars that consume a low-tannin diet take on the catkin form, while caterpillars that consume a high-tannin diet take on the twig form.

Sometimes the means for looking inedible are obtained from the environment. Some species of green lacewings place bits of plant material or other debris on their backs, so they resemble little traveling dust mice. Larvae of *Chrysopa slossonae*, a green lacewing, prey on woolly alder aphids, which are protected by pugnacious ant bodyguards. These ants consume the honeydew produced by the aphids and zealously protect their food source from marauding would-be predators. To avoid detection by the ants, the larval lacewings pluck the waxy cottony fluff from the backs of woolly alder aphids and affix it to specialized "grappling hooks" on their own backs. Thus disguised, they can wander in among the ants without detection—a real-life example of a wolf in sheep's clothing. Another example of the use of environmental debris for protection hits closer to home (at least, to some homes). The masked bedbug hunter (*Reduvius personatus*) is an assassin bug that eats bed bugs. The immature nymphs are covered with a sticky substance that attracts house dust, which soon covers them from head to tarsus (making, no doubt, for some unnerving confrontations for homeowners unaccustomed to dust that travels on its own power).

One other primary defense is not only to look inedible but actually to be inedible. Bad taste in insects can arise in at least two ways. Insects can either manufacture poisons themselves or they can store up (sequester) poisons from the plants they eat. Almost invariably associated with bad taste is aposematism, or warning coloration. Insects that are distasteful, instead of attempting to conceal themselves, actually advertise the fact with unmistakably bright colors. Curiously, the colors assumed by most aposematic insects are the very same colors used by humans to warn of potential danger—red (like stop signs), yellow or orange, and black (like highway warning signs). Thus, most venomous bees, ants, and wasps (Hymenoptera) are brightly colored, often yellow-and-black striped.

Not all brightly colored insects, however, are distasteful. Some are cheats, that is, perfectly edible species cashing in on the fact that predators are loathe to attack brightly colored prey, particularly after a bad experience or two. Such an arrangement is called mimicry. Batesian mimicry, named for H.W. Bates, a nineteenth-century naturalist, is the situation in which a palatable mimic resembles a toxic model. Harmless perfectly edible insects in many orders mimic the venomous bees, ants, and wasps—everything from flies to beetles to even moths. Hover flies in the family Syrphidae are particularly accomplished in this regard. Not only do they have the coloration pattern down perfectly, they even make their stubby fly antennae seem more wasplike by dangling their thin black forelegs in front of their face, where long, black antennae would be expected to occur. The master mimics tend to appear either early in the spring or in late summer, avoiding the period when naive young fledgling birds who don't know enough to avoid yellow-and-black striped insects are first encountering stinging hymenopterans and learning their lessons.

Many defenses are activated only in the presence of a predator. Retreat and hiding are simple but effective responses to the presence of a predator (it's the approach used by cockroaches when you turn your lights on in your kitchen). There's also a distinctive behavior called thanatosis—playing dead. Many predators can detect and capture only living prey, so by playing dead a potential morsel can make a predator lose interest and look elsewhere. This defense is popularly known as "playing possum" but, considering that most of the 20,000-plus members of the family Chrysomelidae (leaf beetles) resort to this strategy when disturbed, it probably should be called "playing beetle." In this deathlike state, many beetles can remain motionless, even when handled, for several hours.

Many species depend on the element of surprise to disconcert or frighten off a predator. These species tend to be cryptic until disturbed, whereupon they flash a startling pattern in the face of their enemy. One very common flash pattern is an eyespot, a marking that resembles an eye (even down to the glint). A bird, thinking it's about to snag a meal, suddenly finds itself looking into what seems to be the baleful stare of a very large animal. Allen M. Young, a tropical biologist, has an interesting theory to account for the distribution of eyespots in tropical butterflies. He noticed that eyespots are very common in butterflies that feed on rotting fruit on the forest floor but relatively rare among butterflies that feed on nectar. He suggested that the butterflies eating fermenting fruit actually become intoxicated and impaired to the point that, if they became aware of a predator, they would need a few seconds to sober up and coordinate themselves to fly away. The eyespots, flashed in the face of the predator, give the bibulous butterflies the few seconds they need. More abstemious species ostensibly have their wits about them at all times and don't need the extra time to gather their wits about them, as it were.

Even in the insect world the best defense is often a good offense. Many species retaliate when attacked. Since insects are by and large not very large, brute strength is rarely an effective means of fending off an enemy. Most insects resort to chemical weapons for defending themselves against attack. Defensive compounds are usually manufactured and stored in cuticle-lined glands that are equipped with a delivery system allowing the insect to discharge the contents with maximum efficiency. Defensive glands come in an almost endless variety of types. Oozing glands discharge their contents slowly as liquid droplets, which may gradually flow over the body's surface. The aptly named stink bugs (Pentatomidae) produce a mixture of over a dozen different vile-smelling substances from paired oozing glands on their thorax. Eversible glands are concealed inside the body until the insect is disturbed; upon disturbance the gland is everted, or popped out, by hydrostatic pressure. Caterpillars in the family Papilionidae all possess a Y-shaped eversible gland that tucks in behind the head when not in use. When a caterpillar is poked or grabbed, the gland (called an osmeterium) pops out and discharges its odoriferous contents (which in some species include such compounds as isobutyric acid, the chief odor principle of rancid butter).

Perhaps most spectacular of all are reactor glands, actually functional chemical laboratories in which two or more relatively inactive compounds are combined to create a chemical reaction. The bombardier beetle owes its name to its ability, when disturbed, to eject a hot liquid from its abdomen several times in succession with such force that an audible popping sound is produced at each discharge. As described by naturalist John Westwood in 1839, "These beetles on being seized . . . immediately . . . play off their artillery . . . burning . . . the flesh to such a degree that only a few specimens can be captured with the naked hand." The reactor gland of the bombardier beetle is two-chambered; the contents of one chamber are moved into the second chamber, where they combine with other materials in a heat- and oxygen-generating explosion. The heat moves the temperature of the discharged liquids to the boiling point and the oxygen gas liberated propels them outward. The hot liquid is released in a series of rapid spurts, averaging 500 pulses per second, rather than in a constant stream. Rapid muscular contraction controls the flow of materials from reservoir to chamber. Overall, the whole system is eerily similar in design to the V-1 buzz bombs used by the Nazis in London during World War II. The buzz bomb, too, was constructed in a multiple-chamber design. It owed its name to the pulsing sound it emitted when set off; the pulsing in turn resulted from the action of an oscillating flapper valve (analogous to the contracting sphincter muscles) that slammed shut after kerosene from one chamber entered into a second chamber and ignited. The force of the explosion pushed the exhaust gases out of the chamber, propelling the bomb forward and starting the cycle over again.

In some species of insects, toxic substances are not stored in glands but instead circulate freely in the blood or hemolymph. These insects, when attacked, bleed to protect themselves in a process known as reflex bleeding. There are specialized weakened areas of cuticle through which an insect can force its blood, often with sufficient force to send it spraying over several inches. Many beetles are capable of reflex bleeding, ladybugs and oil beetles among them. Reflex bleeders are capable of withstanding losses of up to 25% of their total blood volume without suffering adverse effects.

Physiological defenses are the last resort of an insect after the enemy has penetrated its body. These are generally initiated in response to attack by parasitoids. Hemocytes, or blood cells, of the host congregate around the developing parasitoid larva; more adhere and eventually a capsule consolidates. In one sort of capsule, a black pigment called melanin is deposited, cutting off oxygen to the parasitoid. There are countermeasures taken by parasitoids, however. Some parasitoid fly maggots develop a respiratory tunnel that connects to the host tracheae so that they have a constant supply of oxygen, despite being sealed into an airtight capsule.

Insects as medicine (flies in the ointment)

In case of chronic ulceration with a gaping wound, apply locally some bedbugs,
the heads of which should be removed.
—*A System of Pharmacopoeia,* (1590)

DESPITE THE IMPRESSIVE array of toxins insects produce (or perhaps even because of them), humans have been swallowing insects—deliberately—for centuries. Ever optimistic—and opportunistic—people have throughout the ages reasoned that there must be some way to turn insects to their advantage. Perhaps the most articulate expression of this belief that insects exist for the material benefit of humans can be found in a book titled *Insectotheology, or A Demonstration of the Being and Perfection of God from a Consideration of the Structure and Economy of Insects,* published in 1699:

> Every event in nature proclaims that whatever happens is directed by a Being infinitely wise. From this principle it follows that God hath determined the uses of animals in general and of insects in particular, and that he hath destined them to serve, some one purpose, some another. It is not chance, then, that hath made insects useful, but the eternal decrees of providence; and man hath only turned to his use what was originally intended for his service.

No other virtue coming immediately to mind to those charged with divining these purposes, the most reasonable explanation for the existence of insects more often than not had to do with maintaining human health.

Experimentation with insects as curative agents is one of the more ancient scientific pursuits. The Ebers Papyrus, an Egyptian medical treatise that dates back to the sixteenth century B.C., contains references to insect remedies considered ancient even in those days. Etymologically speaking, the very word "medicine" has entomological roots—it derives from the same root as does the word "mead," or fermented honey wine, to which were ascribed remarkable curative properties.

Divine powers are notorious for working in mysterious ways and it has been the lot of humankind over the centuries to ascertain precisely which medicinal powers were bestowed on which insect. The most optimistic presumed that the Supreme Being in his infinite wisdom not only created all things for human use but invested in all things some physical sign, such as shape or color, to demonstrate its particular usefulness. Thus was formulated what has come to be known as the doctrine of signatures, formalized in

Latin as *Similia similibus curantur*—loosely, "Let likes be cured by likes." Botanical examples of the application of the doctrine of signatures include the use of trilobed liver-shaped leaves of *Hepatica* against jaundice, a liver ailment. Entomological examples, however, are by no means difficult to discover. Conspicuously hairy insects, for example, quite often found their way into hair tonics and baldness cures. In his *Historia Naturalis,* Pliny the Elder quoted Varro as saying:

> The heads of flies, applied fresh to the bald place, is a convenient medicine for the said infirmity and defect. Some use in this case the bloud of flies; others mingle their ashes with the ashes of pear used in old time or els of nuts; with this proportion, that there be a third part only of the ashes of flies to the rest, and herewith for ten days together rubb the bare places where the hair is gone. Some there be, again, who temper and incorporat the said ashes of flies with the juice of colewort and brest-milke; others take nothing thereto but honey.

Bees, too, are noted for hirsuteness and, not surprisingly, appear in various preparations designed to alleviate the embarrassments of hairlessness. Dr. James, in his *Medicinal Dictionary* (1743) writes, "If this powder [bee salts] is mixed in unguents with which the head is anointed, it is said to cure the Alopecia and to contribute to the growth of hair on bald places."

In still another application of the doctrine of signatures, singing insects—crickets, katydids, and some grasshoppers—were frequently recommended for disorders of the throat and ears. Throughout Europe during the Middle Ages, the body juice of crickets was regarded as a sovereign remedy for ear and throat difficulties. The practice may even date back to biblical times, when Jewish women were said to carry *chargol* locusts (the precise taxonomic identity of which is hotly contested among scholars) in their ears to prevent earache. Another singing insect, the cicada, found widespread use in China and Japan as a constituent of earache cures—and, curiously, was used in India among the Santals in ointments mixed with male crickets (which, in most species, are the only ones that can sing) to cure screaming fits.

Earwigs, too, were prescribed for those with hearing difficulties, perhaps due to the recognition of the fan- or ear-shaped hind wing (whence the name "earwig") as a signature. From Dr. James:

> Oil of Ear-wigs is good to strengthen the nerve. . . . These insects, being dried, pulverised, and mixed with the urine of a hare, are esteemed to be good for deafness, being introduced to the ear.

Often, signatures or signs are derived not from appearance but rather from behavior. For example, the widespread use of all manner of insects to enhance fertility or virility may have resulted simply from observation of the tremendous powers of increase of insects. Termites in particular, which can form colonies with populations upwards of 1 million individuals, have caught the attention of individuals seeking to restore lost virility or strength; the debilitated Surjee Rao, prime minister of Scindia, chief of the Mahrattas, was purported to have diligently sought out these insects to restore his lost vigor in the last century. Caterpillars of the genus *Cossus* were pulverized and prescribed for stimulating lactation in pregnant women—possibly because the caterpillar, on being disturbed, secretes a whitish, oily substance.

Perhaps the most imaginative application of the doctrine of signatures involved the use of field thistle galls, thin veinlike excrescences along the plant stem caused by the presence of a minute wasp larva, in the treatment of hemorrhoids. Such powerful medicine was *Carduus hemorrhoidalis,* as it was called by the medieval medical community, that simply carrying a gall on one's person was deemed an effective prophylactic.

The kindest thing to be said for most "sympathetic medicines" is that, while almost assuredly not doing anyone any good, they probably caused no one any lasting damage. But, as such, sympathetic medicines make up only a small portion of entomological materia medica. The vast majority of insect curatives cannot be explained in terms of real or imaginary signs or resemblances. For many remedies, it certainly is the case that insects or insect products have some definite therapeutic value. Botanical medicine, though for the most part as hit-or-miss as insect medicine, has justified itself in the public mind in the form of such indisputably effective drugs as reserpine, aspirin, quinine, opium, steroid hormones, digitalis and the like—all of which were first discovered as powerful folk medicines. There is really no objective reason to assume that "bug drugs," in the grand scheme of things, are any less effective on the average than are the botanicals.

As for the qualifications that make insects worthy of consideration in the manufacture of pharmaceuticals, they are by no means wanting. When it comes to the synthesis of chemicals, insects are, as is their wont in most endeavors, tremendously prolific. Not only do they synthesize a wide assortment of their own chemicals—defensive sprays, mating and alarm pheromones, venoms and toxins— but many can incorporate into their bodies chemicals from the plants they eat. Suffice it to say that the enormous assortment of chemicals associated with insects includes compounds that are demonstrably emetic, vesicant, irritating, cardioactive, or neuro-toxic—in short, the insect world is a veritable drugstore of pharma-codynamic agents.

Although medieval physicians did not have at their disposal the technical jargon that doctors today dispense freely, they nonethe-less used a bewildering assortment of equally obscure terms and phrases, many of which related to the classification of ailments and restoratives. The system that dominated medicine for more than five centuries was the one proposed by Galen, a physician who lived from A.D. 129 to 200. The publication of his text *De Methode Medendi* so profoundly affected the science of medicine that for centuries afterward certain drugs were referred to as "galenicals." Galen's theory was that good health was the result of a proper bal-ance of the four elements, characterized as hot, cold, wet, and dry. When one of the elements came to predominate over the others, illness resulted. Treatment consisted of the administration of an agent with the opposing element in predominance. Thus, kermes berries, galls of the scale insect *Coccus ilicus,* were possessed of "cooling and drying"—i.e., astringent—properties, no doubt at-tributable to their high content of organic acids (up to 60% tannic acid for a related medicinal species, Dyer's coccid, on oak trees). Galls formed by a completely unrelated group of insects, the cynipid gall wasps (such as the Aleppo gall), are similarly rich in tannic acid and are prescribed for many of the same ailments as are the coccid galls. Due to the presence of many free hydroxyl groups, tannic acid and other similar compounds have an extraordinary ca-pacity for binding with proteins—a property that may be related to vasoconstriction, or narrowing of the blood vessels. Tannic acid is listed currently both in the British and U.S. pharmacopoeia and is recommended for hemorrhage, gastric bleeding and ulcers, and, until recent times, was produced for commercial use by extraction of various insect galls.

Many insect preparations were deemed "warm and dry"—these were prescribed to assist those afflicted with an overabundance of the cold and wet elements—catarrhs, listlessness and, most of all, lack of interest or ability in matters sexual. Dr. James writes:

> Ants are warm, dry and aphrodisiac and their acid smell wonder-fully enlivens the vital spirits. . . . To use them they must be dis-solved with a little salt and the diseased part anointed with the

liquor. . . . It is preferable to all sorts of apoplectic and strengthen-
ing waters, particularly in the cure of catarrh.

Used for everything from tapeworms to deafness, apoplexy, leprosy, and "the itch," ants won medicinal sanction and official status
as *Spiritus Formicarium* (infusion of crushed ants in alcohol) or
Oleum Formicarium (crushed ants in oil). In infusion, ants were
used to eradicate facial hair in children and to counteract various
skin disorders such as flora, leprosy, and lentigo (freckles).

Many of these preparations may have appeared to be of use in
the treatment of such disorders due to the rubefacient (reddening)
properties of formic acid, principal constituent in the sting of many
ants. Sprayed on human skin, formic acid in aqueous solution may
induce reddening, itching, and irritation. Depending on one's mo-
tivational state and personal proclivities, itching and irritation may
well have had "aphrodisiac" effects. Thus according to Dr. James,
"ants . . . incite to venery."

Perhaps the most celebrated of topical irritants or aphrodisiacs
irrespective of origin is cantharidin, a secretion of the renowned
Spanishfly. The principal source of cantharidin is *Lytta vesicatoria,*
a brilliant metallic-blue beetle in the family Meloidae (and thus not
a fly) that occurs throughout southern Europe eastward to Siberia
(and thus not particularly Spanish). The ancient Greeks knew of
the ability of the beetle's secretion to raise blisters externally and
stimulate the mucosa of the bladder and stomach internally. Pliny
reported that people sleeping under a tree full of them suffered
from their effects and that merely holding one in the hand was
sufficient to stimulate the bladder. Such powerful stuff was, of
course, in great demand. Cato of Utica, disposing of royal property
at auction, sold off his personal stock for 60,000 sesterces, a considerable sum in those days. Throughout the centuries, known
officinally as *Musae Hispanicae* or simply *Cantharides,* the beetle
has been prescribed for hydrophobia, rheumatism, gout, wounds,
buboes, carbuncles, snakebite, leprosy, catarrh, lice, earache, lethar-
gies and internally for dropsy, and as an emmenagogue, diuretic,
and aphrodisiac. By Galen's standards, this insect was "extremely
hot."

The efficacy of Spanishfly as a curative was abundantly extolled
by one John Buncle, Esquire, after a visit to his physician in 1734:

So wonderfully has the great Creator provided for his creature,
man; in giving him not only a variety of the most pleasing food,

but so fine a medicine, among a thousand other, as the Spanish fly, to save him from the destroying fever and restore him to health again. . . . The benefit is entirely owing to that heating, attenuating, and pungent salt of this fly, and this fly only, which the divine power and goodness has made a lymphatic purgative, or glandular cathartic for the relief of man, in this fatal and tormenting malady. Vast is our obligation to God for all his providential blessings. Great are the wonders that he doth for the children of men.

Although John Buncle found no cause for complaint, many of his contemporaries may not have fared so well. Cantharidin, the active principle of *Musae Hispanicae,* is not only "extremely hot," it is also extremely toxic even at low dosages. Taken internally, cantharidin can cause serious gastroenteritis and nephritis; as little as 30 milligrams can be lethal. In fact, in 1698 (some forty years before Buncle penned his panegyric), a London physician named Greenfield was committed to Newgate Prison in a sensational malpractice trial for prescribing an excessive oral dose of *Musae Hispanicae* for a stubborn case of strangury and inadvertently poisoning his patient. While in prison, though, the doctor published a series of papers detailing his more successful experiments with cantharidin in the treatment of bladder and kidney diseases and was eventually released from prison to resume his practice. The reputation of Spanishfly as an aphrodisiac gained considerable notoriety in France in 1772, when the lieutenant général criminél Jean-Pierre Chomel of Marseilles prosecuted the notorious Marquis de Sade for poisoning several prostitutes while attempting to inflame their passions with the covert use of cantharidin.

Cantharidin was an enormously popular medicine in this country until the turn of the century. During the Civil War, when supply ships from Europe were cut off due to naval engagements, physicians resorted to using a native beetle in the same family as the Spanishfly, the "old-fashioned potato beetle," *Epicauta vittata,* so dependent had they become on cantharidin in their practice. Despite the dangers involved in its use, cantharidin as a pure extract still is prescribed, mostly in veterinary medicine, as a rubefacient, counterirritant, and vesicant; it has official recognition in the British, Italian, Swiss, Belgian, and Spanish pharmacopoeias.

Recently, considerable attention has been focused on the natural role of cantharidin in the life of insects that produce it. Its toxicity and distastefulness suggest that it has a defensive function. Indeed, when disturbed, Spanishfly and its relatives in the Meloidae release prodigious amounts of the substance by reflex bleeding at the joints between leg segments (earning them the common name of oil beetles). As a feeding deterrent to insect predators, it is effective at vanishingly low concentrations. The distribution of cantharidin within the body of the beetles suggests yet another function, however. Cantharidin was first isolated and identified in *Lytta vesicatoria* in 1810; Beauregard in 1890 found that in Spanishfly the largest quantities could be obtained from the male seminal vesicles and female ovaries. Biosynthesis seems to be restricted to the male accessory glands. Since males transfer copious amounts of cantharidin to females at mating, the substance may actually serve an aphrodisiac function in the insect.

Yet another beetle, *Dermestes typographicus,* is mentioned by both Dioscorides and Galen for raising blisters, and is recommended by El-Scherif for opening abscesses and furuncles. In the case of *D. typographicus,* otherwise known as the common larder beetle, the ability to raise blisters is not due to the presence of any toxin but by the presence of a thick coat of fine hairs on the larvae that, on contact with sensitive parts of the skin, can cause intense itchiness, redness, and all-around discomfort. Other urticating insects, notably caterpillars of many families, have often been prescribed in the treatment of skin disorders—although Thomas Moffett, prominent seventeenth century purveyor of misinformation regarding the natural history of insects and other animals (and father of the little miss who sat on a tuffet eating curds and whey), has to say only that "the Germans know that the hairy Caterpillar dried and powdered stops the flux of the belly . . . [and] some downy and hairy Caterpillars, by tradition are held to cure children, when they cannot swallow their meat for straightness of their jaws."

By acting on skin and mucous membranes, urticants and especially vesicants may have indeed been of some small service in the treatment of urogenital complaints; there existed, however, a number of remedies for such ailments that may have had at their foundation an entirely different mode of action. Insect blood, or hemolymph, is a remarkable assortment of dissolved inorganic ions, sugars, amino acids and other organic molecules. The precise electrolytic balance of the hemolymph varies tremendously along taxonomic lines. In one group of insects, including mayflies, stoneflies, dragonflies, grasshoppers, and true bugs, the principal electrolytes in the hemolymph are sodium ions and chloride ions (at the other extreme, butterflies, moths, bees, ants, and wasps regulate blood osmolarity principally by varying concentrations of amino acids in solution). These ions, particularly sodium, play an important role in regulating water balance in the bodies of human beings—in fact, in the human kidney, the reabsorption of water is in part a function of the sodium concentration of the fluid.

Far and away, the vast majority of insects used in specific cures for bladder and kidney dysfunction are those that have hemolymph characterized by high sodium ion concentrations. Among these is the cicada—Dioscorides and Galen both recommended roasted cicadas for bladder pains. Dr. James, in his *Medical Dictionary* of 1745, quotes Schroeder and Dale as remarking, "The ashes of the Cricket (*Gryllus domesticus*) exhibited, are said to be diuretic," and also mentions that "grasshoppers (*Locusta anglica minor vulgatissima*) in a suffumigation relieve under a disury, especially such as is incident to the female sex." Curiously, many of these preparations involve roasting and burning. While such treatment may have made unpleasant medicine somewhat more palatable (or at least less recognizable), it might also be the case that, in the same manner that sodium salts today find use as diuretics and urinary alkalizers (e.g., sodium citrate and sodium acetate), the ancients found certain insects to be a concentrated source of salts and thus prescribed them, in lieu of the relatively precious commodity mined from the earth, for bladder and urinary complaints.

Insect hemolymph is not only rich in inorganic ions, it is also full of antimicrobial secretions of various kinds. In recent years, numerous investigators have amply demonstrated the bactericidal properties of extracts prepared from all manner of insects. Thus, the prescription of insect preparations, particularly insect blood, for amelioration of many kinds of infectious diseases might have had some, even if slight, empirical basis. The bed bug, *Cimex lectularius,* is a case in point. Dioscorides attributed to bed bugs no fewer than eleven medicinal virtues, ranging from external application (mixed with tortoise blood) for wounds to attachment to the left arm "in some wool that has been stolen from the shepherds" (Hoeppli and Ch'iang 1940) to cure nocturnal fevers. In Chinese medicine, bed bugs find their way into almost as many prescriptions. From "Recipes for restoring life" (1789):

> Bedbugs used for the treatment of lip-turning furuncle. Pound seven bedbugs thoroughly with some cooked rice and apply the paste on the lesion. It will effect a quick cure.

> For stinking and gangrenous ulcers, pound some bedbugs with *Shui Lung Ku* [a mixture of oil and lime used for filling cracks in a boat], mix with sesame oil and apply locally.

> For stye, drop a little blood squeezed from bedbugs on the affected spot. The swelling will soon subside.

And from *A System of Pharmacopoeia* (1590):

> In case of chronic ulceration with a gaping wound, apply locally some bedbugs, the heads of which should be removed.

Not only do many of these prescriptions mention bed bug *blood* specifically, they almost all involve the use of blood to treat forms of bacterial infections or sepsis: stinking ulcers, styes, wounds, and furuncles. That insect blood, or hemolymph, contains antibacterial activity has been known for over 70 years. Virtually all insects are thought to produce a variety of proteins and peptides with pronounced activity against all manner of microbes, including many human pathogens. Thus, the occasional reference to the use of insect blood or headless insects in entomological materia medica, such as Pliny's prescription of headless *Blaps* beetles against king's evil, leprosy, wounds, bruises, and scabs, may represent early recognition of the occurrence of bacteriostatic agents in insect hemolymph.

Perhaps the most scientific use of insect antibiotics arose on the battlefields of war. Harried and underequipped physicians in both the Napoleonic Wars and the American Civil War noticed with alarming frequency that many of the soldiers whom they judged too far gone to save, with deep, gaping, maggot-infested wounds, recovered more quickly and in greater numbers than did the soldiers whom the doctors had elected to treat. During World War I, an American surgeon, W.S. Baer, discovered that maggots of certain species of Diptera, especially those in the genera *Lucilia* and *Wohlfartia,* preferentially consume necrotic tissue and systematically cleanse the wound. Moreover, in the process, the maggots excrete large quantities of allantoin, a nitrogenous material that sterilizes the tissues. For several years, physicians in the U.S. and Canada employed these maggots (raised under stringently aseptic conditions) in cleaning out deep wounds, particularly those associated with open fractures or chronic suppurations; in fact, in 1934, the Entomological Laboratory of the USDA sent out a questionnaire to surgeons who had used the maggots and found that 91.2% of those responding found the method highly satisfactory. With the isolation and synthesis of the antibacterial agent allantoin, followed by the discovery of antibiotics, the practice of maggot therapy has fallen largely into disuse; however, entomology departments around the country still get occasional requests from surgeons for sterile larvae for particularly refractory cases of osteomyelitis (deep bone infection).

Ants have also aided physicians by facilitating wound healing but in an entirely different fashion. In the Amazon, army ants were used by native peoples to assist in surgery. Surgical incisions are closed by allowing the sharp serrated mandibles of the soldiers of these species to close around the incision, thus suturing the wound. Once these ants lock their jaws, they remain forever closed, even when the head is severed from the body. Carpenter ants are used in a similar manner in the Mediterranean region, as well as in parts of Africa and India.

One other indirect use of insects to heal was popular around the turn of the twentieth century, before arsenicals (and later antibiotics) were discovered to be effective at destroying the microbe that causes syphilis. Syphilis, a widespread venereal disease, is, if untreated, a progressive illness that leads eventually to debilitation, in some cases insanity, and death. Without effective medicines, physicians were willing to try all kinds of harsh treatments to halt the progression of this disease (including, among other things, drastic regimens of bloodletting and purging). When it was discovered that the causative agent of syphilis (a spirochete) could not survive temperatures over 40°C either outside or inside the body, physicians experimented with deliberately infecting syphilitic patients with malaria by exposing them to the bite of a mosquito carrier. One characteristic of a full-fledged case of malaria is intermittent, very high fever; such fevers were often sufficient to eradicate the syphilis pathogen from the body. Of course, former syphilitics remained infected with malaria for the rest of their lives (at the time, no effective treatment for the malaria pathogen was known), but the consequences of a lifelong infection with one of the mild strains of malaria were vastly preferable to those of tertiary syphilis.

Sticky or slimy insects have often proved medicinal, due to their very sliminess and stickiness, in the formulation of emollient or demulcent plasters. Most cockroaches, for example, are unlike the majority of insects in that their exoskeleton is coated not with a waxy layer but rather with a greasy layer; among other things, the grease allows the roach to slide into narrow cracks and crevices where human feet cannot pursue them. In the Prussian pharmacopoeia, the cockroach rated officinal status (*Pulvis Tarakanae*) in a number of preparations and was employed in Jamaica, bruised and combined with sugar, to treat ulcers. Whole insects were also given to children to eliminate worms, possibly by just lubricating them away.

Insect venoms, used by insects for everything from paralyzing prey to protecting themselves from predation by vertebrates, contain an amazing assortment of powerful proteinaceous materials, many of which can cause severe pain and systemic effects upon injection, a fact to which anyone who has been stung by a wasp or hornet can attest. Scientific investigations of the properties of insect venoms are centuries old. Perhaps one of the earliest investigators was Cleopatra, who, searching for a painless, fast-acting poison, tested wasp venom on condemned prisoners and found it greatly to her liking; bee venom could not be used for such insidious purposes, due to the sacred status of the insect.

The mode of action of bee and wasp venom is thought to involve blockage of sensory nerves and, due to the presence of the enzyme hyaluronidase, increased capillary permeability. Thus, the long-established practice of administering bee stings or bee extracts (known officially as *Apis*) in the treatment of rheumatism has a possible physiological basis—increased capillary permeability means enhanced blood flow to afflicted areas and ganglionic blockage means reduced perception of neuralgic pain. The therapeutic value of bee venom is becoming more widely recognized by contemporary physicians; not only are a number of commercial preparations of bee venom available today (Apicur, Virapin and Forapin, to name a few), but the Learned Council of the USSR Ministry of Health sanctioned its use in 1957 in the treatment of certain ailments.

Spider venoms are also showing potential as curatives, particularly for nervous system disorders. Spider venoms are frequently complex mixtures, with over a hundred individual components, and are characterized generally by the exceedingly fine site-specificity of action of these components. In 1982, one component of the venom of a Japanese spider was found to block the action of glutamate, the most ubiquitous neurotransmitter substance in vertebrate nervous systems (a neurotransmitter substance being a chemical that carries signals from one nerve cell to another). Specifically, the venom interferes with the passage of ions, or electrically charged atoms, that move in and out of nerve cells to allow them to transmit electrical signals. Because glutamate is thought to be involved in nerve cell death after injury or stroke, spider venoms may be important not only as research tools to be used in elucidating neurological processes but also potentially as curative agents. To supply the newly increased demand for spider venoms from research scientists, several enterprising entrepreneurs have started up businesses devoted to collecting and selling spider venoms; one such firm, located in Black Canyon City, Arizona, is known as Spider Pharm.

The extraordinary toxicity of certain insect defensive compounds has found use throughout the world not in curing disease but in putting food on the table. The people of the Kalahari in Botswana dig up pupae of several species of chrysomelid beetle, including *Diamphidia nigro-ornata,* and apply fluids from these pupae to the shafts of arrows used for hunting. Up to ten pupae may be used to coat a single arrow, which is then dried over fire. Shot at close range, these poison-tipped arrows can bring down game such as giraffe, ostrich, wildebeest, and zebra. The toxin, which is deadly at a dose of 100 to 250 micrograms per kilogram injected into a mouse, causes frantic movement, twitching, and defecation, followed by paralysis and eventually respiratory failure. The active principle has been identified as a protein, dubbed diamphidiatoxin.

Though couched in far more scientific language and divested of its religious overtones, the belief that insects of necessity contain powerful medicaments persists even today in the form of the ongoing investigation of natural products for anticancer activity. Members of the "grind and find" school of investigative pharmacognosy systematically extract and isolate potential cures for cancer from the unlikeliest of places: blue periwinkle flowers, deep-sea sponges, and even insects. To date, these surveys have unearthed a number of highly effective antineoplastic agents from insect sources; cytotoxic and antineoplastic compounds, potentially useful in the treatment of cancer, have been found in the wings of a butterfly, in the legs of a female stag beetle, and in the venom of the common yellow jacket. Although nowadays no one would seriously argue that these potential drugs were placed in insect storage containers to keep until people developed the technology to utilize them, the basic philosophy that there are so many different chemicals produced by insects that at least one of them is going to do something wonderful for people is a sophisticated reaffirmation of the belief that nature has been good to us and a compelling argument for preserving species and habitats in even the most obscure corners of the planet.

Human entomophagy (rustling up some grubs)

For our best and daintiest cheer
Through the bright half of the year
Is but acorns, onions, peas
Okras, lupines, radishes
Vetches, wild pears, nine or ten
with a locust now and then.
—ALEXI, third century Greek poet

THE REACTION OF most Americans to eating insects is a mixture of distaste and disbelief. The official word for insect-eating, entomophagy, even sounds like some sort of horrible disease. Technically speaking, however, it's a commonplace occurrence in the United States, officially sanctioned by the federal government. The Food and Drug Administration puts out a list of maximum permissible levels of insect infestation or damage, that is, a maximum number of insect adults, eggs, immatures, droppings, or fragments with which a food can be sold. For example, in every 100 grams of Brussels sprouts, there can be no more than 40 aphids and/or thrips; a half-cup of raisins can contain no more than 10 insects or equivalent in pieces and 35 fruit fly eggs. Just about every food imaginable is on the FDA list, and for those interested in tallying their daily insect consumption, you can send away for it for free. For example, there may be as many as 56 insect parts in every peanut butter and jelly sandwich.

The reason these standards exist is, as the FDA brochure states, is that "it is not now possible and never has been possible to grow in the field, harvest, and process some crops that are totally free of natural or unavoidable defects. . . . Even with modern technology all defects in foods cannot be eliminated." The levels of maximum permissible infestation are set not so much for health reasons but for aesthetic reasons; usually the limits are at the lowest level of detectability. Nine leaf miners per 100 grams of spinach, for example, may be no less dangerous than 19—they're just a bit less conspicuous. It has been argued that, since entomophagy is widely practiced worldwide, there should be no legal restrictions on the occurrence of insect parts in food—on the contrary, that the label should simply indicate their presence and the consumer be free to decide whether or not to buy the product.

The point of all this is, though, that insect contamination doesn't seem to do anybody any harm (except maybe psychologically, after it's been pointed out to them, and physiologically to that proportion of the population with an allergy to insects or insect parts). In fact, insects may be contributing a small but significant fraction of our daily protein intake. Basically, there's nothing inedible about insects from the human perspective. A quick examination of the composition of insects reveals that they're not so different from beef, pork, or fish. The chitinous exoskeleton is by and large indigestible, but then again, so is apple skin, and the exoskeleton makes up only a small part of the total biomass (about 4% in caterpillars) and doesn't affect the nutritive value of the insects as food. Analysis of the protein content reveals that insects contain all the essential amino acids (those required by humans for synthesizing proteins), and so are superior as a protein source to most plant proteins. Most insects have an overall higher protein content than do plant foods—crude protein content of caterpillars ranges from 30 to 80%. Insects are even rich in vitamins and minerals (so maybe it's the worm a day and not the apple it's eating that keeps the doctor away) (Table 6.1).

Not only are insects nutritionally suitable as a food source, they're economically feasible as well. An important consideration in raising livestock is a measure called the efficiency of conversion of ingested food (ECI), the amount of weight an animal gains per weight of food an animal eats. Chickens are very good at converting the food they eat into food we can eat; for every 100 pounds of chicken feed, 38–40 pounds of chicken meat are produced. Beef cattle produce only about 10 pounds of meat for every 100 pounds consumed (this is one reason that chicken is cheaper than beef). Another way of looking at this ratio is that 90% of what a steer eats is wasted as far as the consumer is concerned (though probably not from the steer's perspective). ECI measurements for insects are not all that easy to make nor are the ones that have been made completely accurate (they vary with food plant, condition of food plant, age, sex, and temperature, among other things, and there's a lot of error involved in weighing caterpillar droppings). On the average, however, insect ECI values are quite respectable. Caterpillars such as silkworm and tent caterpillar, with ECI values of 19–31 and 14–22 respectively, put sheep, with a paltry 5.3, to shame. What is also important is that insects can be fed food that humans cannot consume directly, such as mulberry leaves and cherry tree leaves; cows and sheep, which are often fed grains such as corn and wheat, are in a sense competing with people for food.

Table 6.1
Nutritional content of edible insects and other animals (amount/100g)
(based on data from Defoliart 1989; Watt and Merrill 1975)

Nutrient	Energy (kcal)	Protein (g)	Calcium (g)	Phosphorus (mg)	Iron (mg)	Thiamine (mg)	Riboflavin (mg)	Niacin (mg)
Daily Requirements	2850	37	1	1	18	1.5	1.7	20
Macrotermes subhyalinus (termite)	613	14.2	0.04	0.438	7.5	0.13	1.15	9.5
Imbrasia ertli (caterpillar)	376	9.7	0.050	0.546	1.9	0	0	0
Usta terpsichore (caterpillar)	370	28.2	0.355	0.695	35.5	3.67	1.91	5.2
Rhynchophorus phoenicis (weevil)	562	6.7	0.186	0.314	13.1	3.02	2.24	7.8
Beef (lean ground)	219	27.4	0.012	0.230	3.5	0.09	0.23	6.0
Chicken (roasted white meat)	166	31.6	0.011	0.265	1.3	0.04	0.10	10.7
Pork (thin class roasted loin)	333	25.8	0.011	0.270	3.4	0.96	0.28	5.8
Fish (broiled cod)	170	28.5	0.031	0.274	1.0	0.08	0.11	3.0

Insect-harvesting may actually be a more energy-efficient way to accumulate protein and calories than hunting for game. David Madsen collected sun-dried migratory grasshoppers (*Melanoplus sanguinipes*) that accumulate on the shores of the Great Salt Lake in Utah in enormous numbers (and which were used historically by many native Great Basin inhabitants). Wave action had deposited numbers of grasshoppers in rows ranging upwards of 6 feet wide and 9 inches thick (up to 10,000 grasshoppers per foot). He found that it was feasible for one person to collect some 200 pounds of grasshoppers in an hour; since the grasshoppers contained approximately 1,365 calories per pound (versus 1,240 per pound for beef), a single gatherer could conceivably accumulate some 273,000 calories per hour of effort invested (dried and salted, the grasshoppers also would require no elaborate cooking or preserving procedures). In contrast, hunting for big game returns only about 25,000 calories per hour of effort invested. While such literal windfalls of insect protein are undoubtedly unusual, Madsen and his colleagues managed to collect 18.5 pounds of crickets in an hour simply by gathering them up from the bushes. Since dried crickets contain about 1,270 calories/pound, their hour of work yielded the caloric equivalent of forty-three McDonald's Big Mac sandwiches.

Not only does reason dictate in favor of entomophagy, so, it appears, does instinct. Insect-eating has been a common theme throughout human evolution. Humans may have been retained the habit from simian forebears; many primates regularly consume insects as part of their diet and some, like the Malayan spectral lemur, exist almost exclusively on insects. Chimpanzees use twigs as tools for extracting termites from their mounds, and such activities can make up over 30% of their foraging time. Gorillas, too, have been observed deliberately partaking of insect (particularly ant and termite) meals. Moreover, there is some evidence that certain species of primates actually produce chitinase, an enzyme that would allow them to digest the exoskeleton.

Worldwide, the most commonly consumed insects are those that are available in large quantities—specifically, migratory locusts or social species such as termites. In virtually all entomophagous cultures, however, certain rare species are consumed not so much for their nutritive value but for their flavor. Many true bugs, which produce pungent secretions from defensive glands, are used in condiments; honeydew-producing homopterans are valued for their sweetness. The earliest known human records documenting entomophagy as a regular practice (aside from cave paintings depicting honey hunts) is a bas-relief executed about 700 B.C. showing servants of King Sennacherib of Assyria carrying locusts skewered on sticks in preparation for a feast. Near Ninevah, in the palace of Asurbanipal there is a contemporary bas-relief depicting a similar scene. The Old Testament makes recommendations as to which insects should or shouldn't be eaten. From Leviticus 11 (21–23): "You may eat of all winged creeping things that go upon all fours which have legs above their feet, wherewith to leap upon the earth. . . .

These you may eat; the *arbeh* after his kind, the *sal'am* after his kind, the *chargol* after his kind, and the *chagav* after his kind" (all names of various developmental stages of locusts—scholars love to dispute this point). The Mishnah Torah, a collection of commentaries on the Bible, interprets the passage to mean "among locusts, these are clean; all that have four legs, four wings, and jointed legs, and whose wings cover the greater part of the body." In the Koran, the prophet Mohammed is said to have eaten locusts while discoursing with his disciples. Locusts are permissible food according to Moslem dietary laws. Mohammed states, "Lawful for us are two dead (animals) and two blood—liver and spleen, fish and locusts." While locusts may have been technically permissible, they were not considered choice fare: there's an old Arabic proverb that goes, "A date is better than a locust." The ancient Greeks were avid insect-eaters, too. One of the earliest Greek references to entomophagy comes from a work by no less a personage than Aristotle of Athens. In his *Historia Animalium* he reports, "the larva of the cicada on attaining full size in the ground becomes a nymph; then it tastes best, before the husk is broken. . . . [as for adults] at first the males are better to eat after copulation than the females, which are then full of white eggs." Grasshoppers of various descriptions were sold in the marketplaces to be ground into flour for cakes or simply served as hors d'oeuvres. Aristophanes refers to "four-winged fowl" at the markets for the poorer classes. Greek historians tell of a people in Ethiopia whom they called the Acridophagi—the locust-eaters (Diodorus, second century A.D., remarked that they were a short-lived sickly bunch).

The Romans, with typical Roman exuberance, indulged in a few unusual insect delicacies. Chief among these was the *cossus,* a grub or caterpillar of uncertain identity that in Rome was considered a delicacy after it had been fed with flour. Under more natural circumstances they bore into the trunks of oak trees. After the Romans, though, insect-eating among Westerners was largely a matter of fending off starvation (as was the case during a few European locust plagues) or of personal eccentricity. It's hard to say why and how western culture lost its appetite for insects. Perhaps more energetically effective methods for raising domestic livestock supplanted the need for supplementary insect protein.

On other continents, however, insect-eating was or is a fact of life. Many of the cultures that eat insects, living in hostile and unproductive environments, even rely upon insects as their major protein source. In many preagricultural societies, fast disappearing under the encroachment of Western agricultural practices, hunting and gathering are the means by which food is secured, and nothing even remotely edible is overlooked. Plagues of locusts never caused the despair in Africa that they have in Europe for the simple reason that, in the absence of their normal food plants, people of the African bush have historically just as happily eaten the locusts.

"Bushmen's rice" is actually a kind of ant with a long body and black head; Hottentots like them, too. The Pedi of South Africa, the Shona of Rhodesia, and the Hottentots of South Africa are locust lovers, as are the Bushmen, and rise before dawn to snag the insects before they warm up and begin moving about. The locusts are customarily boiled, cooled, dried, cleaned, and salted; the legs are ground into flour and cooked with salt and peanut butter. Termites are a popular item—roasted, dried, or fried in fat—throughout Africa, as is the soil from termite mounds. Grubs that tunnel in the trunks of palm trees are eagerly sought and are located by leaning against the tree and listening for them. In Cameroon, the Pange eat a number of insect foods, including no fewer than twenty-one kinds of caterpillars. Almost three dozen species of caterpillars are consumed in the Shaba region of Zaire, and, in the Kwango District, the production of dried caterpillars for consumption exceeds 280 tons per year. In South Africa, roasted mopane worms (the immature stage of a large moth) are bagged and sold as snacks, much as cheese puffs are in the United States.

In Asia, insect-eating is widespread, even today. In Thailand, *maeng dana*, a giant water bug, is available for sale in markets (Fig. 6.3) and in fact has made an appearance in California in import stores (it's used in preparing a spicy sauce called *namphala*) (Fig. 6.4). In Japan, until only the

Figure 6.3
Belostomatid bugs on display at a marketplace in Nakhon Patham, Thailand (Alan Schroeder).

last century, eating land mammals was not regarded with favor; accordingly, insects were consumed with great frequency. Silkworm larvae and pupae even today can be purchased in cans in supermarkets. Ricehoppers are served cooked in soy sauce and sugar (with legs and wings removed) and are so popular that they are processed and sold in supermarkets to be served as cocktail snacks (called *tsukudani*). According to the National Food Composition Tables, compiled by the Resources Council of Japan, *tsukudani* contain 22.5% protein (and are probably more nutritious than beer nuts). Mass production of ricehoppers is centered in Nagano prefecture, where insect eating is very popular—at least ten orders are consumed on a regular basis. In the southern part of the prefecture, insects may contribute in an important way to the mineral nutrition of the inhabitants. Nanshin district, for example, is surrounded by mountains and is subject to heavy rains, making transport of salt into the area difficult. Insects may thus be an important source of sodium and other minerals in the diet.

In North America, most insect-eating has occurred west of the Mississippi, since the aboriginal inhabitants of the regions to the east were, due to favorable climatic conditions, agriculturists early on. Entomophagy was most commonly practiced in the arid Southwest and Mexico. Rocky Mountain locusts, which underwent periodic population explosions and destroyed everything green for thousands

LTD.,PART.
MAHACHAI MARKETING
117/2 MOO 1 TAMBOL THACHEEN
SAMUTSAKORN, THAILAND
TEL. 034-421545

of acres prior to their extinction in the early part of the twentieth century (see Chapter 4), were eaten with great gusto by the Assiniboine and Shoshone. These Native Americans used to go on locust drives, like cattle roundups, in which they beat through the bushes and drove the locusts into an open pit. The Paiutes of the southwestern U.S. ate certain species of pine-feeding caterpillars, which they smoked out of

Figure 6.4
Belostomatid "essence" can be purchased in many Asian grocery stores in the United States.

trees by building small brush fires under the canopy, thus capturing and cooking them simultaneously. The Digger Indians of California caught ants by spreading damp skin or bark over anthills and collecting what accumulated there. The Cheyennes, the Snakes, and the Utes ate lice, ants, grasshoppers, and cicadas. Brine flies, which also undergo periodic population explosions and mass in enormous numbers on lakeshores, were favorites of many tribes in California, Utah, and Nevada.

South of the border, the use of insects as foods in Mexico (although primarily as condiments or special treats rather than as staples) continues today. Agave caterpillars (*gusanos de maguey*) of several species are served as appetizers or as filling ingredients for tostadas or tacos. These same insects can be found floating in the bottom of bottles of mescal, a liquor brewed from agave; the presence of the worm is presumably a guarantee of authenticity, since the caterpillars can be found only in agave plants (Fig. 6.5). *Jumiles,* or stink bugs, are used to season chili, and eggs of waterboatmen (*abuahutl*) are ground into flour for cakes.

Throughout the history of Western civilization there have been but few enlightened individuals who have promoted the use of insects for culinary purposes. Linnaeus, the celebrated taxonomist, recognized that such practices existed—in fact, of the palmworm weevil grub he wrote "*larvae assatae in deliciis habentur*" ("roasted or broiled larvae are considered a favorite food"). Ludolph cooked the locusts that were plaguing Germany in 1693 like crayfish, pickled them in vinegar and pepper, and served them to the Council of Frankfurt, summoned to discuss the plague problem. The notorious Barbary pirates included locusts as regular items on their bill of fare and David Livingstone, of "I presume" fame, ate locusts in Botswana when other food supplies ran short. C.V. Riley, head of the U.S. Entomological Commission at the time of the great locust plague of 1873–76, prepared a number of locust recipes and in fact served an entire locust dinner (soup to nuts) to members of Congress to dramatize his feelings that insect-eating might alleviate much of the suffering brought about by the insects. The great Charles Darwin himself, in his book *Phytologia,* commented on the Chinese habit of entomophagy and mentioned that he himself had sampled some caterpillars of a culinary variety (a sphingid, or hornworm) and found them delicious.

Perhaps the most passionate and convincing advocate of Western entomophagy was one V.M. Holt, who in 1885 published a book titled *Why Not Eat Insects?* In this book, Holt persuasively argued that plant-feeding insects in particular were clean, nutritious, easy to prepare, tasty, and in every other way superior to other arthropods, such as shrimp or crab, which are consumed without a qualm. The book is reprinted periodically, probably for the amusement of the reading public rather than for the benefit of the culinary arts.

While the advocates of entomophagy per se are few (but they do exist—R. Kok suggested in 1983 that insects on board spaceships could provide important protein for space travelers), there are more people interested in using insects as a source of protein for other food animals. The ability of many insect species to develop on material that is otherwise unusable is an additional benefit (e.g., farming chironomid midge larvae on manure to feed to shrimp not only produces shrimp but cleans up polluted areas). The negative psychological impact of consuming insects is greatly attenuated when it is one link removed on the food chain, too. According to extensive studies, people cannot detect the difference between meals prepared with chicken raised on chicken feed and those prepared with chickens raised on Mormon crickets.

There is of late increasing interest, particularly in third world countries, in exploiting insect protein commercially. In South Korea, for example, some rice farmers no longer spray insecticide to

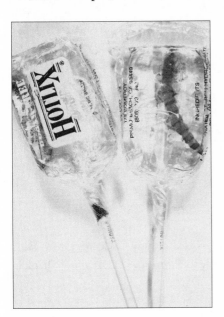

kill the rice-field grasshoppers known locally as *metdugi* but instead harvest them and sell them in addition to or instead of the rice itself; in Mexico, scientists at the National Autonomous University have been developing methods to increase yields of agave worms, a cherished and costly delicacy not only in Mexico but in Japan as well. Dr. Gene Defoliart, at the University of Wis-

Figure 6.5
"Eating the worm" has become easier now that they can be found in tequila-flavored lollipops as well as bottles of mescal.

consin at Madison, founded a periodical, dubbed the *Food Insect Newsletter,* to promulgate scientific information on the history, culture, and feasibility of entomophagy. Other pockets of interest exist as well. One case in point is the Insect Club—a nightclub/bar/restaurant in northwest Washington, D.C., that specializes in, along with insect decor, insect menu items such as mealworm wontons, cricket peanut butter cups, and cricket brittle with ginger essence, prepared with insects reared pesticide-free and shipped fresh regularly from California. At present, however, unless American attitudes undergo a shift of unparalleled proportions, such establishments are likely to remain interesting oddities, frequented perhaps by people who have accepted a dare or had the misfortune to lose a bet.

References

Meat-eating for fun and profit

Hallander, J., 1985. *The Complete Guide to Kung Fu Fighting Styles.* Hollywood: Unique Publications.

Jackson, R.R., and R. J. Brassington, 1987. The biology of *Pholcus phalangioides* (Araneae: Pholcidae): predatory versatility and aggressive mimicry. *J. Zool.* 24: 227–238.

McMahan, E., 1982. Bait-and-capture strategy of a termite-eating assassin bug. *Insectes Sociaux* 29: 346–351.

1993. Praying mantis, kung fu's most effective style? *Black Belt* 31: 87–99.

Price, P., 1976. *Insect Ecology.* New York: J. Wiley and Sons.

Stowe, M.K., J. H. Tumlinson, and R. Heath, 1987. Chemical mimicry: bolas spiders emit components of moth prey species sex pheromones. *Science* 236: 964.

Turnbull, A. L., 1973. Ecology of the true spiders (Araneomorphae). *Annu. Rev. Ent.* 18: 305–348.

Parasitoids

Clausen, C.P., 1940. *Entomophagous Insects.* New York: McGraw-Hill.

Price, P.W., 1980. *Evolutionary Biology of Parasites.* Princeton: Princeton University Press.

Vinson, S.B., 1976. Host selection by insect parasitoids. *Annu. Rev. Entomol.* 21: 109–133.

Biological control of insects

Caltagirone, L.E., 1981. Landmark examples in classical biological control. *Annu. Rev. Entomol.* 26: 213–237.

Caltigirone, L.E. and R.L. Doutt, 1989. The history of the vedalia beetle importation to California and its impact on the development of biological control. *Annu. Rev. Entomol.* 34: 1.

Clausen, C.P., ed., 1978. *Introduced Parasites and Predators of Arthropod Pests and Weeds: A World Review.* Agric. Hdbk. U.S. Dept. Agric., No 480.

Graham, F. Jr., 1984. *The Dragon Hunters.* New York: Dutton.

Huffaker, C.C. and P.S. Messenger, eds., 1976. *Theory and Practice of Biological Control.* New York: Academic Press.

Insects as prey (studies in bad taste)

Blum, M.S., 1978. *Chemical Defenses of Arthropods.* New York: Academic Press.

Burr, M., 1939. *The Insect Legion.* London: Nisbet.

Chapman, R.F., 1971. *The Insects—Structure and Function.* New York: Elsevier.

Dean, J., D.J. Aneshansley, H.E. Edgerton, and T. Eisner, 1990. Defensive spray of the bombardier beetle: a biological pulse jet. *Science* 248: 1219–1221.

Edmunds, M., 1974. *Defense in Animals. A Survey of Anti-Predator Defenses.* Harlow: Longman.

Eisner, T., 1970. Chemical defense against predation in arthropods. In: *Chemical Ecology.* E. Sondheimer and J.B. Simeone, eds. New York: Academic, pp. 157–217.

Greene, E. 1989. A diet-induced developmental polymorphism in a caterpillar. *Science* 243: 643–646.

Kettlewell, H.B.D., 1973. *The Evolution of Melanism.* Oxford: Clarendon Press.

Klausnitzer, B., 1981. *Beetles.* New York: Exeter.

Nakamine, M. and Y. Ito, 1985. "Predator-foolhardiness" in an epidemic cicada population. *Res. Pop. Ecol.* 22: 89–92.

Waldbauer, G. and J. Sheldon, 1971. Phenological relationships of some aculeate Hymenoptera, their dipteran mimics, and insectivorous birds. *Evolution* 25: 371–382.

Wickler, W., 1968. *Mimicry in Plants and Animals.* New York: McGraw-Hill.

Young, A.M., 1979. The evolution of eyespots in tropical butterflies in response to feeding on rotten fruit: an hypothesis. *J. N. Y. Entomol. Soc.* 87: 66–77.

Insects as medicine (flies in the ointment)

Brandt, H., 1960. *Insekten als Rohstofflieferanten.* Munich: Oldenbourg.

Britton, E., 1984. A pointer to a new hallucinogen of insect origin. *J. Ethnopharmacol.* 12: 331–333.

Cowan, F., 1865. *Curious Facts in the History of Insects.* Philadelphia: Lippincott.

Dalton, S., 1975. *Borne on the Wind.* New York: Dutton.

Eisner, T. Chemical defenses against predation in arthropods. In: *Chemical Ecology.* E. Sondheimer and J.B. Simeone, eds. New York: Academic pp. 157–212.

Gudger, E.W., 1925. Stitching wounds with the mandibles of ants and beetles. *J. Am. Med. Assoc.* 84: 1861–1864.

Himan, E.H., 1933. The use of insects and other arthropods in medicine. *J. Trop. Med. Hygiene* 35: 128–134.

Hoeppli, R. and I.-H. Ch'iang, 1940. The louse, crab-louse and bedbug in old Chinese medical literature with special consideration of phthiriasis. *Chinese Med. J.* 58: 338–362.

Howell, M. and P. Ford, 1985. *The Beetle of Aphrodite and other Medical Mysteries.* New York: Random House.

Hultmark, D., 1993. Immune reactions in *Drosophila* and other insects—a model for innate immunity. *Trends Gen.* 9: 178–183.

Ioyrish, N., 1974. *Bees and People.* G.A. Kozlova, trans. Moscow: Mir.

James, R., 1743. *A Medicinal Dictionary.* London: T. Osborne.

Kritsky, G., 1987. Take two cicadas and call me in the morning. *Bull. Entomol. Soc. Am.* 333: 139–141.

Kunze, R., 1893. *Entomological Materia Medica.* Chicago.

Leclerq, M., 1969. *Entomological Parasitology.* New York: Pergamon.

Lesser, F.C., 1799. *Insectotheology.* Edinburgh: W. Creech.

Metcalf, C.L. and W.R. Flint, 1939. *Destructive and Useful Insects.* New York: McGraw-Hill.

Moffet, T., 1967. *The Theater of Insects: The History of Four-footed Beasts and Serpents and Insects.* Vol 3. New York: Da Capo Press [1553–1604].

Pettit, G.R., R.M. Blazer, and D.A. Reierson, 1977. Antineoplastic agents. 51. The yellow jacket *Vespula pennsylvanica. Lloydia* 40: 247–52.

Pettit, G.R. and R.H. Ode, 1976. Antineoplastic agents. 41. The beetle *Allomyrina dichotomus. Lloydia* 39: 129–133.

Phillips, K., 1991. Spider Man. *Discover.* 12(6): 48–53.

Pierce, W.D., 1915. The uses of certain weevils and weevil products in food and medicine. *Proc. Entomol. Soc. Washington.* 17: 151–155.

Posey, D.A., 1987. Ethnoentomological survey of Brazilian Indians. *Entomol. Gen.* 12: 191–292.

Sherman, R.A. and E.A. Pechter, 1988. Maggot therapy: a review of the therapeutic applications of fly larvae in human medicine, especially for treating osteomyelitis. *Med. Vet. Entomol.* 2: 225–230.

Singleton, V.L. and F. Kratzer, 1969. Toxicity and related physiological activity of phenolic substances of plant origin. *J. Agric. Food Chem.* 17: 519–532.

Valenstein, E. S., 1986. *Great and Desperate Cures: The Rise and Decline of Psychic Surgery and Other Medical Treatments.* New York: Basic Books.

Weiss, H.B., 1945. Ancient remedies involving insects. *J. New York Entomol. Soc.* 53: 244.

Weiss, H.B., 1930. John Buncle's panegyric on the Spanish Fly. *J. New York Entomol. Soc.* 38: 49–51.

Weiss, H.B., 1947. Entomological medicaments of the past. *J. New York Entomol. Soc.* 58: 155–165.

Waterston, J., 1929. The Bushman's arrow poison beetle and its parasite. *Nat. Hist. Mag.* 2: 74–80.

Wollard, J.M.R., F.A. Fuhrman, and H.S. Mosher, 1984. The bushman arrow toxin, Diamphida toxin: isolation from pupae of *Diamphidia nigro-ornata*. *Toxicon* 22: 937–946.

Human entomophagy (rustling up some grubs)

Bodenheimer, F., 1951. *Insects as Human Food.* The Hague: Junk.

Bradon, H. 1987. The snack that crawls. *International Wildlife* 17: 16–21.

Castello Yturbide, T., 1986. *Presencia de la Comida Prehispanica.* Mexico City: Banamex.

Cornelius, C., D. Dandrifosse, and C. Jeuniaux, 1976. Chitinolytic enzymes of the gastric mucosa of *Perodicticus potto* (Primate: Prosimian): purification and enzyme specificity. *Int. J. Biochem.* 7: 445–448.

Defoliart, G., 1991. Forest management for the protection of edible caterpillars in Africa. *Food Insects Newsletter* 4: 12.

Defoliart, G.R., 1975. Insects as a source of protein. *Bull. Entomol. Soc. Am.* 21: 161–163.

Defoliart, G.R., l989. The human use of insects as food and as animal feed. *Bull. Entomol. Soc. Am.* 35: 22–35.

Gope, B. and B. Prasad, 1980. Free amino acids in the body tissues of some edible insects of Manipur, India. *J. Adv. Zool.* 77–80.

Gorham, J., 1979. The significance for human health of insects in food. *Annu. Rev. Entomol.* 24: 209–224.

Harris, M., 1985. *Good to Eat: Riddles of Food and Culture.* New York: Simon and Schuster.

Holt, V.M., 1885. *Why Not Eat Insects?* Hampton: Classey.

Ichinose, K., 1989. More on insect eading. *Nature* 337: 513–514.

Kantha, S.S., 1988. Insect eating in Japan. *Nature* 336: 316.

Kok, R., 1983. The production of insects for human food. *Can. Inst. Food Sci. Tech. J.* 16: 5–18.

Krajick, K. 1993. Waiter, there's a fly in my soup and I ordered the cricket salad. *Newsweek* 122: 59E.

Landry, S.V., G.R. Defoliart, and M.L. Sunde, 1986. Larval protein quality of six species of Lepidoptera (Saturniidae, Sphingidae, Noctuidae). *J. Econ. Entomol.* 79: 600–604.

Lindroth, R.L., 1993. Food conversion efficiencies of insect herbivores. *Food Insects Newsletter* 6: 1, 9–10.

Madsen, D.B., 1989. A grasshopper in every pot. *Nat. Hist.* 98 (July): 22–24.

Pemberton, R., 1988. The use of the Thai giant waterbug *Lethocerus indicus* (Hemiptera: Belostomatidae) as human food in California. *Pan-Pacific Entomol.* 64: 81–84.

Shaw, P.C., and K.-K. Mark, 1980. Chironomid farming—a means of recycling farm manure and potentially reducing water pollution in Hong Kong. *Aquaculture* 21: 155–164.

Sokolov, R., 1989. Insects, worms, and other tidbits. *Nat. Hist.* 98 (September): 84–86.

Strezelewicz, M., D. Ullrey, S. Schafer, and J. Bacon, 1985. Feeding insectivores: increasing the calcium content of wax moth (*Galleria mellonella*) larvae. *J. Zool. Anim. Med.* 16: 25–27.

Sutton, M.Q., 1988. *Insects as Food: Aboriginal Entomophagy in the Great Basin.* Ballena Press Anthropological Paper No. 33. Menlo Park, CA: Ballena.

Taylor, R., 1973. *Butterflies in My Stomach or Insects in Human Nutrition.* Santa Barbara: Woodbridge.

Tutin, C.E.G., and M. Fernandez, 1992. Insect-eating by sympatric lowland gorillas (*Gorilla g. gorilla*) and chimpanzees (*Pan t. troglodytes*) in the Lope Reserve, Gabon. *Am. J. Primatol.* 28: 29–40.

Watt, B.K. and A.L. Merrill, 1975. *Handbook of the Nutritional Content of Foods.* New York: Dover Publishing, Inc. (reprint of USDA Agric. Handbook No. 8, Composition of Foods, 1963).

Chapter **7**

PARASITES AND HOSTS

Insects parasitic on humans (what's eating you?)

Fleas! Lice!
A horse peeing By my pillow.
—BASHO, seventeenth century Japanese poet, on staying at a roadside inn.
(Translated by R. Toby)

THE WORD "PARASITE" is almost never used as a compliment; *Webster's New International Dictionary* defines a parasite as "one who eats at the table of another, repaying him with flattery or buffoonery . . . One frequenting the tables of the rich, or living at another's expense, and earning his welcome by flattery; a hanger-on; a toady; a sycophant." Actually, the figurative meanings are accurate interpretations of the biological meanings. A parasite is an animal that obtains its nutrition at the expense of another organism (usually a different species). Unlike predators, parasites generally feed on what are effectively nutritional reserves of their hosts (as opposed to vital organs); thus they do not usually kill their hosts. Host specificity varies enormously: while some parasites are associated with literally hundreds of different host species, others are restricted to only a single species.

Taxonomically, arthropod parasites are a grab-bag group. They are relatively uncommon at the ordinal level among insects with gradual development (that is, without a pupal stage), but the biting lice and sucking lice (which some experts regard as the single order Phthiraptera) are entirely parasitic. A few earwigs and true bugs are parasitic as well. Species with gradual development are almost without exception ectoparasites—parasites that live externally on the bodies of their hosts. Among the orders of insects with complete development, the fleas are wholly ectoparasitic on birds and

191

mammals. Parasitic forms occur sporadically among the beetles—members of the family Platypsyllidae, for example, live a low-profile life as external parasites of beavers. Some moths and butterflies lead an ectoparasitic existence as well; perhaps most unlikely of these are the bloodsucking noctuid moths of Malaysia and Thailand. The order Diptera (flies) contains both ectoparasites and endoparasites of birds and mammals. Most endoparasites are protelean (from *pro,* meaning "before," and *telos* meaning "end")—that is, parasitic only in the larval stages. Included in this category are the bot flies, which spend their formative days in the stomachs, intestines, nasal passages, and other personal spaces of unhappy birds and mammals. Most parasites that are parasitic as adults are ectoparasites; reproduction and dispersal are difficult enough objectives to accomplish without the added handicap of being confined within the body of another organism. However, there are some parasitic mites that complete all life stages inside the bodies of their hosts (in some cases, in such narrow confines as a single hair follicle).

Irrespective of their systematic position, ectoparasites tend to share certain anatomical features. Much of the external structure of permanent ectoparasites (those that rarely, if ever, leave their hosts) is designed for protection against their major enemy—their irritated host. Generally speaking, permanent ectoparasites are effectively two-dimensional, either compressed laterally (side-to-side, like fleas) or flattened dorso-ventrally (top-to-bottom, like lice). Such a body shape (or lack of one) faciliates movement through fur and feathers and hinders the host's ability to remove the parasite. Most ectoparasites are also equipped with a thick, tough cuticle, which can withstand efforts at crushing by the host. Many ectoparasitic forms have wings that are greatly reduced in size, or altogether absent—once a host is found, there is little incentive for a parasite to leave. Ectoparasitic bugs and beetles, for example, are totally wingless; the ectoparasitic sheep ked, a fly, even lacks halteres, the little knoblike structures that replace the hind wings in all other flies. Finally, many ectoparasites have backwards-pointing spines or combs in strategic places on their bodies. These structures, which are not known in nonparasitic species, are found in fleas, polyctenid bugs, a parasitic beetle, and in bat flies. Backwards-pointing spines help the ectoparasite to anchor itself in place and resist host efforts at removal.

It's not so much the case that endoparasites share features as it is that they share an astounding lack of features. Insects that live inside the bodies of other animals tend to lose locomotory appendages and sensory organs. Bot fly maggots, for example, lack eyes, antennae, legs, and most other recognizable body parts. In addition, many endoparasites have highly modified tracheal systems, adapted for what is essentially an aquatic environment inside the bodies of their hosts.

The predominance of parasitic species in the advanced orders with complete development is not altogether surprising in that the principal hosts of these insects, birds and mammals, are a relatively new evolutionary development. The earliest fossils of marsupials and insectivores appear in the late Cretaceous, some 165 to 135 million years ago. In contrast, the forerunners of today's cockroaches and grasshoppers were making their presence felt some 200 million years earlier, in the Carboniferous era (280–345 million years ago). Many insect orders had already undergone substantial diversification prior to the appearance of the birds and mammals. Mammals and birds proved to present an extraordinary evolutionary opportunity for parasitic arthropods because of their reproductive habits. Not only do many of them build nests and shelter their offspring, thereby providing a cozy secure environment for a parasite, they are also relatively long-lived and mate more than once, providing a parasite with opportunities to move to another host with relative ease while their hosts are otherwise pleasantly distracted.

Parasitism does crop up occasionally in orders that are predominantly free-living; in these cases, certain behavioral and anatomical features are easily co-opted for a parasite lifestyle. The true bugs, for example, include in their ranks seed-sucking plant-feeders and insect predators as well as obligate ectoparasites such as the bed bugs, the bat bugs, and some of the kissing bugs. The sucking mouthparts common to all members of the order, with piercing stylets, hypopharyngeal salivary canal, and a muscular pump, are ideally suited not only for piercing plant tissues to suck up sap and for piercing cuticle to suck up hemolymph, but also for piercing mammalian or avian epidermis to suck up warm vertebrate blood.

Humans need not feel unduly persecuted for having a large burden of arthropod parasites. Virtually all mammals are victims of arthropod parasites, the only conspicuous exceptions being cetaceans (whales and dolphins), sea cows (manatees), and scaly pangolins. Possibly the marine habits of the whales, dolphins, and sea cows protect them from parasites to some extent, although a marine existence is no absolute safeguard; there are lice that live in the nostrils of elephant and Weddell seals, for example. In the case of the pangolins, their habits of feeding exclusively on ants may confer some protection at least against ectoparasites, which might easily be killed by angry ants swarming over the body of a pangolin enjoying a meal. The scaly skin of pangolins, which protects them against ant bites, may serve as an impenetrable barrier to ectoparasites as well. Many bird species, particularly those that nest communally, are afflicted by arthropod parasites, and even lower vertebrates, including reptiles, must endure arthropod parasitization. The immature forms of certain trombiculid mites are called chiggers when they inflict themselves on humans; their preferred hosts, however, are snakes (which of course have no hands to scratch the unrelieved itching caused by an infestation of these mites). When it comes to feasting on body fluids, arthropods do not discriminate against their own; many tiny flies in the family Ceratopogonidae, aptly called no-see-ums, suck hemolymph out of the wing veins of larger insects.

It's one of life's great ironies that the major enemies faced by the human species throughout its evolution have been not large, ferocious predators like lions, tigers, and bears but rather tiny insects, on occasion less than one-millionth as large as their target. There's no question that insects are people's worst enemies as far as health matters go; insects have literally plagued humanity since before recorded history. It has even been suggested that the opposable thumb, an anatomical feature distinguishing humans and a few other primates from the rest of the animals, developed in response to the necessity for social grooming—picking parasites off the bodies of close friends and relations. Some of the earliest examples of recorded history include references to insects. At many as five of the ten plagues of Egypt mentioned in the Bible are directly or indirectly attributable to insects (the plagues including flies, lice, murrain, locusts, boils, and possibly darkness). Also mentioned is one of the less endearing gods of the Philistines, Beelzebub, or Lord of the Flies. The term has since come to be synonymous with the devil. On the other side of the world, pre-Columbian pottery dating back to A.D. 1200 and possibly earlier clearly depicts swarms of flies. Insects both as a source of direct injury and as vectors of disease have literally altered the course of human history on occasions too numerous to count.

Insects are injurious to humans in a variety of ways, not all of which are the result of a parasitic association. Among the most direct forms of injury (not to mention insult) is that insects feed on human fluids and tissues. Representatives of about half a dozen orders view humans as a free lunch, with the dish most frequently on the menu being blood (although there are also insects that feed on sweat and even tears). Parasitic insects take two basic approaches to blood feeding. Vessel-feeding insects take blood directly from small veins or venules. Generally among vessel feeders, mouthparts are highly sophisticated and operate on the principle of a hypodermic syringe. Mosquitoes, for example, have mouthparts composed of six stylets. Two elongated sharpened pieces pierce the skin and the vessel and the remaining four form a channel through which blood is drawn up. The bed bug *Cimex lectularius* is also a vessel feeder. The bed bug has a four-part proboscis—a pair of barbed mandibles and a pair of maxillae forming the food tube. Salivary secretions of vessel feeders often contain an anticoagulant, or anticlotting agent, so that the delicate apparatus doesn't get clogged and blocked by a clot. Anticoagulants have been identified in the saliva of kissing bugs and bed bugs, both vessel feeders. The saliva of vessel feeders may also contain anesthetics, or painkilling substances, so that the insect can feed undisturbed by the host (who is unaware he or she is being eaten), and vasodilators, to increase the amount of blood flow in a capillary.

Pool feeders take a much cruder approach to blood feeding. Their mouthparts are not so much like a delicate syringe as they are a can opener or stiletto—the operating principle in extracting blood is brute force. The mouthparts, usually strong and stout, simply rip and tear through tissue, causing profuse bleeding. The insect then leisurely laps up the blood as it flows freely (with no danger of clotting since the blood is moving so fast). The stable fly (*Stomoxys calcitrans*) is a prime example of a pool feeder. Its mouthparts are held out by muscular action like a bayonet and are driven by force into the victim's flesh by a powerful thrust of head and body. Blood-feeding (or rather hemolymph-feeding) on invertebrates occurs as well.

Although consuming blood poses some nutritional challenges, few other body fluids are as readily available and plentiful. Nonetheless, many insects manage to make a living on such unlikely diets as mucus secretions of the face or urogenital region (favorite dining spots of face flies). Even tears from around the eyes of vertebrates serve as arthropod meals; species from several families of moths, mostly from tropical regions, collect around the eyes of cattle, sheep, and even people to imbibe tears. Among ectoparasites of invertebrates, braulid flies (called bee lice) lap up nectar from around the mouthparts of their bee hosts, and some mosquitoes steal nectar and other fluids out of the stomachs of foraging ants.

In terms of body fluids, blood is fairly nutritious, containing almost 20% protein, although much of the protein is present in the form of hemoglobin, which is difficult to digest. Blood does tend to be relatively low in vitamins, however. Most strict blood-feeders avoid vitamin deficiencies by housing in their bodies mutualistic microorganisms that produce the vitamins that are in short supply. These mutualistic associations are most common in obligate, or exclusive, blood-feeders; species that feed as larvae on other food (larval fleas, for example, feed on nest debris) or have alternative sources of food as adults (such as mosquitoes, which feed on nectar as well as blood) often lack these microbes. Transmission of these symbiotic microbes, often housed in special cells called mycetomes, from parent to offspring takes place in a number of different ways. In some species, feces laden with the beneficial microorganisms are smeared over the egg and are consumed by the hatching nymph or larva; in sucking lice, the microbes are incorporated directly into the eggshell wall. Perhaps the most elaborate mechanism of transmission is employed by tsetse flies (*Glossina* spp.). A female tsetse fly produces only a single egg at a time, which hatches inside her body; the growing maggot develops internally and is nourished through a "milk gland," which contains not only nutritive substances but the vital symbiotic microbes as well.

Insects and related arthropods can cause direct injury to vertebrates by getting under their skin, literally. This condition is known as dermatosis. The principal species involved in dermatosis are mites. One such mite is *Sarcoptes scabei*, the scabies mite, also known as seven-year itch, Norwegian itch, mange and a few other more colorful names. Scabies mites burrow into the skin and then hack out long thin sinuous paths, particularly in moist areas: between fingers, behind knees and elbows, ankles, toes, and private places. The major problem with scabies is that after one or two months humans develop an allergic reaction to eggs, mites, and mite droppings deposited in their skin, and start itching madly. Scabies are highly contagious and can be transmitted by contact as casual as a handshake. Scabies epidemics are consequently not uncommon (particularly inasmuch as it may be several weeks before a person realizes he or she is harboring a population). Among the other denizens of human skin are follicle mites, notably *Demodex folliculorum*. These tiny, elongate mites spend almost their entire lives ensconced inside the hair follicles on human faces, particularly those around the eyebrows, nose, and mouth. They're thought to feed on the sebaceous secretions produced by glands near the opening of the hair follicles. Despite the fact that an estimated 75% of humans house follicle mites on their faces, the vast majority of human hosts are unaware of the company they keep, inasmuch as these mites rarely cause any pain or injury.

Unlike dermatosis, some infestations are more than skin-deep and occur inside body cavities. Some insects that feed on dung, instead of waiting for delivery, go directly to the source and establish residence inside the intestine, causing all kinds of problems. Such dung-feeders include beetles; internal infestation by beetles is referred to as canthariasis. Internal infestation by dung beetles is now more or less restricted to places in the world where cleanliness and hygiene are not practiced regularly. Also causing direct injury in a comparable manner are certain flies, such as blow flies and screwworm flies, that normally feed on dead animals or decaying flesh. On occasion, they will infest wounds in living hosts, causing a condition know as myiasis. The human bot fly, or torsalo, is a species that is, however, exclusively associated with normal healthy subcutaneous tissue. *Dermatobia hominis,* as the scientific name implies (*hominus* for "human" and *derma* for "skin"), eats humans. The female fly seeks out female mosquitoes and other blood-feeders in the jungles of Central America and plasters her eggs on the mosquito's abdomen. When the mosquito finds a human to feed on, she pokes a hole with her proboscis; the heat of the human skin causes the bot fly eggs attached to her abdomen to hatch and the maggots crawl down into the preperforated hole to set up residence. Over their 2 to 3 month course of development, the maggots can reach a length of an inch or greater. Attempts to remove these maggots generally cause more damage than allowing them to complete their development undisturbed (whereupon they emerge on their own from under the skin). Recent advances in the treatment of this form of myiasis involve laying several strips of raw bacon over the lesion housing the maggot; in most cases, the maggot moves out of the wound and into the bacon in a matter of hours. Intense discomfort or pain, as well as the disquieting feeling of never being alone, are the major consequences of infestation by human bot flies, although in small children migration by maggots into the brain has proved fatal on some occasions.

Many species of insects, while not parasitic per se, nonetheless cause injury to humans. Insects can on occasion set up residence inside the human body causing accidental injury—cockroaches, for example, occasionally end up lodged in orifices such as ears. This sort of infestation can cause permanent damage to eardrums, not to mention substantial psychic trauma. Thus, there is a premium on developing methods for extricating insects lodged in ears. In a 1988 study reported in the *New England Journal of Medicine,* a patient presented with a cockroach in each ear. The opportunistic physician used this patient to compare two different extrication procedures—lidocaine and mineral oil. Other methods include shining a light in the orifice in order to lure out positively phototactic species, and using suction devices to extricate them manually.

Envenomization is a major source of injury, probably the leading cause of human mortality through direct injury by arthropods. People usually encounter arthropod venom as a result of being stung, not because the arthropod is using venom to bring down big game but because the human unwittingly infringed on its personal space. At least six orders of insects are known to produce venoms, as are a number of noninsect arthropod orders. Venoms are used by many carnivorous arthropods as a means of obtaining prey; a very small animal can bring down a much larger one by injecting paralytic secretions. Such predators include scorpions, spiders, centipedes, assassin bugs, and many wasps. These venoms can also be used, however, in defense against enemies.

Of deaths due to venomous animals in the United States, arthropods account for well over half (the much feared venomous snakes account for only about 30% of deaths by envenomization). Among the arthropods, stinging hymenopterans (bees, ants, and wasps) alone account for about half of deaths due to envenomization. This death toll is at least partly due to the ubiquity of these social insects in places frequented by people; hymenopterans are more likely to run into people than are, say, poisonous jellyfish. Despite their reputation for deadliness, scorpions account for fewer than 2% of envenomization deaths in the U.S; spiders account for about 10 to 15%. Considering the large number of potentially deadly species of venomous arthropods, death due to exposure to arthropod venom is still a pretty unlikely event, statistically; deaths from bee sting, the form that claims the greatest number of lives, number fewer than fifty per year in the U.S. In the case of the honey bee *Apis mellifera,* venom use is strictly for defensive purposes; honey bees consume only pollen and nectar and never use their venom for prey capture. Honey bees possess a large venom sac that is connected to a dartlike stinger (which is actually a modified ovipositor). Because the stinger is barbed, worker honey bees can sting only once; the stinger remains where it has been inserted along with the attached venom sac, which is torn from the worker's body when she crawls or walks away.

Not all venoms derive from bites or stings. Many insects protect themselves with urticating, or stinging, hairs. Urticating hairs are found in sixteen families of Lepidoptera; in all cases, these hairs are used strictly for defensive purposes. The saddleback caterpillar *Sibine stimulea*, for example, is covered with hairs that connect to glands producing histaminelike substances. When the tip of the hair breaks off, as after rough handling, the sharp edge can insert itself under the skin and deliver the venom. Both larvae and adults of species in the Megalopygidae, the flannel moth family, are covered with urticating hairs; contact with adult moths can cause death in small children.

Avoidance of intimate contact with venomous arthropods is not necessarily a safeguard against envenomization. Some arthropods are capable at delivering venom over considerable distances. An assassin bug can spit its protein-destroying saliva over distances of three feet or more, at a rate of about sixteen spits per second. Whip-scorpions have a taillike appendage that revolves like a tank turret while spraying out a defensive secretion consisting mainly of acetic acid, a major constituent of vinegar (and source of the common name "vinegaroon" for these animals). Even some wasps are capable of spraying their venom over several inches.

Venoms generally contain enzymes that break down proteins and cell membranes, as well as compounds that mimic the action of neurotransmitter substances (see Chapter 5). Typical constituents include, among other things, enzymes such as phospholipases and hyaluronidases, which tend to break down cells and tissues. Venoms can also contain other proteins; one such protein, melittin, comprises up to 50% of honey bee venom. Amines such as histamine provoke itching and swelling. In some cases, reactions to venoms are directly attributable to the physiological activity of venom constituents—the peptides, enzymes, and amines that can disrupt cell and organ function. Venoms that affect nerve cell function are called neurotoxic; hemotoxic venoms break down blood cells. Generally, neurotoxic venoms are more life-threatening because nerve cells control such vital physiological processes as respiration. When the nerve cells controlling vital processes cease to function, life—in the absence of an effective antivenin—ceases soon afterward.

A physiological reaction to the injection of venom does not always follow; many factors determine whether or not venoms can cause direct injury to people. Despite the fact that most spiders produce venom, few pose any health risk to humans—either their fangs are not strong enough to penetrate the skin or the venoms are not sufficiently potent to disrupt function. There are exceptions, however. Of sixty-five fatal spider bites in the United States over a ten-year period, sixty-three were attributed to *Latrodectus mactans,* the black widow spider, which produces a highly neurotoxic venom; the remaining two were due to *Loxosceles reclusa,* the brown recluse, which produces a hemolytic venom that causes massive ulceration and tissue necrosis at the site of the bite, but which causes death usually only in small children or emaciated adults. Fatalities can also occur as a result of the physiological effects of exposure to multiple injections of venom—as is the case in some instances of death due to hymenopteran stings. Depending upon the individual, a hundred or more bee stings may bring about cardiovascular failure and death.

Deaths due to envenomization may also be due not to the actions of the venom constituents but due to the body's inappropriate response to their presence in the body—in other words, allergic reactions to injected materials. Allergies are, from the body's perspective, a good idea gone awry. The immune system is designed to protect the body from invasion by foreign bodies, particularly disease-causing organisms. There is a sophisticated and complex array of defenses that are mobilized when a foreign substance invades. Among these responses, certain cells, called lymphocytes, produce antibodies, carbohydrate-linked proteins that, due to unique variations in structure, have a remarkable ability to bind to any number of irritants, or allergens. Once the antibody and antigen are bound together, due to their close molecular fit, the irritant is effectively rendered harmless, taken out of circulation. Chief among these antigen-fighting antibodies is immunoglobulin E, which, like all antibodies, is synthesized and secreted by cells called B lymphocytes. Immunoglobulin E binds to the surface of specialized cells called mast cells and acts as a receptor for its cognate antigen. Formation of an antibody-antigen complex stimulates the mast cells to continue to secrete their contents, including histamine. The next time that particular allergen is encountered, a larger number of mast cells have the matching immunoglobulin E attached, at least in part because among the cells that are produced after the first encounter are so-called memory B cells, which can persist in the body for decades maintaining a record of the unique structure of the particular antigen that led to their formation.

In some cases, certain so-called hypersensitive people respond to exposure to an antigen with an overabundance of mast cells, which release chemicals, such as histamine, that affect smooth muscles of the vascular system. Histamine is the agent responsible for the swelling, itchiness, redness, and respiratory symptoms associated with allergies (mast cells line the respiratory tract and skin by the millions). In severe reactions, the vascular effects of histamine are so pronounced that there is a systemic reaction: first blood vessels narrow, then become more permeable, causing a precipitous drop in blood pressure. This reaction is referred to as anaphylactic shock and, if not treated immediately, can lead to death. Approximately 2% of the population in the United States becomes sensitized to bee venom after a single exposure, and the vast majority of the fifty deaths every year attributed to bee stings are the result of allergic reactions of hypersensitive individuals.

Generally less dramatic than hypersensitive reactions to insect venoms are hypersensitive reactions to inhaled or ingested arthropod body parts or products. Most cases of house dust allergy, for instance, are really allergies to dust denizens. *Dermatophagoides pteronyssinus,* the house dust mite, probably lives in every house that needs dusting. Nobody notices them because they're only 1/100 inch in diameter. Their existence is literally dust to dust—all life stages from egg to adult live in dust (particularly in mattresses). They eat about anything that makes up dust: skin scales, fibers, pollen, fungi, food crumbs, algae, and other such debris. Dandruff, though, is a big favorite, particularly after it has been overgrown by fungus. Cockroach debris, including excrement and body parts, is another major allergen in house dust. At least 75% of asthmatic children display sensitivity to cockroach parts. Due to the frequency with which cockroach parts contaminate food, some food allergies (notably chocolate) may actually be manifestations of cockroach allergies.

The direct injury caused by insects and related arthropods, while substantial, pales into insignificance in comparison with the indirect injury caused by arthropods that are vectors of disease-causing organisms. Much like a gun delivers a fatal bullet, insects deliver pathogens to their hosts. Some of the world's most intractable health problems are arthropod-borne diseases.

Lice, typhus, and war (a lousy situation)

> Her ladyship said when I went to her house
> That she did not esteem me three skips of a louse
> I freely forgave what the dear creature said
> For Ladies will talk of what runs in their head.
> —THEODORE HOOKE, on the third countess of Holland

LIFE FOR US HUMANS has probably been lousy ever since we first appeared on the planet; that is to say, human beings have been prone to periodic infestation by lice for thousands of years. Although there are no fossil lice recovered from caves to confirm this long association, close inspection of the nearest primate relatives of humans reveals that louse infestation, as well as social grooming, are the norm rather than the exception. Perhaps the earliest physical evidence of louse infestation comes from Egyptian tombs, where lice and nits (the eggs of lice) were found mummified along with their human hosts. Hair combs recovered from a number of archaeological sites in Israel dating from as long ago as 100 B.C. almost invariably contain lice at various stages of development. In fact, these combs are double-sided, one side with teeth widely spaced for combing hair and the other side with teeth close together for removing lice and nits. Ectoparasites have been picked out from Viking-age deposits in the western end of Greenland not only from remains of humans but from remains of the sheep they kept as well.

The lice that associate with humans are all sucking lice, distinguished from their biting louse relatives by their haustellate, or sucking, mouthparts, which retract into a little pouch when not in use. Three stylets pierce the skin and capillary blood pressure moves blood into the mouth. Anoplurans, lice with sucking mouthparts, are all bloodsucking ectoparasites of mammals, as opposed to other warm-blooded vertebrates. Their closest relatives are the biting lice, most of which infest birds. While some of the biting lice feed only on hairs, feathers, and other solid foods, others actually ingest blood; this fact, along with the mammal-frequenting habits of some of the biting lice, has led some entomologists to classify all lice into one order, the Phthiraptera. Intermediate forms between the two orders infest, of all things, elephants or warthogs.

Lice in general conform to the gestalt of ectoparasites: they all lack wings and their heavily sclerotized bodies are flattened (top to bottom). In addition, their legs are generally modified for grasping onto fur or feathers. Lice more than most parasites tend to be highly host-specific, perhaps a consequence of gradual development in which every developmental stage is spent on a single host. Specificity can be staggeringly narrow; the glossy ibis is host to no fewer than five feather lice, each of which infests only a particular body part or region. Humans are relatively rare among mammals in having the dubious distinction of serving as host for more than one louse. Probably the most innocuous, but least welcome, of the human lice is *Phthirus pubis,* known as the crab louse due to the presence of relatively massive claws on the second and third pair of legs. *P. pubis,* as the name suggests, infests the pubic region but it can also be found on other parts of the body with coarse thick hair (such as beards, mustaches, or eyebrows). The diameter of the claw is approximately the same as the diameter of a pubic hair and the claw can lock around a hair in a vise-grip fashion. The crab louse is extraordinarily stationary, often remaining more or less in the same spot for its entire development (which is completed in about a month). Crabs generally spread (travel from host to host) during intimate contact, although they can survive for short periods off the host. Thus, there is some basis for concern about toilet seats in unsavory public lavatories, although the ability of lice to travel when off their host is greatly exaggerated in the bathroom grafitto:

> "Don't bother to stretch
> Or stand on the seat,
> The crabs in this place
> Can jump thirty feet."

The only other species in the genus, *P. gorillae,* is a parasite of (true to its name) gorillas.

The genus *Pediculus,* with three or four species (depending on entomological opinions), is, like *Phthirus,* associated exclusively with primates. *Pediculus humanus* is the species that parasitizes people, although for many years the two subspecies of *P. humanus* were given species status. *Pediculus humanus capitis* is the head louse. Elongate in shape, and approximately 2 to 6 millimeters (1/12 to 1/4 inch) in length, it has a tendency to resemble in color the hair of its host. The eggs, or nits, rather large in comparison with the size of the female, are glued to the base of the hair, usually on the scalp. They hatch in about a week to ten days, by which time the hair has grown several millimeters. Development through egg and three nymphal stages takes about three weeks. Head lice tend to infest children (particularly long-haired ones) and can reach epidemic proportions in elementary schools. Frequent contacts

Figure 7.1
Lice (Department of Defense, Armed Forces Pest Management
Board): (above) body lice (*Pediculus humanus humanus*); (below)
delousing, pre-World War I; (right) unit delousing Naples, 1944.

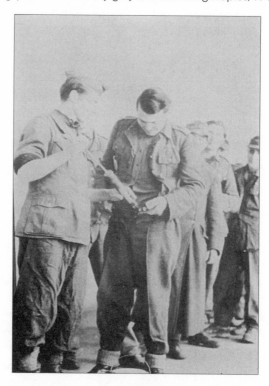

between children at play most likely facilitates their spread. Treatment involves shampooing with insecticidal shampoo and combing hair with fine-tooth comb to eliminate the eggs (from which comes the term "nit-picking").

The body louse, *Pediculus humanus humanus* (Fig. 7.1), is also known as the cootie, seam squirrel, and by a number of other unflattering epithets. Unlike *P. h. capitis*, *P. h. humanus* prefers to reside in the clothes of hosts rather than on their bodies and is capable of surviving a long time off the host. It's considerably more fecund than either *P. h. capitis* or *P. h. pubis*; a female is capable of laying over 300 eggs in a lifetime (in contrast, crab lice lay only about one-tenth that number).

The habits of *P. h. humanus* predispose it to serving as a better disease vector than its cohabitants of human bodies. Not only is the crab louse sedentary, rarely moving more than a couple of centimeters during its life on a host, it has an opportunity to switch hosts only when both parties are in a romantic mood (copulation is, in most human populations, not that frequent an occurrence). In infested clothing, *P. h. humanus* is far more tolerant of life off the host than either of the other human parasites, and clothing can circulate relatively efficiently among prospective hosts. Moreover, with its superior powers of increase, the body louse can build up in epidemic proportions in no time at all. One shirt taken from an infested individual was reported to have contained over 10,000 lice and 10,000 nits (James and Harwood 1969).

Body louse infestation, or pediculosis, in and of itself is anything but pleasant. The bites cause intense itching (which can be socially embarrassing to relieve) and a heavy infestation can bring about a discoloration of the skin known as vagabond's disease or *morphus errorum.* Swelling and skin eruptions can form at the site of the bite. But the major hazard posed by lice is produced at the opposite end of their bodies. Lice act as vectors for exanthematous typhus, a disease caused by *Rickettsia prowazeki,* a rickettsia (a tiny microorganism intermediate in its biology between bacteria and viruses). Like a virus, a rickettsia can develop only inside living cells. The infectious organism was described by Hennaves da Rocha Lima in 1916 and named in honor of two men, H.T. Ricketts and Stanislav von Prowazek, both of whom died of the disease while attempting to isolate and identify the pathogen. The onset of the disease is marked by high fever, excruciating headache, mental disorientation, and a "besotted expression." After five to six days, dark red spots appear on the body. Delirium and a decline in strength follow. Mortality ranges from 10 to 100%, increasing with the age of the patient.

Unlike mosquito vectors of diseases, lice do not inject pathogens along with saliva as they feed—humans after a fashion infect themselves when they accidentally crush a louse or contaminate a feeding site with louse feces. Lice encounter the pathogen that causes typhus in humans when they take a blood meal from an infected human host. While typhus is bad news for people, it's worse news for lice; typhus is always fatal to a louse. The rickettsiae reproduce inside the cells of the louse gut, eventually causing the gut to disintegrate—but not before thousands of infective pathogens are passed out in the feces as the louse feeds on subsequent hosts. Since the feces dry very quickly upon exposure to the air, the fine dust easily gains entry into the human body through louse bites, the lungs, or even the membranes of the eyes, nose, or mouth.

As a disease of people, typhus has a distinctive epidemiology that reflects the habits of its vector. Typhus most often strikes when people are forced to live together in crowded conditions, as during military campaigns or in prison. Thus, typhus is sometimes known as war fever, or jail fever. A form of the disease was described under the name Brill's disease among immigrant eastern European Jewish children living in New York slums in 1898. Typhus, though, has been most important historically as a factor in deciding the outcome of military campaigns. The first clearly recognizable typhus epidemic to be documented in written record struck in 1489, when soldiers arrived in Spain from Cyprus to assist the forces of Ferdinand and Isabella in securing Granada from the Moors. In the siege of Granada, while about 3,000 soldiers died as the result of wounds inflicted by the enemy, 17,000 died of typhus. Cyprus had long been known as a place where this disease was entrenched and was almost certainly the source of the Spanish outbreak.

The next major epidemic in a way brought a delusive resolution to political turmoil in Europe. Since the turn of the sixteenth century, Italy had been a battleground between France and Spain. On February 24, 1525, Spanish forces defeated the French and captured Francis I of France. Two years later, the imperial troops of Spain met up near Naples with superior French forces, now allied with the Turks and backed by the Vatican. About 11,000 imperial troops were surrounded by 28,000 French soldiers. A French victory seemed a certainty until, on July 5, 1527, an epidemic of typhus broke out. Over 21,000 soldiers died of typhus, the siege was lifted, and the remains of the French army beat a hasty retreat (although not hasty enough to escape the imperial troops, which inflicted further casualties). In 1530, Charles V of Spain declared himself sovereign of the Roman Empire.

The historical influence of typhus was felt in both eastern and western hemispheres; the chronicle is long and sad. In 1592, Joachim of Brandenburg lost 30,000 men to typhus in Hungary, which as a result became known as the "graveyard of the Germans." Shortly thereafter, 1618 marked the beginning of the Thirty Years' War, influenced throughout its lengthy duration by typhus. In 1632, Gustavus Adolphus and Albrecht Wenzel von Wallenstein both attempted to capture Nuremburg, but instead lost 18,000 troops to typhus and left to avoid further mortality. Including townspeople, 29,000 people died in seven weeks, by one account. The civilian population was repeatedly hard hit—in 1628 alone, 60,000 died in Lyons. In 1643 the deposed British monarch, Charles I, attempting to recapture London, gave up when typhus crippled both his army and the parlimentary forces.

Perhaps the greatest military victory of typhus, rather than battle strategies or weapons, was the defeat of Napoleon's army. The imperial dreams of Napoleon Bonaparte included conquering the vast reaches of Russia and he set out to accomplish this goal in 1812. Troops began massing in June in east Germany. Counting reserves, Napoleon's forces numbered over 600,000, vastly outnumbering Russia's 250,000 troops. The French made excellent progress until typhus broke out on June 24, 1812, when 500,000 French troops crossed the Nieman River between Prussia and Poland, en route to Moscow via Vilna. This was a new epidemic for the French army. By the start of the battle of Ostrovna, 80,000 had contracted typhus and were either dead or unable to march. As the French troops advanced, the Russian troops continued to retreat, drawing them deeper into the country. Desertions along the way by demoralized soldiers reduced French troop strength even further. When Napoleon's army arrived in Smolensk on August 17, they found it had been torched and abandoned. A confrontation with the Russian army at Valutino two days later did little to advance the French cause. By August 25, French troops departed for Moscow, 200 miles away, with only 160,000 men. By September 5, the number had declined to 130,000. Just outside Moscow in Borodino, in a skirmish with the Russians, the French lost another 30,000 men. Moscow, essentially evacuated before the arrival of the French and systematically cleared of food, clothing, and reasonable shelter, was occupied by the French on September 14. The city was abandoned October 19, with only 80,000 men remaining of the original 600,000; most of the staggering losses were due to the combined effects of hunger and disease rather than to confrontations with the elusive opposing army. The retreat began just as the weather turned cold. Napoleon had not intended to remain so long in Russia, so his troops were ill-prepared for winter weather; in a futile effort to keep warm, soldiers took the clothes of their fallen comrades and put them on, thereby increasing the probability that they would infest themselves with typhus-carrying lice. By the time Napoleon's advance men reached Smolensk on November 8, only 40,000 remained; when the army returned to Vilna in December, only 20,000 were left alive. By the following June only 3,000 of the original half-million troops had survived the ill-fated campaign, and most of those survivors were sick with typhus. Amazingly, Napoleon managed to recruit 500,000 more soldiers upon his return to France in 1813, but the numbers quickly fell to about 170,000 after several skirmishes in Poland; an estimated 219,000 had died of typhus.

Typhus was also decisive in the outcome of the 1854 Crimean War. Statisticians and historians record far more deaths from typhus than from wounds for all warring parties (French, English, and Russian). For the French, the disparity was greatest; while 20,356 died of wounds, 49,815 died of typhus. While the British lost 4,947 to wounds, 7,225 fell victim to typhus. Florence Nightingale, a British nurse sent to minister to the sick and wounded, greatly reduced mortality figures by introducing basic sanitation as a standard hospital practice. Although she had no idea that lice were involved in transmission of the disease, her strict regimen of cleanliness greatly reduced louse populations and hence disease transmission.

Between 1919 and 1923, 3 million people died of typhus worldwide. However, the discovery in 1907 by Charles Nicolle of the transmission of the rickettsia by lice allowed for the implementation of preventative measures (in part earning him a Nobel prize in 1928). These discoveries were made in time to ameliorate the impact of typhus in World War I. This is not to say that typhus didn't exact a terrible toll during that conflict. The assassination of Archduke Ferdinand by an unbalanced Serbian nationalist precipitated a war between Austria and Serbia (a war that later expanded to encompass much of the rest of Europe). In November 1914, typhus broke out in Serbia; at the height of the epidemic, 2,500 new cases were reported each day. Contributing to the morbidity and mortality was the fact that over a quarter of the 400 doctors in Serbia died of typhus, seriously compromising the ability of health care professionals to minister to the sick. Raging unchecked, typhus killed 150,000 people in six months (including half of the Austrian prisoners of war confined in Serbian prison camps). Fortunately, the International Red Cross stepped in and promptly launched a massive delousing campaign facilitated by the donation of almost 1.5 million pounds of insecticide by John D. Rockefeller to Prince Alexis of Serbia. Although in total some 300,000 people died during the epidemic in its entirety, the actions of the Red Cross, precipitated by the recently publicized findings of Nicolle and others, helped to contain the outbreak.

World War II is unique in the annals of military history in that it was the first war in which typhus claimed fewer lives than did the warring parties themselves. This bittersweet victory was made possible by another timely discovery, that of Paul Muller working in a J.R. Geigy chemistry laboratory in Switzerland in 1939. Testing various sorts of known chemicals for their insecticide properties, he came across one, called dichloro-diphenyl-trichloro-ethane, or DDT, with phenomenal toxicity to, among other insects, clothes moths and potato beetles. This material was brought to the attention of the U.S. Department of Agriculture in 1942 by the American representative of J.R. Geigy and extensive testing began, by both the Bureau of Entomology and Plant Quarantine and the War Food Adminstration. Testing proved so promising that in less than two years over 2 million pounds of DDT were being produced for U.S. military use. The first convincing field test of the new miracle insecticide was staged in Naples, Italy, where a typhus epidemic loomed large. In July 1943, after a few cases of typhus had been diagnosed, U.S. and British occupation forces ordered an extensive delousing program, initially with pyrethrum (the botanical powder used back in Serbia in World War I) and subsequently with the newly available DDT. For the first time in military history, a major epidemic was averted, and was attributed to DDT—even though probably equally, if not more, important in halting the epidemic were the rapid mobilization of military health personnel, the extensive field efforts made to ensure all potential victims received treatment, the new and improved instrumentation for dusting people with insecticide, and the efficacy of the old standby botanical insecticide pyrethrum. Since World War II, however, medical and entomological surveillance have been given extremely high priorities by U.S. military officials engaged in any kind of hostilities. Although humans seem sadly doomed to kill each other for political and military objectives, at least in future years it seems unlikely that lice and typhus will be drafted to assist in the process.

Fleas and plague

> He that lies with the dogs, riseth with fleas.
> —G. Herbert, *Tacula Prudentum*

Fleas are probably among the most easily recognized of insects; as the insect morphologist R.E. Snodgrass, a man never noted for overstatement, remarked: "No part of the external anatomy of an adult flea could possibly be mistaken for that of any other insect. The head, the mouthparts, the thorax, the legs, the abdomen, the external genitalia, all present features that are not elsewhere duplicated among the hexapods." They are in fact so unique that entomologists have difficulties determining which other insects fleas are closely related to, since they bear little resemblance to all likely candidates. There are also few insects that present people with so many opportunities to possibly mistake them for any other insects; one or another species of flea attacks humans in just about every country on every continent.

Worldwide, there are about 1,900 species of fleas, or members of the order Siphonaptera. Adults of all species suck blood from birds or mammals. The name of the order, Siphonaptera, is quite descriptive. They are all indeed apterous, or wingless, and are all equipped with siphonlike sucking mouthparts. The mouthparts of fleas are built according to a plan unique in the class Insecta. Piercing the skin is accomplished by a pair of slicing structures. In between these sharp slicers is the epipharynx, a structure that closes off the food channel. These three structures are enclosed by the labium to form an interlocking unit or fascicle. The two maxillary laciniae saw their way through the skin independently and pierce a blood vessel; the flowing blood is drawn up by a pair of muscular pumps in the head. Fleas inject saliva while feeding, which in most hosts provokes an allergic response. Flea bites are quite distinctive, usually clustered in groups of two or three, marked by a central puncture, and capable of producing intense and agonizing itchiness.

There are two other anatomical features shared by fleas in general that reflect not adaptations to exploit a host as food but rather adaptations to avoid the prying, preening claws, nails, or teeth of an irritated host who doesn't appreciate serving as a main course. Fleas are flattened from side to side (laterally compressed); this distinctive shape allows them to move freely in among the hairs or feathers of their host. Coupled with the thick strong exoskeleton (described by Robert Hooke in 1665 as a "polish'd suit of sable Armour, neatly jointed, and beset with multitudes of sharp pinns, shap'd almost like Porcupine's Quills, or . . . conical Steel-bodkins"), it also reduces the probability of a host gaining a firm grasp on their bodies to crush them. The other distinctive anatomical feature is the modification of the third pair of legs for jumping, which fleas do with remarkable agility. The power of the jump comes in part from the rapid downward swing of the muscular femora of the hind legs, but the flea is greatly aided in getting airborne by the presence of resilin, an elastic protein, in the thorax. This protein, like rubber, can store and release energy; when muscular contraction releases the stored energy in the protein, fleas can achieve leaps of over 30 centimeters (12 inches) and accelerations of 135 g. Surprisingly, not all fleas are jumpers; several species that attack birds or squirrels in their nests in trees crawl to make their way around (jumping under such circumstances might lead to a precipitously hard landing).

While adult fleas are distressingly familiar to most people, the larvae come as a bit of a surprise. Flea larvae are unlike their parents in appearance in almost every conceivable way; bristly, cylindrical, legless, and eyeless, they present an overall wormlike impression. Unlike their parents, larval fleas are not blood-feeders, at least not in the strict sense. The preferred food in most species is actually the dried digested blood excreted by adults frequenting the same neighborhood, which they consume with chewing mouthparts. Some flea larvae actually use their mouthparts to attach to the anus of adult fleas as they feed on the host, thus ensuring that fecal material does not go to waste.

In general, female fleas lay their relatively large white eggs in the nest of the host; while eggs are laid occasionally among the hairs of the host, these are not cemented in place (as are nits, or louse eggs) and as a result usually fall off into the nest to accomplish the same end. The larvae feed on dried blood and other offal that can be found in their surroundings; some fleas, like the larvae of *Pulex irritans,* the human flea, can survive entirely on debris other than flea droppings. Development through three larval instars can take anywhere from a week to almost seven months, depending on food availability and temperature. When ready to pupate, the larva spins a cocoon with sticky silk, to which local dirt and debris can adhere. Pupation can last from about a week to about a year; it's been suggested that at least one of the environmental cues that prompt emergence of the pharate adult (which rests fully developed inside the pupal case) are the vibrations associated with the footfalls of an approaching host. Adults of some species can survive for over two years without food.

Unlike some ectoparasites (like lice, for example, which spend all life stages on their hosts), fleas spend comparatively little of their life cycles actually on the bodies of their hosts; by some estimates, cat fleas (*Ctenocephalides felis*) spend as much as 90% of their adult life and 100% of their larval life off the host. Consequently, they are not extraordinarily host-specific, although species tend to have favorites among their many suitable hosts. Humans share their fleas with many other mammals. The cat flea not only feeds on cats but will bite humans with great enthusiasm and is actually the predominant flea on dogs in many parts of the U.S. The Oriental rat flea *Xenopsylla cheopsis,* as the common name implies, is found with greatest frequency on black rats (*Rattus rattus*) but it freely feeds on humans as well. On the other hand, while *Pulex irritans* is known as the human flea, it is not averse to feeding on cats, dogs, rats, goats, skunks, coyotes, pigs, spiny anteaters, and even ducks and owls.

One characteristic of the hosts of fleas, though, is that they are sedentary over at least part of their life cycle; flea larvae need a nest in which to grow. Thus, mammals that travel in herds with relatively mobile young, like horses or bison or giraffes, aren't usually bothered by fleas. Although there are fleas that infest birds (such as the hen flea and the sticktight flea), they are far less successful parasites of birds in general than they are of mammals, perhaps because bird nests are inhabited for such short periods of time and are often located on tree limbs, a locality that is both inaccessible and inhospitably arid for the moisture-loving wingless fleas.

In general, those mammals that do live in nests tend to do so when raising their young. Fleas have developed some remarkable physiological tricks for synchronizing their development with that of their host. The female rabbit flea, *Spilopsyllus cuniculi,* for example, does not begin to develop eggs until she feeds on a pregnant female rabbit. When baby bunnies are born, the flea feeds on them and only then can attract a male for mating. A few days after mating, the female flea lays eggs and flea and rabbit raise their young in the same nest.

The association between humans and fleas has a definite dark side, due to the association of fleas with the bacterium *Yersinia pestis*. *Y. pestis* is basically a disease of rodents. Fleas ingest the bacteria by feeding on an infected host. The bacteria are sucked up along with the blood into the pharynx, through the esophagus, past the proventriculus (a filterlike arrangement consisting of seven circlets of spines surrounding a narrow passage) and into the stomach. Both in the stomach and in the proventriculus, the bacteria can grow and multiply. They multiply to such an extent that they can block the proventriculus altogether, preventing the flea from ingesting blood. The starving flea, however, continues to attempt to feed, regurgitating masses of bacteria into the host. For many rodents, this is not a major problem since they do not manifest symptoms of disease even when infected by the bacteria. However, for other less fortunate species, including mice, rats, squirrels, gerbils, and prairie dogs, infection by *Yersinia pestis* can mean swift and agonizing death. When the disease affects populations of wild animals, it is referred to as sylvatic plague.

Plague would just be one of many causes of mortality in rodents and of no great concern to humans except for one thing— when rodent hosts start dying off, many rodent fleas, in particular the Oriental rat flea, have no qualms about feeding on humans. Given the close association between humans and certain species of rats (willingly or unwillingly), the consequences of this change in diet have been enormous for the history of Western civilization.

Infection in humans by *Y. pestis* takes three forms. In bubonic plague, after about six days' incubation, the first symptoms appear: a gangrenous pustule at the site of the flea bite and tenderness and swelling of the lymph nodes. These symptoms are followed by the appearance of purplish regions of discoloration caused by subcutaneous hemorrhaging. The lymph nodes become hard and swollen, ranging from about almond-sized to orange-sized. These are the buboes characteristic of full-blown infection and are the source of the name "bubonic plague." This form of the disease is also known as Black Death, "black" due to the discoloration caused by massive hemorrhaging and "death" due to the 50 to 60% mortality rate.

In septicemic plague, *Y. pestis* enters the bloodstream, either through a flea bite or by an accidental break in the skin (trappers who unwittingly skin plague-infested animals have contracted the disease after cutting themselves with their skinning knives). Death can occur in less than a day. Septicemic plague is transmitted by fleas and is characterized by enormous numbers of bacteria in the system, overwhelming the defenses of the liver and spleen. Because of its rapid course, septicemic plague is almost invariably fatal.

Pneumonic plague, unlike the other forms of plague, is not transmitted by an insect vector. Instead, it travels directly from person to person by way of airborne sputum. After a two- or three-day incubation period, the body temperature drops and a cough, characterized by the production of bloody sputum, results from the presence of infection in the lungs. Mortality from pneumonic plague is, as from septicemic plague, close to 100% if untreated.

If diagnosed in time, plague can be cured with the use of drugs, including sulfanilamide and antibiotics (namely, streptomycin, chlortetracycline, and chloramphenicol). Moreover, vaccines have been developed from killed bacteria or avirulent strains that can confer immunity. Such was not always the case. Prior to the development of antibiotics and vaccines, the only thing that could stop an epidemic of plague was a winter cold enough to kill the fleas.

Thus, plague characteristically comes in pandemics—worldwide epidemics. Pandemics tended to be of the bubonic variety, it being the only form of plague that permits its victims to live long enough to pass on infection to others (the particularly virulent septicemic plague often extinguishes itself by killing off its victims before they have contact with others). *Y. pestis* is a permanent resident of the Middle East, Central Asia, and east Africa (the so-called inveterate focus), where it is endemic or always present at low levels in local rodent populations, many of which harbor bacteria without manifesting symptoms of disease. Throughout history at intermittent intervals, ecological and environmental conditions change so that plague moves out of its normal circles and enters human society.

Probably the first recorded pandemic struck in A.D. 541 during the reign of the the Byzantine emperor Justinian I, who was in the midst of attempting to recapture parts of the Roman Empire held by the Germans. Although written records are sketchy, the disease appeared to have begun in Egypt and soon spread through Asia, North Africa and Europe. In Constantinople, 200,000 died in four months. After it had run its course, an estimated 20 to 25% of the population of southern Europe had been wiped out. Needless to say, the plague put an irreversible crimp in the grandiose plans of Justinian and may have contributed to the downfall of the Byzantine Empire the following century.

Once plague had established a beachhead in Europe, it continued to reappear in 10- to 24-year cycles for the next two centuries. The next major pandemic didn't begin until the mid-fourteenth century. The legacy of Genghis Khan, the Mongol Empire was still strong in the fourteenth century. Mongol horsemen probably contracted plague in Yunnan, China, and transported it throughout the empire. At Caffa, a settlement of Italians from Genoa was besieged by Tatars bent on conquest. Their numbers were soon decimated by plague. Not content to retreat in utter defeat, the Tatars hurled the bodies of dead plague victims behind the fortified walls of the city. The terrified Genoese vacated the city as soon as the Tatars removed themselves and opportunity presented itself. By sailing back home to Genoa, they introduced bubonic plague to Europe. When they arrived at home, sick and dying, they created mass panic among the residents who responded to their arrival by firing burning arrows at the three ships, thus dispersing them to other ports and introducing the infection throughout the area. By 1348, the pandemic had begun.

About a third of Europe's population died in a relatively short period of time. Plague reached France within months of its appearance in Italy; from there it established itself in Germany and then across the Channel in England. In London, 30,000 of the 70,000 residents perished. The consequences of the Black Death (a name by which the disease was known starting only in the seventeenth century) were enormous. In the population at large, fanatical pursuit of pleasure soon became widespread, possibly due to the realization that death could strike at any time. Not only were the common people afflicted by hedonism, even the clergy showed signs of moral breakdown. At least partly due to the evident corruption within the Church (and also partly due to the total lack of influence exerted by the Church in halting the progress of the disease), the postplague era was characterized by the breakdown of clerical influence and the proliferation of religious sects. The most bizarre of these were the flagellants, who marched from town to town for public displays of self-abuse. They were insidious contributors to the death rolls not only by preaching religious intolerance and instigating mass murder of Jews (accused of poisoning the wells despite the fact that they too were dying in droves) but also by traveling from town to town delivering *Y. pestis* to fresh hosts faster than rats could hope to do.

Preoccupation with death and misery dominated the arts for years to come. One representative work of literature from the era, Boccaccio's *Decameron,* is a chronicle of experiences of different individuals during the plague. A change in social order followed the death of both clergy and serfs—wages went up, due to worker shortages, and landowners were compelled to offer their tenant laborers greater inducements. The working class prospered while the aristocracy never recovered its former strength and influence, and the Black Death contributed significantly to the decline of the manorial system in Europe. A cultural change was brought about by substantial losses among the clergy, who were the principal educators of the era. Replacement by lay instructors changed the nature of education in England. This empowerment of the masses led to numerous popular uprisings throughout Europe. The Jacquerie (peasant revolt) in France in 1358, the Ciompi Revolt in Florence in 1378, and the English Peasant's Revolt, instigated by a proposed poll tax, in 1381, were all manifestations of the new political order desired and expected by the lower classes after the plague.

One of the last gasps of the second pandemic was the Great Plague of London in winter 1664–1665, which may have been precipitated by crop failures in the years before and the concomitant movement of hungry rural rats into the city. Excellent detailed (and heartrending) accounts by Samuel Pepys while the plague raged, and after the fact by Daniel Defoe, describe the mortality and misery. At the beginning of the plague, London's population was estimated at 460,000; less than a year later, an estimated 100,000 people had died. The city, overcrowded and unsanitary before the introduction of the disease, was ripe for an epidemic. As many as 7,000 deaths in a week were not uncommon. So many people died that cemetery space was totally exhausted; plague pits were dug, and endless streams of carts carried the dead from their homes to their anonymous final resting-places. Outside the walled city of London, at least 100,000 other deaths were attributed to plague.

Out of ignorance and desperation, many draconian measures, designed to protect the as-yet uninfected, were enacted. The lord mayor of London unwittingly contributed to the spread of the disease by ordering the extermination of all cats and dogs in the city; rat populations were noticeably larger thereafter. Plague doctors peddled charms, amulets, and herb bouquets whose odor was said to drive away the poisonous airs. One curious legacy of the Great Plague is the child's nursery rhyme "Ring around the rosies," a grim reference to the disease ("rosies" being the buboes), its unsuccessful treatments ("posies" were sniffed to ward off deadly miasmas), and its all-too-common outcome ("all fall down"). Mysticism abounded; amulets inscribed with the word "abracadabra" repeated many times were thought to possess protective powers. Perhaps most inhuman (but nonetheless understandable in the context of the times) was the practice of locking people into their homes if a family member or resident was known to have died of plague in the house. City watchmen patrolled the streets to make sure that no one attempted to escape; inside their homes, many perished of thirst or starvation rather than plague.

It's a matter of contention as to what ended the Great Plague in London. Deaths had already begun to decline by January 1666, possibly due to the effects of cold winter temperatures on fleas. Some people believe, however, that the death knell for the Great Plague was the Great Fire, a fire that began in a bakery in the heart of London and raged unchecked for four days, destroying 373 acres within the city. All in all, it was a rough time for London.

The third pandemic traces its beginnings to Yunnan, China in 1892. Reaching Bombay in 1896, it ended up killing some 6 million people in India alone. Deaths from plague were reported in Thailand in 1904, Burma in 1905, and Tunisia in 1907. It arrived in Hawaii by way of two ships from Hong Kong docking in Honolulu in 1899. On December 12, a Chinese bookkeeper died of bubonic plague in Chinatown and that entire section of the city was promptly quarantined. By December 29, despite house-to-house inspections and disinfection, nine more people were reported to be suffering from the disease. The Board of Health decided that their only option for controlling the spread of the disease was controlled burning of infected houses. Thousands of Chinese were evacuated from their homes and relocated in quarantine camps. The first of forty-one fires was set on December 31. On Janury 20, 1900, however, things went awry. During a controlled burning, the wind shifted and fire raged out of control, burning for seventeen days and destroying thirty-eight acres acres and over 4,000 homes. Four months later, the epidemic was officially deemed over and the Chinese allowed to return to what was left of their homes.

The third pandemic progressed to South America, reaching Paraguay in 1899; from there it spread overland. That same year, it reached U.S. shores when, on June 27, the SS *Nippon Maru* docked in San Francisco from Hong Kong by way of Honolulu. Two people had contracted plague on board ship (which was infested with plague-infected rats). Although the ship was quarantined, nobody knew about the eleven Japanese stowaways, two of whom jumped ship and ended up dying in San Francisco Bay. The rats, oblivious to quarantine laws, successfully reached shore complete with fleas and nine months later San Francisco was hit. By February 29, 1904, 126 people had been infected and 122 died.

One major difference between the third pandemic and the first two was that people had a much clearer idea as to the cause and prevention of plague. Miasma, or poisonous air, was thought to be responsible for the spread of contagion during the first and even second pandemic. In the nineteenth century, however, the germ theory had led to spectacular successes in medicine and epidemiology. One factor working in favor of medical research was that bubonic plague is caused by a bacterium, which, unlike the virus that causes yellow fever, for example, is easily visualized with a microscope. In 1894, Shibasaburo Kitasato, working with plague patients in Hong Kong, discovered what he called *Bacterium pestis*; at more or less the same time, also in Hong Kong, Alexandre E.J. Yersin described the same organism as *Bacillus pestis* (the nomenclature has a long and confusing history; in 1923, the name was changed to *Pasturella pestis* and was finally changed again in 1970 to its present name, *Yersinia pestis*). Three years later, Masanori Ogata ground up fleas from infected rats and injected them into healthy ones, who contracted plague, suggesting that plague may be transmitted by crushing fleas. A year later, one P. L. Simond in Bombay demonstrated that plague bacteria entered the host via flea bites and suggested ways to check the spread of plague by ship. He found the bacterium in the rat and in the gut of the flea, and in all other ways clinched the story. The British Indian Plague Commission however, in its infinite wisdom rejected his findings (releasing a statement in 1899 that "there is absolutely no evidence that the disease has ever been carried from one country to another by plague-infested rats in ships").

Yet another breakthrough came in 1903, when Charles Rothschild, a wealthy Englishman with a passion for collecting fleas, described the hitherto unknown Oriental rat flea *Xenopsylla cheopis* from Egypt (collected in the shadow of the pyramid of Cheops), the chief vector for the plague bacterium. In 1905, Simond was finally vindicated by the new British Commission for the Investigation of Plague in Bombay. W. G. Liston, in charge of the new investigations, indicated *X. cheopis* as the vector.

Today, plague, once diagnosed, is treatable with antibiotics, developed during the 1930s. The specter of pandemic, however, remains. Bubonic plague is far from an historical curiosity, even in the United States. Almost 300 cases were diagnosed between 1956 and 1987, half of these in New Mexico, where plague is widely established in populations of feral rodents. Bacteria of all description have been known to develop resistance to antibiotics; if multiply resistant strains of *Y. pestis* were to develop, mortality rates would undoubtedly rise again, although never to the levels of previous centuries, during which neither the causative agent nor its insect collaborator were known.

The buzz on mosquitoes

MOSQUITOES ARE HARDLY the only insects that nourish themselves at the expense of human corpuscles. But mosquitoes, over and above any other group of insects, are more than just a minor annoyance. Because of the pathogenic microorganisms they not only carry around but actively culture in some cases, mosquitoes are an outright threat to human health. They are particularly adept at transmitting diseases caused by viruses—the list includes yellow fever and the encephalitides—but they also act as vectors, or carriers, of disease-causing nematodes and protozoans. Even today, malaria, a disease caused by protozoans vectored by mosquitoes, afflicts between 200 and 300 million people and kills at least 2 million every year.

It's a little surprising that the scourge that annually lays low a significant fraction of the world's population is on average less than half a centimeter (0.2 inches) in length. Worldwide, there are about 2,500 known species of mosquitoes, or members of the dipteran family Culicidae. Mosquitoes range from tropical jungle to arctic tundra and have been found at elevations of 14,000 feet in Kashmir and at depths of almost 4,000 feet below sea level in gold mines in Africa. The taxonomic characters uniting the many species in the family include the distinctive mouthparts (the proboscis used for bloodsucking) and the presence of scales on the veins and posterior margin of the wings. The distinctive proboscis consists of six piercing stylets, made up of the labrum, the mandibles, the maxillae, and

Figure 7.2
(Counterclockwise from top left) *Anopheles* adult, *Anopheles* eggs, *Anopheles* larva, close-up of *Anopheles* egg, *Culex* adult in biting position.

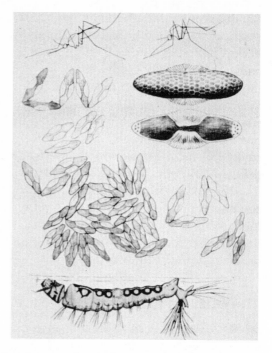

the hypopharynx. Food is taken up from the wound in the channel formed by the grooved labrum and the hypopharynx, and various anticoagulant and anesthetic substances may enter the wound through the salivary pore in the hypopharynx.

Despite the wide variety of habitats frequented by mosquitoes, their lives all have a certain number of features in common (Fig. 7.2). The larvae of all mosquitoes, for example, are aquatic. The larvae, wormlike creatures with a well-defined head and a bulbous thorax, are known as wrigglers. A wriggler out of water is as uncomfortable as a fish out of water; it leads a totally aquatic existence, obtaining oxygen by way of a siphonlike arrangement at the tip of the abdomen. Mosquito *eggs*, on the other hand, can often survive for a long time on dry land. Some species lay their eggs on dry land (usually on the edges of ponds or wet spots) and the embryo remains dormant until the egg is immersed (accounting for the sudden emergence of certain species—called floodwater mosquitoes—after heavy rain). Other species lay eggs either singly or in rafts directly on the water surface.

Most wrigglers are filter feeders that consume microbes and other bits of plant and animal debris carried into the mouth by means of currents generated by the oral brushes, two fanlike clumps of hairs on the maxillae. The larvae of *Toxorhynchites*, however, buck the trend and are predaceous on other mosquito larvae. Irrespective of species, there are four larval molts. Mosquito pupae are known as tumblers by virtue of the fact that, unlike most insect pupae, mosquito pupae are capable of considerable independent movement (most of which consists of tumbling head over heels as they hang near the water surface). The pupa characteristically possesses respiratory structures called trumpets on the thorax, which project up through the surface film to provide abundant supplies of oxygen. Generally, the pupal stage is quite brief, often only two or three days in duration.

While it's safe to say that many species of mosquitoes feed as adults on human blood, it is even safer to say that the majority of individual mosquitoes in the world never feed on human blood. First of all, many species of mosquitoes find consuming human blood a thoroughly distasteful prospect, preferring instead to concentrate on other vertebrate species. Even in species that are anthropophilous (associated with humans), only the females of the species are equipped for blood-feeding—they generally need a blood meal in order to develop and lay viable eggs. Males generally feed either on nectar or not at all. Both sexes of *Toxorhynchites* feed on nectar and there are even species in the genus *Harpagomyia* that feed exclusively on the regurgitated food of ants in the genus *Crematogaster.* However, a disproportionate amount of attention has focused on those species that are associated with humans, for obvious reasons.

Aedes (a mosquito, a plan, a canal)

> . . . the time has come when the mosquito with its little trumpet
> will sound my hymn of glory.
> —CARLOS FINLAY, January 2, 1901

The species that are probably the closest associates (if not the closest friends) of humans are in the genus *Aedes*. About half of the 150 species of mosquitoes in North America are in the genus and are generally recognizable by their habit of resting butt-end down as they feed and by their intensely painful bites (the name of the inland floodwater mosquito, *Aedes vexans,* in fact means "irritating pest"). Eggs are laid one at a time in or out of water. Species in this genus infest salt marshes—the Jersey mosquito *Aedes sollicitans,* which occupies the extensive salt marsh areas of New Jersey, is sometimes called the convict mosquito because of the conspicuous stripes all over its body. These same stripes inspired the common name of the tiger mosquito, *Aedes albopictus,* a relative newcomer to North America.

In terms of human health and happiness, the most important species of *Aedes* is *A. aegypti.* It's found worldwide between 40° north and south latitudes. A close and constant associate of humans, it breeds preferentially in containers that catch rainwater: pots, tin cans, gutters, cisterns, rain barrels, and even in baptismal fonts. Unlike other container breeders, it is rarely found in naturally occurring catch basins, such as treeholes. *A. albopictus,* a recent introduction into this country with similar breeding habits, gained entry into the U.S. in used tires imported from Asia; not only do tires catch water, they also retain heat throughout the winter months and provide idyllic accommodations for developing larvae.

As the specific epithet *aegypti* implies, *A. aegypti* is not native to North America. It is thought to have been introduced from West Africa in slave ships in the late sixteenth century. Unfortunately, *A. aegypti* did not travel to the New World alone; it was accompanied by a virus that causes the disease yellow fever in humans. The first documented case of yellow fever in the New World was in Brazil in 1642.

Yellow fever usually begins from three to six days after exposure to the virus through the bite of an infected mosquito. Initial symptoms include headache, dizziness, fever, nausea, vomiting, and generalized aches and pain; serious cases progress to a second stage (often after a seeming remission of symptoms) characterized by widespread hemorrhage, vomiting of black material, delirium, and jaundice, or yellowing of the skin and eyes caused by accumulation of bile pigments in the blood; this yellowing is the source of the name, "yellow fever." Untreated, about 10% of cases end in the death of the patient. Survivors of a bout with yellow fever, however, will enjoy lifelong immunity from subsequent infection.

Seaports and coastline cities have always had more than their share of yellow fever epidemics not only because of the steady influx of visitors who may be harboring an infection but also because of the seaworthiness of the vector *A. aegypti,* which frequently travels on board ships in barrels containing fresh water supplies. The sugar trade established foci of disease in Havana, Boston, New York, Philadelphia, and New Orleans. In Philadelphia, one outbreak was particularly devastating; almost 5,000 people (about 10% of the population) died during the epidemic. The epidemic seems to be traceable to the immigration of about 2,000 refugees from Santo Domingo into Philadelphia in 1793; these refugees were fleeing both the slave revolts and the yellow fever epidemic raging in their home ports. Dr. Benjamin Rush, an esteemed Philadelphia physician, signer of the Declaration of Independence and arguably the father of modern psychiatry, diagnosed the first case in August. On August 24, in the beginning of the epidemic, 17 deaths were recorded in a twenty-four-hour period from yellow fever; by October 11, the daily tally was up to 119. In the month of September alone, 1,400 yellow fever deaths were recorded.

Dr. Rush was to play a critical role in the progress and outcome of the yellow fever epidemic. Virtually nothing was known of the etiology of the disease and Rush fervently attributed it to the stench arising from a pile of rotting coffee dumped on a dock near where the first case had been reported. He himself contracted the disease and recovered and grew adamantly enthusiastic about the regimen to which he attributed his cure—massive bloodletting and purging with jalap and calomel (mercurous chloride). In hindsight, this regimen was exactly the wrong prescription for yellow fever patients, what with their already substantial blood loss through hemorrhage and vomiting. Nonetheless, Rush's tremendous enthusiasm, energy and reputation combined to impress the public into demanding an opportunity to undergo the rigorous treatment. It's entirely likely that Rush unwittingly expedited the death of a significant number of his patients. To his credit, however, it must be said that he rightly believed that the disease was not infectious in the sense that people could not contract it from contact with an infected person and he bravely and tirelessly visited the sickbeds of hundreds of patients in an era when many lesser physicians shunned contact with the sick completely. Rush died years later without ever suspecting the significance of the sizable mosquito infestation that coincided with the summer outbreak of yellow fever. The epidemic, exacerbated by a long hot summer conducive to mosquito breeding, finally came to a halt with the first killing frost of October. Ironically, in 1814, the year after Rush died, a less-celebrated physician named David Hosneck made an astute and vital observation—that a period of eight to fourteen days passes between waves of infection, suggesting that a developmental period is involved in maintaining cycles of infection.

Between 1668 and 1893, there were at least 135 outbreaks of yellow fever in the United States, called "yellow jack" in reference to the yellow quarantine flag that was displayed in afflicted towns. An

Figure 7.3
Colonel William Crawford Gorgas, who spearheaded the campaign to eradicate mosquito-borne disease from Cuba and Panama in 1899-1908 (Abbot, 1915).

epidemic in New Orleans in 1878, which resulted in the loss of 6,000 lives, was attributable to a single ship from Havana, departing during a yellow fever outbreak in that city. Havana at the time was renowned as a city of filth and contagion. A desire to move in and clean up the city, and thereby reduce health risks to American cities engaged in trade, may have been at least partly responsible for U.S. participation in the Spanish-American War of 1898.

American troops occupied Havana in 1899 and Army Major William Crawford Gorgas was commissioned to eliminate yellow fever from the city (Fig. 7.3). In 1898, during the Spanish-American War, approximately 2,450 American servicemen died. Only 385 of them, however, died in battle; yellow fever accounted for some 230 deaths and other diseases the rest. Major Gorgas believed that yellow fever, like typhoid fever, dysentery and so many other urban ailments of the day, was caused by filth (*fomites,* invisible particles of filth) and so undertook to eliminate yellow fever by improving sanitation. Although a massive cleanup of the city did reduce the incidence of other diseases, the worst outbreak of yellow fever in two decades struck in 1900, prompting the army to send a team of four physicians to investigate the situation.

Major Walter Reed was accompanied to Havana by James Carroll, Jesse Lazear, and Aristides Louis Agramonte. Once there, they contacted Carlos Finlay, a Cuban doctor who had for years insisted that mosquitoes, specifically *Stegomyia* (now called *Aedes*), were involved in the transmission of yellow fever. Although he had claimed to have effected mosquito transmission under laboratory conditions, his experimental results were equivocal and he was generally not taken very seriously by the medical establishment. However, Lazear had spent several years working on mosquito transmission of malaria in the southern United States and convinced Reed to pursue the idea. Also lending credence to the theory was the counsel of Henry Carter, U.S. Public Health Service quarantine officer assigned to Havana in 1898. He had arrived there after an investigation of yellow fever in Mississippi convinced him of the existence of a two-week "extrinsic incubation" period for yellow fever. Such an incubation period would coincide with the development of the pathogen in the mosquito vector.

Reed, on assignment from the Army Medical Museum, was aware of Ronald Ross's recent findings that malaria is vectored by mosquitoes and initiated an experimental test of Finlay's theory (although he shortly afterward had to return to Washington to complete a report on typhoid fever). The first tests were disappointing, largely because the team was unaware that yellow fever is infectious only in the first three days of illness. Carroll, however, successfully (?) contracted the disease from a mosquito that had fed on a patient in his second day of illness. Tragedy struck, though, when Lazear, either bitten accidentally or as a result of unauthorized experimentation on himself, died of a particularly violent case of yellow fever in October.

In order to meet the rigors of scientific proof, a systematic experiment was designed. Away from heavily populated areas, an experimental area, named "Camp Lazear" in honor of their fallen comrade, was set up by the team. Camp Lazear consisted of two cabins in close proximity to one another. One cabin was occupied by seven American army volunteers, who lived with the fluid-drenched clothing and sheets from infected soldiers. The cabin was heavily screened to prevent the admission of mosquitoes. That none of these soldiers fell ill was a demonstration that yellow fever could not be transmitted by contact with body fluids. In a second cabin were thirteen paid Spanish immigrants who had clean sheets and clothing but who were exposed to mosquitoes that had fed on yellow fever patients. Eleven of the thirteen fell ill with yellow fever, demonstrating convincingly that the bite of an infected mosquito was a necessary component to contracting yellow fever.

In the face of the compelling evidence that mosquitoes were vectors of yellow fever, Gorgas redoubled his efforts to sanitize Havana, but focused his attention on eliminating the breeding grounds of mosquitoes: bodies of standing water. Soldiers systematically swept through the city, emptying water jugs and fining citizens ten dollars (a substantial sum in those days) for violations. Gutters were flooded with oil (which clogs the respiratory apparatus of larvae as they come to the surface for air) and standing water was virtually eliminated. In short order, so was yellow fever. In only a year, yellow fever was effectively eliminated from Havana, where it had held sway for over 200 years.

Inspired by its Havana experience, the United States government set about to repeat its success in Panama. For seven years, between 1881 and 1888, the French had attempted to construct a canal across the Isthmus of Panama and were scuttled in that attempt by massive deaths of laborers, from yellow fever among other ailments. Ferdinand de Lesseps, the canal company chief, called it quits in 1889 (after an estimated loss of as many as 20,000 lives). The U.S., recognizing the economic benefits of a canal across Panama, commenced a canal project in 1904. Gorgas was dispatched as chief sanitary advisor accompanied by staff and $50,000 in supplies. He fumigated the place with 120 pounds of the botanical insecticide pyrethrum, the equivalent of a year's supply for the entire United States (Fig. 7.4), much to the annoyance of the populace, particularly in view of the fact that an epidemic erupted the following year anyway. Gorgas was severely criticized from all sides (including the Canal Commission, several of whose members still subscribed to Rush's stench theory) but nevertheless survived several dismissal attempts (among others by then Secretary of War William Howard Taft). Sanitation prevailed, however, and the 1905 epidemic was the last. Gorgas was the only high-ranking officer to remain in Panama throughout the entire project, witnessing the first ship through the canal in 1914.

The rest of the yellow fever story is largely immunological rather than entomological. Early on, attempts to isolate the "germ" that caused the disease were unsuccessful and abandoned; unknown to these early workers, heady with success at identifying many bacterial and fungal disease agents, yellow fever was caused by a filterable agent far smaller than even the smallest bacterium. It was eventually identified as a virus and scientists from The Rockefeller Institute developed a vaccine to confer lifelong immunity. The yellow fever vaccine was actually the first successful vaccine to be developed in the twentieth century (the first, in fact, since Louis Pasteur developed his rabies vaccine in 1885). Max Theiler, of Rockefeller Institute, adopted Pasteur's technique of cycling live virus through the brain tissue of nonhost experimental animals; at each cycle, the virulence of the virus is attenuated until the virus is so transformed it cannot cause disease. It can, however, stimulate the body to produce antibodies, which then remain in the body to

Figure 7.4
Conquering mosquitoes
in Panama (Abbot, 1915):
(above) pouring oil on
stream surfaces to destroy
larvae; (left) spraying
insecticide.

protect an individual from any subsequent exposure to the natural, virulent virus. By 1936, Theiler had perfected his vaccine, known as 17D live yellow fever vaccine. It was soon called into service by the U.S. military, to inoculate soldiers and other personnel being shipped overseas during World War II; action in many tropical regions meant that many Americans would be exposed to *Aedes aegypti* and yellow fever.

The success of the yellow fever vaccine was not without setbacks. To scale up vaccine production, the virus was grown in tissue culture; this tissue culture was then injected into chicken eggs and allowed to incubate. After incubation, the infected chick embryos were finely processed and combined with human blood serum. After centrifuging and filtering out particulates, the filtrate served as vaccine. The problem with this method was made glaringly apparent when, during a Rockefeller Institute-sponsored effort to eradicate yellow fever in Brazil, over a million people inoculated with the vaccine contracted hepatitis. The virus that causes this disease circulates in human blood serum and was too small to be filtered out during vaccine production. Despite the fact that the drawbacks of this method of vaccine production were known, the Institute, when asked by the U.S. military to supply large quantities of vaccine on the eve of the second World War, elected to continue to use human blood serum. As a result, in 1942 (when the vaccination program was undertaken) 50,000 cases of hepatitis were reported at military installations in the United States and overseas; over eighty people died. By the end of 1942, a new, serum-free method of vaccine production was in place at the Institute and has been used, with few complications, ever since.

Since the vaccine was developed, an estimated 200 million doses have been administered throughout the world. Even though there is a vaccine to prevent the disease, yellow fever continues to cause suffering and even death at least in part because people at risk are not receiving the vaccine. Between early 1988 and late 1990, some 8,685 cases were reported, resulting 2,643 deaths. The number of cases reported is but a fraction of the total—it has been estimated that as many as 200,000 cases and 30,000 deaths occur worldwide every year. In Africa, where yellow fever runs rampant in over 30 nations, mortality rates can reach 80% or higher.

Other arboviruses (arthropod-borne viruses) associated with *Aedes* species include dengue, or breakbone fever; eastern and western encephalitis; Venezuelan equine encephalitis; St. Louis encephalitis; chikungunya (from a native African word for "doubled up"); oropouche; and bunyamidera. In view of this panoply, there is justifiable concern over the recent introduction of *A. albopictus,* the first disease vector to be introduced into the U.S. in 300 years. This species is a proven vector of dengue and a suspected vector of a number of forms of encephalitis. The introduction of Asian tiger mosquitoes into the United States is yet another example of the remarkable capacity of human beings to create problems for themselves. *A. albopictus* is native to tropical and temperate Asia and India. In 1985, however, it was discovered breeding in a landfill in Harris County, Texas. Soon afterward, breeding populations were reported in many other localities; by 1992 it had been found in in more than ten states, ranging north to Maine and New Hampshire and west to Illinois and Iowa. Its prevalence around tire piles led to the realization that its likely mode of entry into the country was in imported tires. Despite the fact that used tires constitute the largest landfill problem in this country, Americans import several million used tires every year from Asia, particularly from Japan (where cold-tolerant populations of *A. albopictus* abound). These Asian tires are used extensively for producing retreads. Used tires are fabulous breeding sites for Asian tiger mosquitoes and it's likely that a few of the imported tires housed living larvae, which happily completed their development in their new American homes. Asian tiger mosquitoes are a potential health threat of major dimensions—in addition to being vicious biters, they are competent vectors of several devastating diseases, including dengue; hemorrhagic fever; yellow fever; and eastern, western, Lacrosse, and St. Louis encephalitis. With over 3 billion used tires available in dumps all across the country, this species stands to expand its range tremendously and every neighborhood it moves into is a potential site for an outbreak of disease.

Aedes mosquitoes are unquestionably capable of spreading all kinds of diseases, but their prowess as disease vectors is not without limits. The notion that mosquitoes may be involved in the transmission of human immunodeficiency virus, or HIV, the virus that causes AIDS, first gained credence when an unusual cluster of AIDS cases with no apparent risk factors was reported in 1985 in Belle Glade, Florida. The high frequency of mosquitoes in this area suggested to some investigators that insects may have played a role in transmitting the disease. Subsequent epidemiological analysis, however, failed to confirm the connection; among other things, the HIV-positive population of Belle Glade included no children under 10 or adults over 60, groups that are certainly at risk of being bitten and that therefore should have been at risk of exposure to the disease. Rather, the epidemiological findings were entirely consistent with sexual transmission as the principal cause of disease. Moreover, laboratory experiments confirmed that the HIV virus does not proliferate inside the body of mosquitoes and that the viability of the virus inside the body of mosquitoes (and other bloodsucking insects) is poor. All told, given that the amount of HIV virus circulating in human blood is generally very low, that mosquitoes ingest and transfer very small volumes of blood, and that the virus does not persist for very long inside the mosquito's body, while mosquito transmission of AIDS cannot be categorically ruled out, it can certainly be considered a highly unlikely event, probably one of the few nice things ever said about a mosquito.

Anopheles (to the vector go the spoils)

Considering all the havoc they have wreaked with human affairs, it's remarkable that very few people would recognize an anopheline mosquito on sight. They can be distinguished from other sorts of mosquitoes only by a set of rather esoteric characters; the length of the palpi (about as long as the proboscis), the shape of the scutellum (evenly rounded or strap-shaped), and the absence of scales on the abdomen. Conspicuous characters more easily recognized by the layperson (or prospective meal) are the very long legs, even for a mosquito, the spotting on the wings, and the habit of adults when taking their blood meals to stick their abdomen way up in the air. Worldwide, about 300 species lay claim to membership in the genus *Anopheles* (of which 15 live in North America).

In North America, adult female *Anopheles* (already fertilized during the fall) are about the only ones to survive the winter. Spring thus begins with egg-laying. Eggs are generally laid one at a time either while the female is floating on the surface of the water or as she is hovering above the water surface. The eggs of anopheline mosquitoes are quite distinctive; most have the same general ovoid shape tapered at both ends, vaguely reminiscent of a Viking ship. They are as seaworthy as a Viking ship as well, equipped with air-filled floats to keep them up near the surface. Anophelines tend to prefer quiet neighborhoods, away from currents and wave action, and thus occupy seepages and bodies of impounded water with great frequency. There's tremendous variety in the group, however, and some species breed in tree holes, flowing sunlit streams, and even in the catchwater of bromeliad epiphytes in the tropics.

Eggs generally hatch in about two days; unlike the eggs of *Aedes,* anopheline eggs cannot tolerate desiccation. Larvae develop, eating microbes and debris, in about two weeks and pupation lasts about three days. All told, development takes about three weeks, actually a fairly long time by culicid standards. Mating generally involves the formation of a male mating aggregation; males hover together in a group above or around the places where females are likely to emerge. After mating, females in most cases must have a blood meal in order to develop eggs and the blood required is generally of the warm variety (that is, from birds or mammals).

By sucking the blood of a series of warmblooded hosts, certain anopheline mosquitoes can act as vectors of pathogenic organisms that circulate in the bloodstream. Among these are protozoans in the genus *Plasmodium,* which cause in humans the disease malaria. Humans are afflicted by only four species: *P. vivax, P. ovale, P. malariae,* and *P. falciparum.* Plasmodia actually depend on the mosquito not only for transportation but for completing part of the life cycle. In general, mosquitoes ingest the gametocytes or sex cells of the pathogen that circulate in the bloodstream of humans infected with malaria. There are two forms of gametocytes: microgametocytes and macrogametocytes. Each microgametocyte matures to produce about eight microgametes, each with whiplike extensions or flagella (in a process called exflagellation); these swim to fuse with the macrogametocytes. Once these fuse, a zygote is formed, which undergoes development to form an ookinete, or moving egg. Ookinetes then form cysts in the midgut wall of the mosquito, where they develop, explode, and release long thin cells called sporozoites into the body cavity. The active sporozoites then swim their way into the salivary glands of the mosquito. When the mosquito takes a blood meal on a mammalian (frequently human) host, it injects saliva into the puncture wound and with the saliva pass out sporozoites. Once in the human bloodstream, the sporozoites head for the liver, where they undergo a transformation into merozoites; after some time, some merozoites develop into an asexual stage called a schizont, inside liver cells. The merozoites can invade red blood cells, to take on yet another form, the trophozoite. The red blood cell is converted into a self-destructing factory that releases either more merozoites inside a human being or more gametocytes inside a mosquito; the merozoites invade more red blood cells in the human to prolong the disease and the gametocytes inside the mosquito continue the cycle of infection. Gametocytes that happen to form inside a human being are doomed to die unless they are sucked up by a mosquito, in which they can complete development.

The roundabout life cycle of the pathogen is responsible for one of the most distinctive characteristics of malaria: intermittent fever, or regularly alternating paroxysms of fever and chills. Most of these paroxysms have three distinct phases: chill, followed by fever, followed by copious perspiration and return to normal temperature. Different forms of malaria are marked by the different frequency of occurrence of these paroxysms. In quotidian malaria, the attacks come at 24-hour intervals; in tertian malaria, every third day; and in quartan malaria, every fourth day. These episodes correspond to the release of the parasites and their metabolites, as they exit from the decimated red blood cells.

Aside from the distinctive pattern of fever (which is responsible for several of the more colorful names of malaria, including intermittent fever, chills and fever, remittent fever, and *Wechselfieber*), there are other clinical manifestations of malaria, including loss of function of spleen and liver. The parasitized blood cells cause complications by blocking capillaries in the brain and elsewhere. Yet another diagnostic feature of malarial infection is the presence in macrophages (cells derived from white blood cells) of two pigments that are actually breakdown products of hemoglobin. The pigments, hemozoin and hemosiderin, are released when red blood cells burst and are engulfed by the macrophages. One of the most severe forms of malaria, caused by *Plasmodium falciparum,* is frequently fatal.

Malaria has had a major impact on the course of human history by causing massive mortality. Among the events whose outcome may have been determined by the pattern of malarial infection was Alexander the Great's attempted eastern conquests, and the outcome of the Crusades. It's difficult to say with any certainty, however, since the causative agents of malaria have been known for less than a hundred years. Intermittent fevers are specifically mentioned in Babylonian and Assyrian medical works, and in the Old Testament (Leviticus 26:16), reference is made to "burning ague." In India, the Ayurveda, part of the Atharvana and a major medical treatise, refers to the "King of Diseases," one characterized by alternating hot and cold fevers every day or every third day.

Perhaps the clearest and most comprehensive early accounts of malaria was provided by Hippocrates of Cos (460–377 B.C.), the so-called "Father of Medicine." He not only described in minute detail the periodicity of fevers and the overall physiological consequences of the disease, he also associated the incidence of the disease with proximity to swamps and marshes. In his work *Airs Waters Places* he wrote, "If there be no rivers, and the water that the people drink be marshy, stagnant, and fenny, the physique of the people must show protruding bellies and enlarged spleens" (Russell 1955).

Malaria was even more a problem in ancient Rome than in ancient Greece, as evidenced by the existence of temples dedicated to the goddess *Dea Febris* (the less than comely goddess of malaria, described as a bald old hag with protuberant veins and stomach). Marshes and swamps were again recognized as sources of contagion. Marcus Terentius Varro (116–27 B.C.), for example, wrote in *Rerum Rusticarum,* "Note also if there be any swampy ground . . . because certain minute animals, invisible to the eye, breed there, and, borne by the air, reach the inside of the body by way of the mouth and nose, and cause disease which are difficult to be rid of." In general, however, marshes and swamps were thought to cause disease by way of "bad air"–or as Italians would say, *mal'aria* (rather than, as some physicians thought, *malaqua,* or bad water). The word malaria doesn't really show up with any frequency in the English language until the nineteenth century. "Ague" was the term in vogue (much to the annoyance of French-speaking Voltaire, who couldn't understand why it was pronounced as two syllables while the cognate word "plague" was accorded only a single syllable when pronounced).

Name notwithstanding, there was little consensus as to what caused the disease. In the late nineteenth century, its periodicity lent itself well to the relatively new germ theory of disease, developed over the last two centuries as a result of the microscopic investigations made possible by innovations in optical technology. In 1847, E. Bartlett summarized the leading contenders as causative agents of malaria: chemical products of marshes, moisture, animal or vegetable decomposition products, and "invisible living animalculae"; he personally ascribed to the notion that "invisible living animalculae" were responsible for the disease. A false hope that the causative agent was found was generated by the isolation and description of *Bacillus malariae* by Edwin Klebs and Corrado Tommasi-Crudeli in Italy, which they claimed to have found in marshy mud and the urine of malaria patients, and which they also claimed caused malaria upon injection into rabbits (unlikely in retrospect since rabbits are immune to malaria).

Microscopic work proceeded apace nonetheless and focused early on the presence of malarial "paludic pigments" as a sign of infection. Charles Louis Alphonse Laveran, a surgeon with the French Army in Algeria, however, was the first to recognize plasmodia in the red blood cells of malaria patients in the military hospital. Upon close inspection of the pigmented bodies, he found signs of life: "I had suspected for a long time the parasitic nature of these bodies when on November 6th, 1880, while examining one of the spherical pigmented elements in a preparation of fresh blood, I noticed with joy at the periphery motile filaments of the animated nature of which there was no room for doubt." His description of *Oscillaria malariae* met with skepticism initially but other colleagues confirmed his observations and the protozoan was established as a malarial parasite. Laveran received a Nobel prize in 1907 for his perseverance.

Laveran's discovery set the stage for further advances. Gerhardt in 1884 demonstrated that injections of blood from malaria patients could cause malaria in healthy individuals and, in 1885–6, Camillo Golgi demonstrated the association betweeen febrile symptoms and the parasite's life cycle and also differentiated among the various species of *Plasmodium,* associating each with its own characteristic form of malaria. The story was far from complete, however; the mode of transmission of the disease was still unknown in 1894, almost ten years after the causative agent had been identified. Mosquitoes had actually been suspected of playing a role in the transmission of malaria many times over the centuries (and as long ago as the sixth century, according to Kalidasa, the Sanskrit poet). Many so-called primitive peoples also steadfastly maintained that mosquito bites could lead to fatal fevers (malaria in parts of Africa is called *mbu,* as are mosquitoes). The European scientific community, however, was slow to embrace the idea.

In 1894, the work of Patrick Manson lent considerable credence to the theory that mosquitoes are vectors of malaria. A Scottish physician working in China, he was absorbed in studies of filariasis and conjectured that mosquitoes were the vectors of filarial worms. Close inspection revealed that filarial worms not only gained entry into mosquitoes when they fed on infected humans but also that the worms grew and developed inside the body of the mosquito. Manson published his observations in a paper titled, "On the development of *Filaria sanguinis hominis,* and the Mosquito considered as a Nurse" in 1879. Manson believed that the filaria entered humans who drank water in which infected mosquitoes died. It was left to Joseph Bancroft in Australia to demonstrate its transmission by insect bite in 1889. Nonetheless, Manson's work established the fact that mosquitoes could harbor and maintain pathogens of humans in their bodies.

The final conclusive proof of insect involvement in malaria transmission was the work of Ronald Ross, a British surgeon stationed in India. Aware of Manson's work, Ross befriended him while on leave in England. Upon his return to India, he set out to prove Manson's theory that mosquitoes are vectors of malaria. He got nowhere for a while at least partly because, not knowing much about insects, he was working almost exclusively with *Culex* and *Aedes* mosquitoes, neither of which act as vectors of malaria. In 1897, he finally got hold of *Anopheles*. Ross successfully induced mosquitoes to feed on a malaria patient and then witnessed the parasites develop in the bodies of the mosquitoes. Ross thus described the sexual phases of the parasite. (William George MacCallum, working in Ontario, described the asexual phase in droplets of crow's blood on a slide in 1897.) Since exflagellation was known to occur in mosquitoes, Ross realized that it represented the sexual phase of the pathogen and that mosquitoes were the definitive hosts for malaria parasites (that is, the hosts in which sexual reproduction took place). After several setbacks, including a government transfer and a widespread reluctance to find human volunteers to test his theories, due to some untimely and unsuccessful experimentation by other investigators on a plague vaccine, on the fourth of July, 1898, Ross saw sporozoites in the mosquito salivary glands. The last link, proof of metaxeny, or host change, was forged when Ross, working with birds, managed to infect a mosquito from a malarial sparrow and inoculate a healthy sparrow with infected mosquitoes. For his efforts, Ross received a Nobel prize in 1902.

The relatively rapid recognition of the men involved in ascertaining the origin and transmission of malaria was acknowledgment of the significance of their findings. Once the cause of malaria had been discovered, its prevention was possible. Malaria had been treated for a long time first with bark of cinchona and then with its purified active ingredient quinine; it cured malaria by killing the protozoan parasite. But by eliminating the vector it would be possible to eliminate the disease altogether. Thus began a series of intensive efforts at sanitation and swamp clearing that were largely successful. Malaria, for example, was eliminated from the Panama Canal Zone by 1913 by judicious application of oil to surfaces of standing water to kill larvae and liberal spraying of pyrethrum to kill adults. Once the success of these techniques was demonstrated in Panama, their adoption in the United States resulted in a decline in the average mortality rate from higher than 10 per 100,000 in southern states to only 3.02 per 100,000 in 1940.

The advent of synthetic organic insecticides also had a tremendous impact on the incidence of malaria. In the United States, the already low incidence of the disease fell even further as a result of the vigorous prevention program launched by the U.S. Public Health Service. In 1948–9, after 21 months of study, only 59 cases of malaria could be found in Alabama, Georgia, Mississippi, and South Carolina after extensive residual DDT spraying. Results of DDT application were even more dramatic in tropical areas. In Brazil, for example, there are no fewer than five species of mosquito capable of acting as vectors. A National Malaria Service was established in 1941 and at a cost of only about 21 cents per person initiated a spraying program in which 3 million homes were sprayed annually, a practice that cut down the incidence of the disease from 14,782 in 1945 to 1,192 in 1949. Similar success stories emerged from Belize, Venezuela, Sri Lanka, Italy, and many other countries all over the world.

Within a decade, however, the success story turned out to have yet another chapter. Indiscriminate spraying of toxic chemicals, with the goal of eradicating mosquito species, has resulted instead in the evolution of resistance to insecticides. In 1956, only five species of anopheline mosquitoes were known to be resistant to any insecticides; by 1968 at least 38 species displayed resistance to at least one type of insecticide. Insecticide resistance in mosquitoes takes on many forms. Some species display biochemical resistances–that is, they possess elevated or altered forms of enzymes that break down the insecticide to harmless metabolites. Other species display behavioral resistance—for example, they are so sensitive to the presence of the insecticide that they avoid contact with it and thereby fail to acquire a lethal dose. In many parts of Africa, the interior walls of huts were routinely sprayed with DDT to kill mosquitoes that habitually rested there; after several generations these mosquitoes were replaced by others that were still susceptible to the effects of the insecticide but that rested instead on the *unsprayed* outer walls of huts, where they never encountered the toxins.

Thus, by the mid-1960s, the incidence of malaria began to increase worldwide despite preventative spray programs. In 1968–9, after preventative spraying was cut back, an epidemic of malaria hit Sri Lanka, affecting over 500,000 people. Renewed intensive spraying cut the number of cases in 1972 to about 150,000, but resistance to DDT in *Anopheles culicifacies* resulted in an upsurge, despite massive spraying, in 1975 to over 400,000 cases. Reinvasions of malaria in areas thought to be malaria-free have occurred in Central America, the Middle East, Africa, India and the South Pacific. Complicating the resurgence associated with the acquisition of insecticide resistance is the fact that populations of *Plasmodium* have developed resistance to chloroquine, quinine, and several other drugs used as prophylatics.

Today, a greater diversity of weapons is in use in the fight against malaria. To combat the pathogen, vaccine development is underway. As for dealing with the vector, no longer is there total reliance on chemical toxins; alternative approaches include draining and filling standing water, screening houses to prevent mosquito entry, use of netting and insect repellents to prevent bites, pouring oil on larval breeding sites—in short, most of the activities that were so effective at eradicating malaria prior to the discovery of synthetic organic insecticides. Despite these efforts, however, malaria still afflicts millions of people every year and is the leading cause of death worldwide for children under the age of 5.

References

Insects parasitic on humans (what's eating you?)

Andrews, M., 1977. *The Life that Lives on Man.* New York: Taplinger.

Brewer, T.F., M.E. Wilson, E. Gonzalez, and D. Felsenstein, 1993. Bacon therapy and furuncular myiasis. *J. Amer. Med. Assoc.* 270: 2087–2088.

Dunlap, T.R. 1981. *DDT: Scientists, Citizens, and Public Policy.* Princeton: Princeton University Press.

Frazier, C.A. and F.K. Brown, 1980. *Insects and Allergy and What to Do About Them.* Norman, Oklahoma: University of Oklahoma Press.

James, M.T. and R.F. Harwood, 1969. *Herms's Medical Entomology.* New York: Macmillan.

Marshall, A.G., 1981. *The Ecology of Ectoparasitic Insects.* New York: Academic Press.

O'Toole, K., P.M. Paris, and R.D. Stewart, 1985. Removing cockroaches from the auditory canal: a controlled trial. *New England J. Med.* 312: 1192.

Lice, typhus, and war (a lousy situation)

Andrews, M.R., 1941. *Fatal Partners*. New York: Doubleday, Dean.

Cartwright, F. F., 1972. *Disease and History*. New York: Dorset.

Mumcuoglu, Y.K. and J. Zias, 1988. Head lice, *Pediculus humanus capitis* (Anoplura: Pediculidae) from hair combs excavated in Israel and dated from the first century B.C. to the eighth century A.D. *J. Med. Entomol.* 25: 545–547.

Roricalli, R. A., 1987. The history of scabies in veterinary and human medicine from Biblical to modern times. *Vet. Parasitol.* 25: 193–198.

Sadler, J.P., 1990. Records of ectoparasites on humans and sheep from Viking Age deposits in the former Western settlement of Greenland. *J. Med. Entomol.* 22: 628–631.

Zinsser, H., 1935. *Rats, Lice and History*. New York: Little Brown.

Fleas and plague

Bowsky, W., ed. 1971. *The Black Death. A Turning Point in History?* New York: Holt, Rinehart and Winston.

Carter, F., 1988. *Exploring Honolulu's Chinatown*. Honolulu: Bess.

Cowrie, L., 1970. *Plague and Fire*. London: Wayland.

Cowrie, L., 1972. *The Black Death and the Peasants' Revolt*. London: Wayland.

Crook, L.D. and B. Tempest, 1992. Plague: a clinical review of 27 cases. *Arch. Intern. Med.* 152: 1253–1256.

Friedman, J., 1986. "He hath a thousand slayn this pestilence": the iconography of the plague in the late Middle Ages. In: *Social Unrest in the Late Middles Ages*. F. Newman, ed. Binghamton, New York: Medieval and Renaissance Texts and Studies, pp. 75–112.

Gottfried, R., 1983. *The Black Death: Natural and Human Disaster in Medieval Europe*. New York: Free.

Gregg, C.T., 1985. *Plague: An Ancient Disease in the Twentieth Century*. Albuquerque: University of New Mexico Press.

Hirst, L.F., 1953. *The Conquest of Plague*. Oxford: Clarendon.

James, M.T. and R.F. Harwood, 1969. *Herms's Medical Entomology*. New York: Macmillan.

Lehane, B., 1969. *The Compleat Flea*. New York: Viking.

Morris, J.T. and C.K. McAllister, 1992. Bubonic plague. *Southern Med. J.* 85: 326–327.

Nohl, J., 1924. *The Black Death*. New York: Harper.

Snodgrass, R.E., 1946. *The Skeletal Anatomy of Fleas (Siphonaptera)*. Smithsonian Misc. Coll. Vol. 104, No. 18, Washington, D.C.: Smithsonian Institution (Pub. No. 3815).

Twigg, G., 1984. *The Black Death: A Biological Reappraisal. London: Batsford.*

Ziegler, P., 1969. *The Black Death.* New York: Harper and Row.

Mosquitoes and disease

Anonymous, 1992. Yellow fever: the global situation. Bull. World Health Org. 70: 667–669.

Busvine, J. R., 1976. *Insects, Hygiene, and History.* London: Athlone.

Chase, A., 1982. *Magic Shots.* New York: Morrow.

Harrison, G., 1978. *Mosquitoes, Malaria and Man: A History of the Hostilities..* New York: Dutton.

Harwood, R. and M. James, 1979. *Entomology in Human and Animal Health.* New York: Macmillan.

Hawley, W. A., 1991. Adaptable immigrant. *Nat. Hist.* 100(7/91): 55–56.

Horsfall, W., 1962. *Medical Entomology—Arthropods and Human Disease.* New York: Ronald.

James, M.T. and R.F. Harwood, 1969. *Herms's Medical Entomology.* New York: Macmillan.

Miike, L., 1987. *Do Insects Transmit AIDS?* Washington, D.C.: Office of Technology Assessment Staff Paper 1.

Powell, J.H., 1949. *Bring Out Your Dead: The Great Plague of Yellow Fever in Philadelphia in 1793.* Philadelphia: University Pennsylvania Press.

Russell, P.F., 1955. *Man's Mastery of Malaria.* London: Oxford University Press, University of London Health Clark Lectures at London School of Hygiene and Tropical Medicine.

Williams, G., 1969. *The Plague Killers.* New York: Scribner.

Rai, K.S., 1991. *Aedes albopictus* in the Americas. *Annu. Rev. Entomol.* 36: 459–484.

Chapter 8

ODD APPETITES AND OUT-OF-THE-WAY PLACES

Coprophagy (dung-eating) and sacred scarabs

I fly among those of divine essence
I become in it Khepra
I am that which is in the bosom of the gods
The Great Beetle.
—EGYPTIAN PRAYER FOR THE DEAD

INSECTS ARE NOTORIOUS for having peculiar appetites, but there are perhaps no stranger appetites from a human perspective than those of coprophagous arthropods—insects that eat excrement (dung, manure, feces, and whatever else it is euphemistically called). It's strange not only from an aesthetic angle but also from a nutritional angle. Presumably, excrement is that part of ingested food that cannot be metabolized or assimilated by its producer to yield energy or materials for growth. The fact that it is subsequently ingested by other organisms suggests that they are either extremely efficient at processing food or that they have substantially different metabolic needs than do other organisms. It's likely that many dung-feeders, or coprophages, actually obtain the bulk of their nutrition not from dung per se but from the microbes associated with it. Nonetheless, the few studies that have been done have shown that almost 90% of what a dung-feeder takes in goes right back out the other end.

It's not only that dung is difficult to eat that makes a life of coprophagy so challenging; dung is also not easy to find on a regular basis. It is generally deposited randomly throughout the environment and, because it can quickly desiccate under the right circumstances, it doesn't remain edible, even by dung-insect standards, for very long. Thus, most species that breed in dung must do so with all deliberate speed in order to complete development before the pat dries up and blows away. Virtually all (if not all) species that spend their life in and around dung are holometabolous, or undergo complete rather than gradual development. Insects with gradual development are less likely to manage the growth rates necessary to pack an entire lifetime into just a few days.

In almost all cases, dung-feeding has cropped up in groups that otherwise engage in scavenging for a living. The most conspicuous member of the dung-feeding guild are the beetles, particularly in the family Scarabeidae, although dung-feeding shows up intermittently in other beetle families as well (e.g., in the Hydrophilidae, the water-scavenger beetles). The other group in which coprophagy is widely established is in the Diptera, where maggots in many families frequent dung. Among these are house flies and latrine flies, the so-called small dung flies, stilt-legged flies, eye gnats, and the black scavenger flies. Outside the Diptera and Coleoptera, coprophagy appears sporadically. One of the more bizarre forms of coprophagy is displayed by a caterpillar. *Cryptoses chalepi* is a pyralid moth that lives in the algae-encrusted fur of sloths, which hang almost immobile for long periods of time from branches of tropical trees. For many years, the breeding site of this moth was completely unknown. Patient fieldwork by an entomologist in the jungles of Central America revealed that the sloths climb down to the forest floor once a week to defecate. The moths ride around in the fur of the sloth awaiting just such an opportunity; they debark and oviposit in the fresh pat, which otherwise would be next to impossible to locate in the tropical forest floor.

Finding the dung is the first obstacle for successfully establishing a coprophagous lifestyle and it is a formidable one. Many species, like the sloth moth, associate with the source of their food and simply await the inevitable event. Some species of beetles, in their eagerness, can cause a rather unusual and uncomfortable form of insect infestation known as canthariasis—the beetle enters the body through the anus and establishes residence in the dung in situ. This condition is mercifully rather rare and is almost nonexistent where proper hygiene is practiced.

For long-distance location of dung pats, a rather sophisticated system of chemical detection is used by some species. Irrespective of its source, dung smells more or less the same, due to the presence of chemicals that result from the breakdown of proteins. Studying the chemical factors that attract dung insects to dung requires a certain kind of intellectual fortitude, but fortunately there are scientists who have gamely undertaken the task. Valeric acid was found to be the dung component attractive to *Geotrupes ster-cororus*, a scarabeid found on deer dung in the Belgian Ardenne Forest. Olfactory cells on the antennae of *Geotrupes auratus*, a Japanese dung beetle associated with cow pats, are sensitive to 2-butanone, phenol, p-cresol, indole, and skatole. Of these, indole and skatole, breakdown products of protein, have an odor that is unmistakably fecal from the human perspective. Whereas 2-butanone seems to be the primary long-distance attractant for the beetles, the other constituents aid in close-range orientation.

If there is a fairly efficient long-distance detection mechanism in place for locating dung, a veritable bonanza awaits the co-prophage. Cattle in pastures, for example, provide approximately 12 pats per day per individual; sheep produce 0.24 droppings per square meter per day, for an annual yield of about 480 kilograms (about 10,500 pounds) per hectare. It is difficult to discern a pattern of common anatomical adaptations to dung-feeding—in all probability, adopting dung as a diet involves physiological and behavioral changes more than anatomical ones. Among other things, dung as a foodstuff doesn't fight back, as do vertebrate hosts, insect prey, or even plant food. About the only distinctive anatomical feature exhibited by some but not all dung-feeders are mouthparts that can filter nutritious microbes from the matrix of undigested food particles.

Dung insects have developed a variety of ways of handling dung once it has been located. Beetles in particular display a diverse array of behaviors upon encountering a dung pat. Many species simply lay their eggs directly in the dung where it was deposited and develop on it in situ. Others tunnel underneath and bury it on site; burying the dung not only prolongs its suitability as a foodstuff and as a habitat by reducing its exposure to environmental variation, but it also protects the developing grubs from parasites and predators. Dor beetles (species in the genus *Geotrupes*) excavate underground galleries several feet down. A male and female together construct the gallery by first digging straight down and then adding on side galleries, packing them about four inches deep with dung. A single egg is deposited in each gallery. *Onthophagus*, only a bit over 1/4 inch in length (6–9 millimeters) can tunnel down 4 inches. In some species, adults actually remain in the galleries to care for their offspring. One such species is *Copris lunaris*. The females overwinter underground in a chamber, which is entered in the spring by a male. The male carries out soil while the galleries are under construction and brings in dung from the outside for provisioning.

The female in the chamber packs and shapes the dung into a "dung loaf" weighing as much as 6 ounces. The loaf is later sliced into several pear-shaped brood cells and provided with an egg. The female (as well as the male in some cases) remains with her brood not only after the eggs hatch but until the larvae pupate, a period that can run four months.

Perhaps the most spectacular means of handling dung is rolling it away to bury it far from where it was first dropped. Scarab beetles who engage in this behavior are sometimes called tumblebugs (Fig. 8.1). Adult scarabs in the genus *Scarabaeus* form two types of balls, one for their own use and one for use by their offspring. Dung can simply be sectioned where it lies or it can be compacted from smaller bits. The balls are rolled backwards, with the hind legs moving forward on the ball and the front legs moving backward on the ground. Whereas both sexes construct feeding balls, only fertilized females construct a brood ball. The female excavates a gallery and reforms the spherical ball into a pear shape. One egg is deposited in each chamber and a female can construct on average about six of these chambers over her lifetime.

Ball-rolling is a remarkable example of physical exertion on the part of the beetles. In one study, ball-rolling beetles, themselves averaging about 2 to 5 grams in weight, moved dung balls that ranged from 6 to 244 grams, attaining speeds of up to 20 centimeters per second (and averaging 14.4 centimeters per second). Unlike most insects, quite a number of dung beetles are endothermic, or capable of regulating their body temperature independent of the environment, which may help them meet the energetic demands of their ball-rolling habits. Some of the larger endothermic scarabs, weighing in at 20 grams or more, are up to eight times larger than some endothermic vertebrates (like hummingbirds or shrews). Endothermic activity means that a scarab can build a dung ball much faster (1.5 to 5 minutes at 40°C versus 10 to 35 minutes at 26°C). Speed is essential, not only to make maximum use of a changeable resource but also because other dung beetles in the neighborhood are not averse to swiping already completed dung balls from other individuals. A host of other dung-feeders move in given half a chance, too. Competition is intense—in east Africa, one dung pat deposited by an elephant attracted 4,000 beetles within only half an hour and a dung mass of 1.5 kilograms (over 3 pounds) was cleared away in two hours by 16,000 dung beetles, who themselves weighed collectively more than ½ kilogram (1 pound; Hanski 1987).

Since scarabeid dung beetles are relatively large and relatively conspicuous as insects go, it's not surprising that their unusual and easily observable behavior has attracted considerable attention. The ancient Egyptians, probably beginning with the Ethiopians who first settled the Nile, found in the behavior of *Scarabaeus sacer* much symbolic significance. That round pellets are occasionally rolled from east to west, the same path taken by the sun on a daily basis, evoked the metaphor of world as dung ball. The dung ball was also seen as a symbol of the sun—cuticular projections on the front legs of the beetle could be thought of as resembling rays of the sun and the thirty tarsal segments (five per leg) as representing the thirty days of each month. The moon, too, was invoked by the twenty-eight-day developmental period of the larvae, approximating the lunar month. The dung beetle was conceptualized as "courageous warrior" clad in armor. Contributing to this image was the mistaken notion that all dung beetles were male—perhaps males rolling their food balls were more conspicuous—and thus were entirely self-sufficient. Hence, the appellation "Father of the Gods" for the sacred scarab. Finally, dung beetles symbolized the regenerative

Figure 8.1
Elephant dung beetle (*Heliocopris*), the largest dung-rolling beetle, and sacred scarab (*Scarabaeus sacer*).

Pthah, or the creative power—this from the emergence of larvae from dung and mud (the ancients didn't know eggs were deposited inside) and from the fact that scarabs were among the first animals to reappear after the Nile floods, an annual catastrophic event.

Virtually all throughout Egypt, scarabs were worshipped as deities, especially in connection with resurrection and reincarnation. They appear on the royal sepulchres of Biban el Molu, more ancient than the pyramids. Scarabs were often embalmed and entombed with the dead. In fact, the pyramids may simply be representations of dung beetles on a grand scale, with the pyramid a "deified dung pat" containing the soul in transformation, like a pupa awaiting metamorphosis. Later dynasties grew more sophisticated and instead of using real scarabs used stone icons beginning about the Third Dynasty, 3933 to 3900 B.C. These were often engraved with writings from the *Book of the Dead*. Each king had a distinctive scarab upon which were engraved royal names, deeds, and titles, and the scarabs were often used as seals for royal documents. The beetle itself was incorporated into the hieroglyphic phonetic alphabet in the twenty-sixth dynasty (the "h's" in "Pthah" are written with this symbol).

Although dung does not attract much attention or even polite conversation today, dung and dung-feeders are exceedingly important economically. On the negative side, many dung associates, by also associating with humans, can act as vectors for disease. The house fly is one principal example, implicated in the transmission of typhoid fever, yaws, dysentery, anthrax, cholera, and a number of other diseases. Recognition of the role of *Musca domestica* as disease vector in the early part of the twentieth century led to an intensive publicity campaign to "swat that fly," spearheaded by L.O. Howard of the USDA C. F. Plowman and W.F. Dearden published a "popular and practical handbook" entitled *Fighting the Fly Peril* in 1915, entreating the populace to devote the same energy to eradicating flies that was at the time being devoted to the eradication of malaria-carrying mosquitoes. It's difficult today to appreciate the magnitude of the "fly peril" at the turn of the century; since the introduction of the automobile, one of the chief breeding sites of house flies—horse manure—has been largely eliminated from city streets. It's one of the great ironies of the century that the automobile was introduced as a "nonpolluting" alternative to horse-drawn transportation.

On the other hand, without coprophages, the earth would soon be piled deep with the waste products of its inhabitants.

Coprophages provide an invaluable service by processing and removing the waste of others. This activity in the United States has actually been measured in terms of dollars and cents. Dung takes an extraordinary amount of pastureland out of commission every year. G. Fincher (1981) estimated, based on approximately 96,075,000 head of cattle in the U.S. on farms in January 1979, allowing 0.08 square meters per cowpat and ten defecations per day, and assuming these cattle are on pasture all year, that cows "could theoretically produce enough dung to cover 4,973 hectares of pasture per day, or 1,815,145 hectares per year. Corrections due to feedlots and to dung dropped in woods, feedlots, and barns, and dung buried by beetles still yields a figure of 335,678 hectares of pasture continuously covered by cattle dung every year." The importation of dung beetles, by increasing nitrogen cycling, by reducing parasitism by dung-breeding gastrointestinal parasites (which costs the dairy industry alone some $164 million a year), and by increasing grazing, could contribute an estimated $2,068,360,154 to the economy every year.

The economic importance of dung beetles became patently obvious in Australia, where the only native mammals are marsupials (pouched, rather primitive species not particularly closely related to the rest of the world's placental mammals). Potential entrepreneurs introduced cattle onto Australia's lush rangeland, only to discover that the native dung beetles were incapable of processing the dung of anything other than the marsupial mammals to which they were accustomed. Native dung beetles could return less than 20% of dung nitrogen back to the soil. Accordingly, cattle dung accumulated at an alarming rate and along with this dung grew populations of the pestiferous bush fly *Musca vetustissima*. In 1964, the CSIRO (principal scientific agency of the Australian government) launched the Dung Beetle Project, and arranged for introductions of foreign dung beetles accustomed to placental mammal (particularly cattle) dung. Some forty-four species were introduced in the next fifteen years, and fourteen have settled in to establish residence, with the net result that Australia now has a thriving cattle and dairy industry.

Insects in carrion (of maggots and murderers)

> I heard a fly buzz—when I died.
> —Emily Dickinson, *Poem No. 465*

IT'S BAD ENOUGH knowing that insects pester humans in their daily lives but it's really adding insult to injury to think that they will continue to do so after the humans are dead and supposedly gone. Dead bodies (of any vertebrate) provide nourishment and livelihood to a large number of insects. That such is the case has been recognized since time immemorial and may in fact be at least part of the motivation behind the distinctly human practice of burying the dead with elaborate ceremony. Biblical references to carrion-feeders abound. In Isaiah 14:11, for example, it is written, "Thy pomp is brought down to the grave and the noise of thy viols, the worm is spread under thee and the worms cover thee." If that's not depressing enough, there's a passage in Job (21:26) reminding mortals that "they shall lie down alike in the dust, and the worms shall cover them."

Insects associate with carrion or corpses for a variety of reasons. Necrophagous species are those that actually consume dead flesh. Most of the truly necrophagous species are flies or beetles. Among the flies, members of the family Calliphoridae (the blow flies) and Sarcophagidae (the flesh flies) are particularly conspicuous elements of the dead-animal fauna. House flies, black scavenger flies, brine flies, coffin flies, and, particularly in later stages, cheese skippers are also associated with carrion. Among the beetles, necrophagous families include the burying beetles and the hide or skin beetles; histerids and tenebrionids are associated with later stages of decay. About the only other conspicuous members of the dead animal fauna are lepidopterans, especially tineids, or clothes moths.

From the insect perspective, there are definite advantages to utilizing carrion as a food resource. Unlike many forms of animal protein, carrion does not struggle or run away. There are, however, decided drawbacks as well, notably that carrion is neither an abundant nor predictable resource. Its suitability also changes dramatically over a relatively short time. Life cycles and behavioral traits in the carrion fauna reflect the ephemeral nature of carrion. Species associated exclusively with carrion, particularly in its early stages, tend to have a phenomenal capacity for finding dead animals. Many members of the carrion fauna are exceedingly sensitive to odors of decay and putrefaction at short range; silphids can, for example, detect the presence of skatole, a degradation product of protein, at concentrations in the air of about 9 parts per million (although the effective range over which they are sensitive is only on the order of about 1 meter). House flies and blow flies are attracted to a number of decomposition products, including acetic acid, benzoic acid, butyric acid, valeric acid, indole, acetone, phenol, *p*-cresol, and methyl disulfide.

Aside from having remarkable sensory capabilities, many members of the carrion fauna also have remarkable digestive capabilities. Larval blow flies, for example, are among only a handful of species that can metabolize collagen, the principal protein constituent of connective tissues. Immature clothes moths and hide beetles possess enzymes that can metabolize keratin, the proteinaceous component of hair and skin, which is indigestible to most organisms due to the fact that some of its amino acids are linked by usually unbreakable sulfur-containing chemical bonds.

If there is one word that characterizes the dung fauna, it's "competitive"—it's a dead-dog-eat-dead-dog world out there. Competition comes not only from other arthropods but from vertebrate carrion-feeders as well. Scavengers (such as vultures, wild dogs, or hyenas) may remove from 60 to 100% of the aboveground carcasses before they can be colonized by arthropods at all. High mortality and wide population fluctuations are thus typical of the lives of many arthropod carrion feeders. It often happens that more eggs are laid on a single carcass than can possibly reach maturity; ecologists refer to this sort of competition for food, in which, due to overcrowding, the possibility exists that all individuals fail to complete development, as scramble competition (in contrast to contest competition, after which clear winners and clear losers can be identified).

Many life-history attributes of carrion insects are designed to reduce competition both within a species and between species. Fast growth and flexibility of resource requirements are typical of carrion feeders. Blow flies, for example, can complete development in ten to fourteen days, gaining an average of 5% of their final larval weight per hour. Flexibility is key, too; they can pupate, depending on the availability of food, at a wide range of sizes (with the minimum size only 12% of the standard weight) and over a wide range of time intervals (from 50 to about 3,000 hours). In addition, some carrion feeders actually kill potential competitors. Despite the fact that legless, sightless maggots would hardly seem to be formidable enemies, flesh fly maggots routinely kill blow fly maggots in large numbers.

Also characteristic of carrion fauna is the ability to make quick and efficient use of a corpse. The ability to find carrion quickly is at a premium, since first arrivals can preempt resources and outcompete latecomers. *Sarcophaga* flesh flies practice larviposition (deposition of maggots, rather than eggs), which may provide them with a competitive edge compared to flies that must undergo an egg stage in carrion. *Lucilia* blow flies go to the extreme of laying their eggs directly in wounds of injured animals, not even waiting for them to die before moving in.

Some carrion insect behaviors are aimed at preventing access of the carrion to would-be competitors. *Dryomyza anilis* flies actively defend a body from other members of the same species that attempt to oviposit. Food relocation accomplishes the same end for *Nicrophorus* and other silphids. Silphids are known as burying, or sexton, beetles because they take the bodies of dead mammals and birds and bury them, thereby preventing potential competitors from finding, ovipositing, and feeding on the body. They are aided and abetted in their struggles by an unusual mutualistic association with a mite. The bodies of the silphid beetle *Nicrophorus* are covered with tiny mites (*Poecilochirus necrophori*) found only on the bodies of burying beetles. When a silphid locates a carcass, the mites leave the body of the beetle and move onto the carcass. There, they feed not on the dead remains but rather on the eggs and developing maggots of carrion flies that lay their eggs on carcasses at about the same time that burying beetles begin to prepare a carcass for burial. After processing and burying a carcass, burying beetles lay their eggs in the burial chamber. The larvae are fed by the parents initially, who in a tender example of parental care regurgitate partly digested dead flesh to meet the nutritional needs of their offspring. Eventually, the larvae can feed directly on the carrion. When they are ready to pupate, the adults dig an escape tunnel for them and, after pupation, the newly eclosed adults emerge and seek out fresh carrion. As they emerge, they are boarded by the mites, which accompany them to the carrion and to new sources of fly eggs and maggots. The association between the mites and the silphids appears to be mutually beneficial—the mites, which are flightless, are able to move from one fresh carcass to another, and the silphids are able to keep a carcass free from maggots for the duration of the development of their offspring.

Not every insect found in a dead body is necessarily necrophagous. A sizable community of predators and parasites has arisen that specializes on the insects that are found in carrion. Among the most important predators of carrion-feeding insects are certain staphylinids (or rove beetles). Even some of the necrophagous blow fly maggots turn carnivorous in late instars. There are also omnivorous species in carrion that include dead meat as part of a wide and varied diet. Wasps are among these species and are frequently found feeding on dead flesh. Most of the stories of augury attributed to "bees" in carcasses in all probability involved wasps such as yellowjackets rather than bees. Finally, there are adventive species, species that visit a corpse simply as a bit of topographical relief in their environment. Springtails and other litter dwellers, including mites and millipedes, are often found on or around dead bodies, not so much as a result of deliberate colonization but rather just as a result of their soil-dwelling, debris-feeding habits.

Like dung, dead bodies are habitats that are extremely unpredictable in distribution and extremely variable in terms of size, condition, and general composition. The changes in physical features of a corpse over time, however, are so predictable that distinct waves of colonization by stage-specific species have been described (although there is some controversy as to whether these represent distinct stages or a continuum). The period immediately after death is associated with chemical changes originating within the body, such as the release of body fluids and gases. Bluebottle and house flies visit and lay eggs at this stage. They arrive so quickly and develop so fast that for centuries people believed that maggots appeared spontaneously when meat began to decay. This notion of spontaneous generation was extremely difficult to dispel. Francesco Redi of seventeenth-century Italy finally performed a series of classic experiments, however, that demonstrated that maggots appear in meat only if flies are allowed to lay eggs on the meat first.

Shortly after death, fermentative changes attributable to microbes within the body result in detectable odor emission, and flesh flies move in. Wasps are also frequent visitors of carcasses at this stage. When fats begin to break down, volatile fatty acids are released. These attract hide beetles, *Aglossa* (a pyralid moth), and other species that can feed on decomposing fats. This stage lasts from three to six months. Caseic fermentation—protein breakdown—follows and attracts a number of species specialized at metabolizing protein-breakdown products. In this group is the cheese skipper, *Piophila casei*.

Fermentative changes are accompanied by evaporation of many body fluids. Carcasses in later stages attract silphid and hister beetles, as well as some flies. Eventually, fluids evaporate completely and the body becomes desiccated. At this stage, hide and skin beetles move in as well as clothes moths and their relatives. These are species that can break down and metabolize collagen, the major protein constituent of connective tissue. Mites also colonize at this stage. Finally, only the real dregs remain. Such things as fecal material from the resident insects are consumed by scavengers such as spider beetles (*Ptinus brunneus*) and other such unfussy types.

The progression of species in a decomposing corpse is so precise and predictable that it is actually admissible as evidence in court. Forensic entomology makes use of life-history information and species identification to estimate time of death (postmortem interval), usually of crime victims, so entomologists can be called into court to testify as expert witnesses. Probably the earliest recorded incident in which insects helped to identify a murderer was the one described in a thirteenth-century Chinese manuscript, in which a murder by sickle was under investigation. All of the local farmers were called in and asked to lay their sickles on the ground. Flies were attracted to land on only one sickle; faced with such damning testimony, the guilty party promptly confessed. The modern use of forensic entomology for crime solving dates back to 1855, in which the knowledge of the insect fauna of human corpses was used to solve the murder of a child. A plasterer working on a mantelpiece in 1850 discovered the body of a child while he was working and Dr. M. Bergeret of Jura was called in to conduct the autopsy. He estimated from the presence of mite eggs and certain life stages of a blow fly (*Sarcophaga carnaria*) that the occupants of the house in 1848 were the likely perpetrators.

By the end of the century, a number of landmark papers had been published, including classics by J.-P. Mégnin (*La faune des Tombeaux* in 1887 and *La Faune des Cadavres* in 1894) and the report in 1898 by M.G. Motter, "A contribution to the study of the grave. A study of one hundred and fifty disinterments, with some additional experimental observations." Since that time, entomological evidence has been used in countless murder cases and has played a critical role in obtaining convictions.

One of the most important functions an entomologist has in a murder investigation, particularly in estimating the postmortem interval, is to consider the effects of local variation on the progress of a normal succession. Many environmental (as well as biological) factors influence the rate and nature of faunal succession. The time of day that a body is disposed of affects the timing of colonization; many blow flies do not fly at night, so a body may escape colonization for 12 hours or more, depending on the length of the night at the time. The extent to which a corpse is exposed to the elements influences the probability that it will be discovered by ovipositing flies. A carefully wrapped body may escape detection for a considerable length of time. The geographical location where the body is deposited plays a role in determining which species establish residence. Species of blow fly, for example, vary with the locality, with some more or less restricted to urban areas and others far more common in rural or undisturbed places.

The manner of death also influences the nature of the insect community that eventually colonizes the corpse. Normally, flies prefer to oviposit in any of the nine natural body orifices (the mouth, two nostrils, two ears, two eyes, and two more private places). A knife or gunshot wound allows access to parts of the body that would normally not be colonized until much later in the decay process. Poisons such as parathion or arsenic can slow down the growth rate of flies and beetles, and some drugs, such as cocaine, can speed up the growth rate. In addition, insects can sequester certain fat-soluble compounds in their own tissues such that, while detectable levels of a poison may no longer be present in the corpse itself, they may be present in insect tissues in concentrations high enough to measure.

There are now more than two dozen full-time forensic entomologists in the United States, working in close concert with police and other law enforcement agencies to solve all manner of crimes. While certainly the flashiest of crimes in which entomological evidence is important are the murders and other violent crimes, forensic entomologists can also provide critical insights into other sorts of misdeeds. Neglect of children or elderly invalids is often indicated by flourishing maggot infestations in diapers or in open sores, for example. Molecular biological techniques are currently in development to facilitate the identification and age estimation of insects associated with crime scenes. Career criminals may soon find it necessary to take a crash course in entomology to elude not only law enforcement agents but also the telltale necrophagous insects.

Bodies in the pantry

PEOPLE MAY THINK that coming across a maggot-infested dead body is a rare event, generally associated with unsavory neighborhoods or wrong sides of tracks, but actually it's a fairly commonplace occurrence in the most respectable of homes, if the concept of dead body is stretched far enough. For the human species, civilization meant not only domesticating plants but also domesticating animals; domestication led again to surpluses of products that required storage and that were quickly exploited by opportunistic arthropods. In order to preserve meats for long-term storage, they were dried and cured and, as such, like it or not, became effectively indistinguishable from carrion, dried and cured by exposure to the elements. Some insect species normally accustomed to feeding on carrion quickly incorporated dried meats and cheese (which is a dried form of animal protein cured by microbial action) and became domesticated (or at least associated with humans) right along with the animals providing them with sustenance.

One of the more peculiar partakers of stored animal protein is the cheese skipper, *Piophila casei,* a piophilid fly. In nature, females lay eggs in dried carcasses and the maggots feed by stripping off shards of flesh. In houses, cheese skippers infest ham, bacon, and, as their name implies, cheese. The name "skipper" refers to the remarkable ability of the legless maggots to propel themselves 6 inches into the air by bending into a circle and contracting their abdominal muscles; like a spring, the contraction propels them into the air. As a curious historical aside, "maggoty" cheese was once considered an exquisite delicacy—until it became apparent that these larvae were capable of surviving the trip through the alimentary canal.

Clothing woven from animal wool or hair is another stored product subject to damage by insects normally associated with dead bodies. As they do in corpses, beetles and moths figure prominently in people's drawers, closets, and sweater chests. Many carpet or hide beetles, in the family Dermestidae, feed on dead animals. While *Attagenus megatoma* will, if pressed, eat cereals, dead insects, dust, or animal droppings, they're particularly fond of fur and feather products and consume clothing, curtains, carpets, and rugs. Two species of caterpillars, in the family Tineidae, are partial to woolens, the webbing clothes moth *Tineola bisselliella* and the case-making clothes moth *Tinea pellionella.* Carpet beetles and clothes moths are among the select few that have enzymes that can sever the sulfur-containing bonds in keratin, the principal protein of fur and feathers, and break it down. Thus, these insects are literally capable of "splitting hairs."

Carpet beetles and clothes moths, while pestiferous, are at least fairly temperate in their eating habits. In contrast, there are stored-product species whose eating habits resemble the omnivorous habits of humans. This resemblance does not endear them to humans but rather creates considerably more conflict. Probably the most conspicuous of the omnivorous home invaders are the cockroaches. While there are at least fifty species of cockroaches in North America, only about four are troublemakers in the sense that they move into people's houses and make themselves at home (and hence are known as domiciliary species). All four are cosmopolitan in distribution and probably originated in tropical Africa or Asia. Central heating and a constant food supply mean that people's houses are the next best thing to their homeland.

The American cockroach *Periplaneta americana* probably arrived in the U.S. on slave ships from Africa some 300 years ago and is now firmly established not only here but all over the world. The largest of the domiciliary species, it packages about a dozen eggs in an ootheca, which it deposits on objects in protected places. The immature nymphs take up to a year to complete development. These cockroaches are sometimes called waterbugs due to their predilection for damp basements and sewers. The Oriental cockroach *Blatta orientalis* is dark brown to black in color and is slightly smaller than the American cockroach; females are wingless, but both sexes can walk appreciable distances when the need arises. These cockroaches are also primarily basement-dwellers. The German cockroach *Blattella germanica* is not really German (in fact the Germans call it the Prussian roach) but is probably from equatorial Africa (at least, that's where its near relatives still reside). Unlike the other domiciliary species, German cockroaches carry their oothecae, packed with three to four dozen eggs, wedged in their abdomen until just before the eggs are ready to hatch. German cockroaches are extremely prolific and extremely pestiferous, frequenting kitchens, living rooms, and bedrooms as well as damper areas of the house.

The brown-banded cockroach, or TV roach, *Supella longipalpa*, is a relative newcomer to North America and was possibly introduced into this country by soldiers returning from the Pacific theater at the end of World War II. Of the chief domiciliary species, it seems most comfortable in the heat (and is probably descended from ancestors that lived in drier regions of the African continent). Since *S. longipalpa* frequents warm places, it is frequently encountered in and around household appliances and electrical apparatus (hence the moniker "TV" roach).

While pathogenic organisms, including fungi, bacteria, viruses, parasitic worms, and protozoans, can be found on the bodies of cockroaches, they have never been directly implicated in the transmission of human infectious diseases. They nonetheless do present a health risk. Among other things, their association with filth means that they can physically contaminate food with bacteria, and they have been associated with transmission of *Salmonella* and other foodborne illnesses. As if that weren't enough, their tendency to seek out dark corners occasionally puts them in places where they are decidedly unwelcome, not the least of which are the ears of unsuspecting sleepers. The cost of controlling cockroaches runs over $1.5 billion every year.

Detritivores (the dirt on termites)

ALTHOUGH THE NOTION of eating pizza with mushrooms hardly merits raised eyebrows, the concept of eating pizza topped with slime mold, protozoans, bacteria, or dirt is not likely to be a big money-maker. Humans tend to tolerate decay only in small doses, as far as diet is concerned. This is not to say that humans eschew all forms of rotting compost; among the most highly treasured gourmet treats are cheeses (cured animal protein) marbled with bacteria such as *Penicillium roqueforti* and sauternes produced from grapes riddled with noble rot, *Botrytis cinerea*. Among arthropods, however, detritus is a highly regarded foodstuff. For one thing, detritus (defined as "that which remains after disintegration") is certainly abundant. While in general detritus is low in protein and nutrients, and relatively rich in indigestible polymers such as cellulose, detritivores can more than make up in quantity what their diet lacks in quality.

What exactly is being consumed by detritus-feeders isn't always clear. While some arthropods consume dead plant parts, others actually filter through the soil and feed selectively on microscopic bacteria or fungi, which are actually a fairly good source of protein and vitamins. Detritivores display some remarkable adaptations to compensate for the poor nutritional quality of their food. Some sowbugs, for example, eat their own feces to recover some of the nutritional material that gets passed out the first time. Some aquatic fly larvae (such as *Tipula abdominalis*) that feed on leaf litter actually metabolize carbohydrates produced by the fungi that have colonized the leaves, rather than the leaves themselves; breakdown of leaf litter proteins occurs in the abdomen and is facilitated by symbiotic gut bacteria. In at least four species of longhorn beetles that infest dead or dying trees, digestion of cellulose and other wood components is made possible by the fact that the beetles capture and make use of enzymes produced by fungi growing in the decayed wood.

Probably the most efficient and most spectacular consumers of dead, decayed, or otherwise seemingly inedible foodstuffs are the termites. Termites, members of the order Isoptera, are sometimes called white ants, although they are much more closely related to cockroaches than to the true ants (a relationship that most species would be unwilling to acknowledge). Termites do resemble ants in the sense that they, like many of the hymenopterans, lead a highly social life. All of the 1,900-odd known species of termites (including fossil forms) are social. The nature of their society differs in its particulars from the social life of hymenopterans. Among the reproductive individuals living in a colony are the fertile females, called queens, and fertile males, called kings. Unlike the drones of honey bee or ant societies, termite kings actually have a life beyond inseminating the queen (although their work load is still on the light side in comparison with workers). So-called primary reproductives are winged fertile kings and queens that depart the colony to mate, shed their wings, and found a new colony. Once she lands and settles in, the queen in many species undergoes a grotesque transformation in which her abdomen swells to enormous size (in excess of 5 inches in length in some cases); unable to move, she spends much of the rest of her life in a royal chamber immobile except for pumping out thousands of eggs. Fecundity varies, but, at one extreme, *Odontotermes* queens can release more than 86,000 eggs on a good day.

In contrast with hymenopteran societies, the king and queen of a termite colony are not alone in their sexual prowess; in some species supplementary reproductives are permitted to live in the colony, held in reserve in case the queen should fail to produce a normal egg quota or in case the king should fail to perform his royal duties to the queen's specifications. The remainder of the individuals in the colony are sterile, or nonreproductive. These sterile workers are of two types. Most numerous are the workers, wingless individuals of various sizes and of either gender that take care of all food gathering and dispensation—eggs, very young nymphs, and royals are incapable of feeding themselves so it is up to the workers to make sure they get enough to eat. Soldiers, also wingless, are in charge of defending the colony from intruders. In some species, soldiers are equipped with massive, heavily sclerotized jaws that are used for biting or shredding enemies. In other species, particularly in the subfamily Nasutitermitinae, the head is highly modified in shape so as to resemble an oil can. Through the pointed tip of the cone, these blind and seemingly insensate soldiers are capable of directing a flow of acrid, viscous, fluid at any would-be predators.

As a group, termites are capable of consuming an astounding variety of foods. Perhaps most impressive is their ability to consume living and dead wood. To this end, they are assisted by some symbiotic microbes that have taken up residence in the termite gut. These bacteria and protozoans produce cellulases, enzymes that can break down cellulose, the glucose polymer that makes up the vast bulk of plant structural material; without the appropriate enzymes, cellulose is completely indigestible. The absolute reliance, at least in some of the lower termites, on these gut symbionts has led to the strange practice of anal trophallaxis—deliberate consumption by termites of anal secretions of nestmates. In these secretions are the microbes vital for termite survival; ingesting them, by whatever means, ensures a steady supply of cellulase in the gut. Among higher termites, some species actually produce their own cellulases and thus have abandoned the practice of anal trophallaxis. Because some termites can attack living trees, they can cause damage to a wide variety of economically important plants, including fruit trees and ornamentals. The ability of termites to make a meal out of dead wood, though, means that every year they cause millions of dollars worth of damage to things held precious by humans: walls, fences, roofs, chairs, tool handles, paper products, rugs, and the like.

Some plant-feeding termites actually cultivate an association with symbiotic organisms outside the gut. Termites in the subfamily Macrotermitinae build special chambers in their nests to house "fungus comb." These *Termitomyces* fungi, known only from termite nests, are tended by the termites, who maintain them and provide them with a substrate—termite feces—on which to grow. In exchange, the termites consume the nutritious fungal flesh. Winged reproductives about to found a new colony pack away spores in their digestive tract to ensure that the new colony will have an adequate food supply.

Many of the ostensibly wood-feeding termites actually feed on an assortment of other materials, including grasses, twigs, leaves, plant litter, animal droppings, and many types of stored products. There are a few termite species that actually eat dirt—that is to say, they ingest humic materials found in the soil. This is not a rare, aberrant practice—so-called humivorous termites can be found throughout most of the world (except, curiously, in Australia and parts of Asia). These humivores possess specially modified mouthparts, with flanges designed for stone-crushing. Humivorous termites, like their wood-chewing relatives, also appear to depend on microbial symbionts for processing the organic minerals, polyphenolic substances, microbes, and arthropod and fungal corpses that make up the nutritive part of soil.

Irrespective of how peculiar the termite lifestyle may appear, it is a highly successful one. Termite societies are among the most populous in the animal world; a single colony can contain 20 million or more individuals. To house these hordes, termites construct elaborate structures. Termite mounds run anywhere from only a few inches in height and diameter to over 27 feet in height and 80 to 90 feet in diameter (Fig. 8.2). Mounds can be made of a variety of materials, but most mounds are made up of soil with or without excrement. By virtue of their size and number (in some parts of Africa, there are more than 800 termite mounds per hectare), termite mounds can tie up an enormous amount of soil. In Congo savanna, about 30% of the soil surface is occupied by termite mounds, which have a combined weight of over 2,400,000 kilograms (about 5,300,000 pounds) per hectare.

All of this excavation has enormous economic implications. First of all, soil-dwelling termites are likely to cause damage to underground cables or pipes simply by virtue of their earth-moving and soil-processing activities. More importantly, termite activities greatly influence the structure and composition of soils. Termite activities can accelerate soil erosion by clearing off plant cover and by removing organic material from the soil, thereby increasing the likelihood it can be blown away. The complex gallery system of

Figure 8.2
Termite mound in Australia (A. Berenbaum).

some mounds may serve as a drainage system that undermines the stability of the ground surface. On the upside, however, the mounds themselves are generally highly fertile, high in organic matter and rich in calcium, phosphorus, potassium, magnesium, and other vital plant nutrients. In parts of Africa and Asia, termite mounds are flattened and then used for crop planting. Termites also contribute mightily to the decomposition of organic matter, greatly accelerating the rate at which nutrients locked up in woody litter gets released back to the living world in usable form. Some of the materials broken down by termites, such as lignin and cellulose, are refractory materials that might otherwise remain resistant to decomposition for a long time.

Again, by sheer number, termites have an impact on ecosystem dynamics rivaled only by the activities of another social animal, *Homo sapiens*. As a result of the fermentation processes that are carried out in their guts, termites produce large quantities of methane as a waste gas. Termite contributions to atmospheric methane levels have been measured and in some places are comparable to contributions made by human industrial activities. Because methane is a greenhouse gas, that is, a gas the presence of which in the upper atmosphere may bring about a change in global temperatures, termite flatulence, for want of a better descriptor, has the potential to affect ecosystem function on a global scale.

Adjustments to life in or on the water

ONE OF THE great advances in the evolution of arthropods was the colonization of dry land, an event that led at least in part to the tremendous diversification of the class Insecta. One aspect of that diversification, curiously, has been the recolonization of aquatic environments. Today's aquatic insects differ in several fundamental ways from ancestral aquatic arthropods. For one thing, all aquatic insects are derived from terrestrial forms and their anatomy and physiology clearly reflect their terrestrial origins. In addition, the vast majority of aquatic insects live in freshwater environments; most ancestral aquatic arthropods were saltwater dwellers, as are many contemporary aquatic classes of arthropods (notably the crustaceans).

Adopting an aquatic lifestyle presents several physiological challenges to the basic terrestrial insect body plan. One of the greatest challenges is obtaining oxygen. In terrestrial insects, oxygen is taken in as a gas directly from the atmosphere through spiracles. In some aquatic insects, this system is hardly modified. Many aquatic insects live almost entirely on the surface (e.g., water striders, which glide along on the water's surface) and thus have no need of special respiratory equipment for taking oxygen up from water. Even insects living underwater can exploit atmospheric oxygen directly by way of siphons, or extensible tubes, that allow the spiracles to break through the water surface while the insect is submerged. The rat-tailed maggot (*Eristalis tenax),* the larval stage of a syrphid fly, owes its name to the long extensible siphon connected to its posterior spiracles. A rat-tailed maggot measuring 2 centimeters in length can extend its siphon six times the length of its body to reach the water surface. Rat-tailed maggots customarily are found in polluted waters, where dissolved oxygen concentrations in water are very low. Water scorpions (*Ranatra* species) are equipped with a breathing tube or snorkel that delivers oxygen from the water surface to its spiracles. Respiratory siphons are typically equipped with a fringe of hairs around the opening that are hydrofuge—that is, they are water-repellent. When an insect travels to the surface, this water-repellent fringe prevents entry of water into the siphon; at the surface, the fringe opens up like an umbrella, pushing water away from the opening and expanding the area over which air can be taken up. Other aquatic insects make use of atmospheric oxygen without traveling to the surface to obtain it. Beetles in the chrysomelid subfamily Donaciinae (as well as mosquitoes in the genus *Mansonia*) use a sawlike appendage at the base of the abdomen to cut into the stems of aquatic plants. The spongy interior of these plants is filled with oxygen-rich air, and the insects can thus tap into an air supply underwater.

In order to remain submerged for a long time, however, insects require a system for extracting dissolved oxygen from the water. Very small insects with thin cuticle can absorb dissolved oxygen directly through the cuticle. *Chaoborus,* the phantom midge, has a closed tracheal system that does not open to the outside; oxygen enters via diffusion through the cuticle, which is so thin as to be almost transparent (hence the common name of *Chaoborus,* or "glassworm"). In most permanent underwater dwellers, only certain specialized appendages are equipped with thin cuticle and are responsible for the uptake of dissolved oxygen. These structures are called tracheal gills. In addition to being covered with extremely thin cuticle, they are abundantly supplied with tracheae; feathery in appearance, the increased surface area of these finely divided structures increases the efficiency with which dissolved oxygen can be absorbed. Immature mayflies, damselflies, and stoneflies breathe underwater via tracheal gills. The tracheal gills of mayflies and stoneflies are feathery extensions of the body wall of the abdomen; damselflies, in contrast, are equipped with three paddle-shaped structures projecting from the tail end of the abdomen. The tracheal gills of dragonfly nymphs are contained inside the rectum, in this case called the branchial chamber. Water flow over the tracheated surface is maintained by uptake via the rectum, followed by forcible expulsion of spent water through the same orifice. This forcible expulsion can be used to propel the dragonflies through the water at great speed and serves as an effective escape tactic when they are pursued by predators.

Dissolved oxygen moves into a gill thanks to the physical process of diffusion, the net tendency of molecules to move from areas of high concentration to areas of low concentration. The oxygen content of the water is higher than that of an insect; thus, there is a net movement of oxygen molecules into the insect. The process of diffusion also makes possible respiration via a physical gill. A physical gill is a bubble of air that extracts oxygen out of the water. Many insects (particularly adult water beetles) travel to the water surface and trap a bubble of air, often under a wing cover. This air bubble contains atmospheric gases in the same proportion as the air at the water surface—roughly 20% oxygen and 78% nitrogen. Water, however, is composed of approximately 34% oxygen. Thus, oxygen tends to move from the water into the air bubble trapped by the insect. The oxygen then moves into the insect's tracheal system via spiracles that open into the bubble. Meanwhile, carbon dioxide is "exhaled" or released by a respiring insect into the bubble; since the concentration of carbon dioxide in the bubble is far higher than that of the surrounding water, the carbon dioxide tends to diffuse out of the bubble into the water. In water, nitrogen is dissolved at concentrations of approximately 65%, so nitrogen also tends to diffuse slowly out of the bubble. As it does, the bubble grows smaller, and the beetle must eventually travel back to the surface to replenish its air supply.

Some insects have developed a physical gill that never needs recharging. A plastron is an extremely thin layer of air held fast to the body surface by an area of tiny, densely packed fine hairs (only 10 micrometers or less in length). The hairs are hydrofuge, or water-repellent. Oxygen diffuses into the air layer, which, in contrast with an air bubble, never grows smaller since the hairs (not nitrogen gas) maintain its size and shape. The air layer is never displaced by water pressure and as a result insects with plastrons can remain underwater effectively forever, as long as they reside in bodies of water with relatively high oxygen concentrations. Plastron respiration is used by stages of aquatic insects incapable of moving up to the water surface; such stages include the egg stage and the overwintering, relatively immobile diapause stages of certain species.

In bodies of water with very low oxygen concentrations, some aquatic insects make use of a respiratory pigment, hemoglobin, that can bind to oxygen. This same pigment is used by humans to lock oxygen into red blood cells, which carry the gas via the circulatory system wherever it's needed in the body. Larvae of chironomid midges are called bloodworms because their hemolymph, or body fluid, is full of bright red hemoglobin. These larvae characteristically inhabit polluted, slow-moving waters notoriously low in oxygen content. While hemoglobin can bind to oxygen, it can also release oxygen when ambient concentrations are low, so the pigment may act as an oxygen storehouse for periods of oxygen deprivation in these insects.

Yet another challenge to insects living in aquatic systems is getting around. Water is some 700 times denser than air and acts as a considerable impediment to movement. A number of aquatic insects minimize locomotion problems by simply "skating" on the surface of the water. Water striders are equipped with very long legs, tipped with hydrofuge hairs. Their body weight, spread over a considerable area, is insufficient to break the surface film, on which they can move freely. Unlike most aquatic bugs, water striders (sometimes called "jesus bugs" because of their ability to walk on water) do not have claws at the tips of their tarsi (they are instead offset), where they might pierce the surface film and cause the striders to sink. Aquatic collembolans (springtails) can literally hurl themselves up in the air using the water surface film as a launching pad. Whirligig beetles also skate on the water surface. Their bodies are divided by a "plimsoll" line, above which there is a water-repellent surface and below which there is a wettable surface. The beetles move forward (at great speeds) by paddling with their flattened middle and hind appendages, boat-fashion. A streamlined body shape cuts down on water resistance.

While whirligig beetles row on the water surface, in respectable rowboat fashion, other aquatic insects move underwater by rowing. Waterboatmen and backswimmers both possess hind legs that are elongate, flattened, and equipped with long fine hairs (all of which serve to increase the surface area in contact with the water). On the power stroke, the hairs are held perpendicular to the body; on the back stroke, they lie flat and offer little resistance. The backswimmer actually moves in an upside-down position, with its back forming a keel for stability. Diving beetles also row their way through the water, stroking their flattened fringed hind legs together. The majority of aquatic beetles, however, stroke their legs alternately, rather than together, like a kayaker rather than a rower.

Many aquatic insects have no special adaptations for locomotion underwater and spend most of their time clinging to or crawling along the bottom. A constant problem for bottom-crawlers is buoyancy; being less dense than water, they face a constant challenge to remain submerged. Many simply cling to vegetation when not actively in pursuit of prey or in flight from a predator. Phantom midges can adjust their buoyancy by means of a pair of gas bladders, the contents of which can be regulated depending on water pressure. One group of backswimmers (in the genus *Buenoa*) appear to use hemoglobin not so much for respiration but for regulating buoyancy by controlling the oxygen content of the body; the greater the oxygen gas content, the greater the buoyancy. Caddisfly larvae (caddisworms) can remain on the bottom by virtue of their habit of constructing cases or houses of twigs, stones, pebbles, or sand cemented together with silk. These cases not only protect caddisworms from passing predators but also serve to anchor them to the substrate, where they can forage for debris or for prey.

Some aquatic insects have no apparent anatomical modifications for life underwater. Mymarids, or fairyflies, are tiny wasps that lay their eggs inside the underwater eggs of water beetles and other aquatic insects. The wasps for all intents and purposes simply "fly" underwater, using their wings to propel themselves to their destinations.

Living underwater has brought about alterations in sensory systems of aquatic insects as well. Water striders and other surface dwellers are extremely sensitive to ripples on the water surface, which are used not only to detect the presence of prey but also to signal interest to prospective mates. Whirligig beetles send out ripples across the water as they travel, which they monitor with their antennae held on the water surface. Reflected waves indicate the presence of an obstruction, so an alternate route can be taken if the obstruction is an immobile obstacle. Whirligig beetles are also anatomically endowed for simultaneously detecting visual signals from both the aquatic environment and the air as befits a denizen of the water surface. Their eyes are effectively divided in half; the upper half scans the skies in search of potential predators while the lower half examines the water column for potential prey. Chemical signals are important in aquatic environments as well; this sort of stimulus may be even more valuable than visual signals which, in turbid or silty water, may not be very reliable. Many of the hard-bodied underwater forms (such as adult water beetles) are endowed with noxious defensive secretions, which are released when the insect is threatened by a predator.

Fly-tying and fly-fishing

A man may fish with the worm that hath eat of a king
And eat of the fish that hath fed of that worm.
—WILLIAM SHAKESPEARE, *Hamlet*

FISHING IS IN all probability the world's second oldest profession. Archaeological remains reveal bones and bits of tusks fashioned into crude hooks; as technology advanced, these were replaced with copper and iron. One brilliant inspiration, hit upon by some unnamed genius as long as 3,000 years ago, was not simply to disguise the hook but to fashion it into a semblance of something delectable to attract fish. Thus were the beginnings of fly-fishing.

The first recorded artificial fly proper dates back to the Macedonians, who attached feathers to hooks to fish the Astraeus River. The historian Claudius Aelian, described in great detail the imitation insects in his book *De Animalium Natura* (A.D. 200):

> These fish feed on a fly which is peculiar to the country, and which hovers over the river. It is not like the flies found elsewhere, nor in shape would one justly describe it a midge or bee, yet it has something of each of these. In boldness it is like a fly, in size you might call it a bee; it imitates the color of a wasp, and it hums like a bee. . . . They do not use these flies at all for bait for the fish; for if a man's hand touch them, they lose their color, their wings decay, and they become unfit for food for the fish. For this reason they have nothing to do with them, hating them for their bad character; but they have planned a snare for the fish, and get the better of them by their fisherman's craft. . . . They fasten red . . . wool round a hook, and fit on to the wool two feathers which grow under a cock's wattles, and which in color are like wax

—an apt description of a contemporary fly called a red-hackle (Leonard 1950). The identity of this insect, called *Hipporous,* is not known.

Written accounts of fly-fishing are virtually nonexistent for about 1,200 years; the next appeared in 1486. Juliana Berners was the prioress of the Benedictine convent at Topwell, in the vicinity of St. Alban's, England, where she became intimately familiar with the natural world and developed a consuming passion for insects and fishing. The result of her interest was the publication in 1486 (printed by no less than William Caxton, the man responsible for introducing the movable printing press to England) of her tome *A Treatyse of Fysshynge with an Angle*. The book describes in fine detail a number of flies for which the prioress provided names (some still in use today), including the dun fly, yellow may, stone fly, black louper, dun cutte, maure fly, tandy fly, wasp fly, and drake fly.

The first illustration of a fly did not appear until 1652, when John Denny depicted one in his book *The Secrets of Angling*. The following year was a (no pun intended) watershed year for fishing in general—it was the year that Isaak Walton published his timeless classic, *The Compleat Angler*. In the book, Walton described with style and enthusiasm all aspects of fishing, including the life histories of aquatic insects. He also emphasized the importance of knowing the local insect fauna and the feeding preferences of the fish. His description of the twelve kinds of artificial flies to be used on top of the water may be the first literary reference to dry fly-fishing. His book was so wildly popular that it rapidly went through several editions; the fifth edition, published in 1675, included an extensive classification of flies compiled by Charles Cotton, a Walton disciple.

The scientific revolution of the eighteenth century found its way into fly-fishing in the nineteenth century in the sense that entomology developed as a scientific discipline, and its contributions in the form of an enhanced understanding of insect ecology and life history were incorporated into the fly-fisherman's arsenal. In England, Alfred Ronalds published, *The Fly-Fisher's Entomology*, in which he described flies that closely resembled real and distinctive insect species. Much of the information available today on aquatic insects was obtained as a result of career fishermen and hobbyists (by the turn of the century, sport fishing was firmly established as a national pastime). The tradition continues today; one of the finest entomological texts on aquatic insects—William McCafferty's *Aquatic Entomology*—is designed for use not only by entomologists but by anglers as well.

As for actual fly-tying, there are tremendous differences of opinion and practice but basically two schools of thought exist.

According to some, flies are most effective if they are impressionistic, that is, if they capture the essence of the insect and ignore the details. Size, color, and general *gestalt* are the principal concerns. According to others, flies must be realistic in every detail, even down to feathering pectinate antennae or fashioning abdominal gills on mayfly larvae. All parties agree, however, that the artificial fly must look real to a fish. Obtaining the perspective of a fish is a complex mental exercise. Among other things, phenomena such as countershading and light filtering by the water column must be taken into account. An amazing assortment of materials have been drafted for use in the construction of artificial flies. Feathers are a common feature. They are particularly suitable because, in many species, banding patterns create the illusion of segmentation, a must for simulating anything insect-like. Feathers from literally dozens of species of birds have been incorporated into fly designs. Game birds in particular have been popular (perhaps because in the off-season fishermen are likely to go hunting, too). Fur or hair from a variety of mammals not only has the appropriate translucence from the fish perspective but it is also buoyant and can keep a fly up on the surface of the water. Favorite furs include groundhog, lynx, badger, monkey, squirrel, bucktail, fox, and goat.

As for deciding on a design, the spectrum of choices facing an angler is not quite as broad as that facing an aquatic entomologist. Fly-tying fishermen tend to concentrate their efforts on a select few orders of aquatic insects. Of primary interest are mayflies, stoneflies, dobsonflies, and caddisflies (all of which lead lives closely tied to, if not always immersed in, water), and of secondary importance are true bugs, beetles, moths, bees, ants, and wasps (many of which are only accidentally aquatic). Despite the fact that beetles and true bugs are extremely abundant in all kinds of freshwater habitats, they are not favored as models for flies. It's possible that these primarily predaceous insects, equipped with potent venoms and powerful jaws, are not preferred fish food. Studies of gut contents of trout lend support to the notion that you won't catch too many fish with a beetle (one entomologist found that less than 2% of what was in the stomach of brown trout was beetles).

The importance of intimate knowledge of insect life cycles in successful fishing is illustrated by the proliferation of detail in mayfly flies. No fewer than five types of flies represent mayflies (Fig. 8.3): nymphs, duns (subimagoes or winged immatures), spinners (male and female adults), and spentwings (dead mayflies). Every year, calendars are published with recommendations on which flies

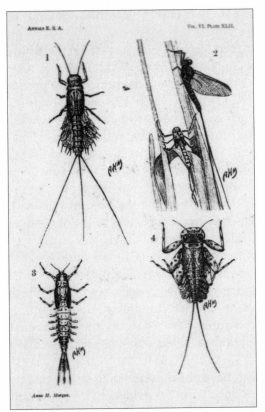

Figure 8.3
Mayflies (Morgan, 1913)
1. Mature nymph
(*Blasturus cupidas*)
2. Adult freshly emerged
from subimago skin
3. Nymph (*Ameletus
ludens*)
4. Nymph (*Epeorus
humeralis*)

to use at what time of year. Timing fly use to coincide with major insect emergences is all part of constructing the most convincing deception. Many fish display prey-switching behavior, that is, they customarily pursue selectively prey species that are most abundant. Thus, a fisherman with knowledge of emergence times and population fluctuations will be able to provide fish with a more enticing lure.

Flies are classified by how they are used. Wet flies are those that are designed to sink beneath the water surface (and are popular for use with trout and salmon). They are constructed to resemble adults emerging from pupal skins at water surface, females ovipositing below the water surface, small crustaceans, or drowned surface flies. In constructing wet flies, absorbent material is used to ensure sinking, and feathers and fur are used sparingly in order to ensure room for movement. Wet flies include divided wing, closed wing, down wing (female caddisfly), hackle tip, palmer hackle (caterpillar), flat wing (female stonefly), and translucent wing (such as mayflies, with females tied so as to convey the impression of egg sacs at the tip of the abdomen, complete with wings made of fish skin).

Dry flies are designed to remain on the surface of the water. Buoyancy is key here and many floatable materials are used in their construction (e.g., cork). Of dry flies, the most popular are probably the upwing types: duns and spinners. Double-paired wings include divided wing, split wing, and hackle tip wing. Downwing types include sedges (caddisflies), alderflies, rolled wings, hairwings, and reverse wing quill. Spentwing types include dead drakes and spinners and are recommended for quiet waters. Flatwing types are the stoneflies, and nobody knows what fanwing types like the Royal Coachman are supposed to be, but they appear to work nonetheless. There are detached body types (with the fly longer than the hook), palmer hackles (caterpillars), bivisibles (conspicuous not only to the fish but to the angler, so he or she can see where they go), parachute, and reverse fly (to simulate upstream flight). There are various nymphs depicted as well, such as the humpback (mayflies), double wingpad (stoneflies), compressed body (damselfly nymphs). Because these nymphs have no wings, they can't technically speaking be considered flies (since, of course, nymphs don't fly).

Although fly-fishing has never been unpopular, it has undergone a recent resurgence in popularity (as evidenced by the recent release of a major Hollywood studio production, *A River Runs Through It,* a paean to the spiritual and restorative powers of fly-fishing). Today, devotees can order all manner of flies from a number of nationally distributed catalogues. Shirts, ties, and even jewelry featuring fly motifs are even available to satisfy the most devoted enthusiasts.

Aquatic insects as bioindicators (it's the pits)

WHILE THE HABITS of aquatic insects have been of intense interest to anglers for at least a millenium, they have in this century become of great importance to society at large for an entirely different reason. Because of their abundance, their diversity, and their relatively low rank in aquatic food webs, aquatic insects are particularly sensitive to changes in the quality of their habitat. Within the last few decades, the sorts of changes to which aquatic habitats have been exposed have largely been the result of human assaults on the environment. Among other twentieth century insults, aquatic habitats have been subjected to pesticides from agricultural areas, spills and runoff from mining operations, acid rain, and wastewater discharges from sewage treatment facilities and nuclear power plants. Oftentimes, the aquatic insects are the first organisms in aquatic systems to suffer the ill effects of such human contact. Since the early 1960s, armies of entomologists, environmental engineers, and conservation biologists have been inventorying and monitoring populations of aquatic invertebrates in order to assess the environmental impact of human activities.

Of particular interest to people who assess aquatic environmental quality are the comings and goings of stoneflies and mayflies. As a rule, these insects live close to sediments and as a result are likely to come in contact with all manner of pollutants. They tend to be extremely sensitive to environmental perturbations and are susceptible to even slight increases in water temperature, acidity levels, or heavy metal contamination. A decline in the relative abundance of these insects in an aquatic community is a definite warning that the entire system has been compromised. Because mayflies and stoneflies are popular prey of game fish, a decline in their numbers is often soon followed by a decline in the diversity and abundance of the fish community.

On the other hand, chironomid midge larvae (including the bloodworms) are environmental indicators of an entirely different nature. These insects are notorious for their ability to thrive in the most horrific of environments. In one acid strip-mine lake in southern Illinois, for example, picturesquely named "Bradley's Acid Pit," population densities of a species of midge (*Chironomus* near *maturus*) approached 50,000 per square meter, despite pH levels as low as 2.7 (on a par with vinegar). An increase in the relative abundance of chironomids in a community is usually a sign of neighborhood decline. The Environmental Protection Agency has published a list of pollution tolerances for some 230 species of chironomids, for use in water quality assessments.

Small size notwithstanding, aquatic insects may play a vital role in facilitating the recovery of aquatic ecosystems following contamination. The burrowing habits of some mayflies, for example, can accelerate the rate at which toxic materials are adsorbed into the sediment. Many herbicides, pesticides, and other organic chemical pollutants are relatively insoluble in water. Once introduced into an aquatic environment, they can either drift down into the sediment or get taken up by living organisms; of course, once buried, they pose less risk to ecosystem function than they do traveling through food chains. Generally, diffusion of these chemicals through the semisolid sediments at the bottom of a lake or river is a slow process. However, many aquatic invertebrates make their living by burrowing through these sediments. The burrowing mayfly, *Hexagenia limbata,* tunnels extensively underneath the sediment surface, flushing out particles that clog the burrows and pulling water through to keep themselves supplied with oxygen. This pushing and pulling has the effect of turning under surface sediment and exposing underlying sediments to dissolved materials. Overall, the quotidian activities of burrowing mayflies can more than double the rate at which water-insoluble pollutants can get adsorbed onto sediments and increase the depth at which these materials get buried.

References

Coprophagy (dung-eating) and sacred scarabs

Bartholomew, G. and B. Heinrich, 1978. Endothermy in African dung beetles during flight, ball making and ball rolling. *J. Exp. Biol.* 73: 65–83.

Cambefort, Y., 1987. Le scarabée dans l'Egype ancienne. *Rev. l'Hist. Rev.* 204-1: 3–46.

Cherry, R.H., 1985. Sacred scarabs of ancient Egypt. *Bull. Entomol. Soc. Am.* 31: 14–16.

Cowan, F., 1865. *Curious Facts in the History of Insects.* Philadelphia: Lippincott.

Fletcher, T. B., 1924. Intestinal Coleoptera. *Indian Med. Gaz.* 59: 296–297.

Fincher, G., 1981. The potential value of dung beetles in pasture ecosystems. *J. Georgia Entomol. Soc. Suppl.* 16: 316–333.

Hanski, I., 1987. Nutritional ecology of dung- and carrion-feeding insects. In: *Nutritional Ecology of Insects, Mites, Spiders, and Related Vertebrates.* F. Slansky and J. Rodriguez, eds. New York: Wiley, pp. 837–844.

Hanski, I., 1988. Are the pyramids deified dung pats? *Trends Res. Ecol. Evol.* 3: 34–45.

Inouchi, J., T. Shibuya and T. Hatanaka, 1988. Food odor responses of single antennal olfactory cells in the Japanese dung beetle, *Geotrupes auratus* (Coleoptera: Geotrupidae). *Appl. Ent. Zool.* 23: 167–174.

Istasse, A., M. Magema, C. Gaspar, J. Wathelet, and M. Severin, 1982. Responses of *Geotrupes stercorosus* (Coleoptera, Scarabeidae) to valeric acid. *Ann. Soc. R. Zool. Belg.* 112: 41–44.

Klausnitzer, B., 1981. *Beetles.* New York: Exeter.

Plowman, C. and W. Dearden, 1915. *Fighting the Fly Peril.* London: Unwin.

Waage, J.K. and G.G. Montgomery, 1976. *Cryptoses chaloepi:* a co-prophagous moth that lives on a sloth. *Science* 193: 157–158.

Insects in carrion (of maggots and murderers)

Abstracts of Forensic Entomology. *Proc. XVIII Int. Cong. Entomol.* Vancouver, B.C., Canada, July 3–9, 1988, pp. 267–268.

Catts, E.P. and N.H. Haskell, eds. 1990. *Entomology and Death: A Procedural Guide.* Clemson, South Carolina: Joyce's Print Shop.

Greenberg, B., 1985. Forensic entomology: case studies. *Bull. Entomol. Soc. Am.* 31: 25–28.

Goff, M.L., A.I.O. Mori, and J.R. Goodbrod, 1989. Effect of cocaine in tissues on the development rate of *Boettcherisca peregrina* (Diptera: Sarcophagidae). *J. Med. Entomol.* 26: 91–93.

Heath, A.C.G., 1982. Beneficial aspects of blowflies (Diptera: Calliphoridae). *New Zeal. Entomol.* 7: 343–348.

Keh, B., 1985. Scope and applications of forensic entomology. *Annu. Rev. Entomol.* 30: 137–154.

Leclercq, M., 1969. *Entomological Parasitology: The Relations between Entomology and the Medical Sciences.* New York: Pergamon.

Milne, L.J. and M. Milne, 1976. The social behavior of burying beetles. *Scientific American* 245: 84–89.

Motter, M.G., 1898. A contribution to the study of the fauna of the grave. A study of one hundred and fifty disinterments with some additional experimental observations. *J. New York Entomol. Soc.* 6: 201–231.

Scott, M.P. and J.F.A. Traniello, 1991. Guardians of the underworld. *Natural History* 100 (June): 32–36.

Smith, K.G.V., 1986. *A Manual of Forensic Entomology.* Ithaca, NY: Cornell University Press.

Springett, B.P., 1968. Aspects of the relationship between burying beetles, *Necrophorus* spp., and the mite, *Poecilichorus necrophori* Vitz. *J. Anim. Ecol.* 37: 417–424.

Underwood, A., 1989. The witness was a maggot. *Science Digest* 97 (November): 24–33.

Bodies in the pantry

Brenner, R.J., 1993. Cockroaches, allergies, and construction practices. *Pest Management* 12: 18–23.

Cornwell, P.B., 1976. *The Cockroach.* London: Hutchinson.

Kopanic, R.J., 1993. Cockroaches and *Salmonella* contamination. *Pest Management* 12: 11–13.

Oldroyd, H., 1965. *The Natural History of Flies.* New York: Norton.

Detritivores (the dirt on termites)

Fraser, P.J., R.A. Rasmussen, J.W. Creffield, J.R. French, and M.A.K. Khalil, 1986. Termites and global methane–another assessment. *J. Atmosph. Chem.* 4: 295–310.

Lee, K.E. and T.G. Wood, 1971. *Termites and Soils.* New York: Academic.

Martin, M.M., 1989. *Invertebrate-Microbial Interactions; Ingested Fungal Enzymes in Arthropod Biology.* Ithaca: Comstock Publication Associates.

Martius, C., R. Wassmann, U. Thein, A. Bandeira, H. Rennenberg, W. Junk and W. Seiler, 1993. Methane emission from wood-feeding termites in Amazonia. *Chemosphere* 26:1–4.

Rouland, C., A. Brauman, M. Labat and M. Lepage, 1993. Nutritional factors affecting methane emission from termites. *Chemosphere* 26: 1–4.

Waller, D.A. and J.P. LaForge, 1987. Nutritional ecology of termites. In: *Nutritional Ecology of Insects, Mites, Spiders, and Related Invertebrates.* F. Slansky and J. Rodriguez, eds. New York: Wiley, pp. 487–532.

Adjustments to life in or on the water

Chapman, R., 1982. *The Insects: Structure and Function.* Cambridge: Harvard University Press.

Merrit, R.W. and K.W. Cummins, 1978. *An Introduction to the Aquatic Insects of North America.* Dubuque, Iowa: Kendall Hunt.

Milne, L. and M. Milne, 1978. Insects of the water surface. *Sci. Am.* 238: 134–142.

Resh, V. and D.M. Rosenberg, 1984. *The Ecology of Aquatic Insects.* New York: Praeger.

Wigglesworth, V.B., 1964. *The Life of Insects.* New York: Mentor.

Fly-tying and fly-fishing

Koller, L., 1963. *The Treasury of Angling.* New York: Golden.

Leonard, J.E., 1950. *Flies.* San Diego: Barnes.

Marbury, M.O., 1988. *Favourite Flies and Their Histories.* Secaucus, NJ: Wellfleet.

McCafferty, W., 1981. *Aquatic Entomology.* Boston: Science Books.

Morgan, A.H., 1913. A contribution to the biology of May-flies. *Ann. Entomol. Soc. Amer.* 6: 371–426.

Schwiebert, E., 1973. *Nymphs.* New York: Winchester.

Trench, C.C., 1974. *A History of Angling.* Chicago: Follett.

van den Broek, G.J., 1984. The sign of the fly: a semiotic approach to fly-fishing in Britain. *Am. J. Semiotics* 3: 71–78.

Aquatic insects as bioindicators (it's the pits)

Clements, W.H,. D.S. Cherry, and J. Cairns, Jr. 1988. Structural alterations in aquatic insect communities exposed to copper in laboratory streams. *Environmental Tox. Chem.* 7: 715–722.

Gerould, S. and S.P. Gloss, 1986. Mayfly-mediated sorption of toxicants into sediments. *Environmental Tox. Chem.* 5: 667–673.

Green, P.C., 1989. The use of Trichoptera as indicators of conservation value: Hertfordshire gravel pits. *J. Environ. Management* 29: 95–104.

Lynch, T.R. and C.J. Popp, 1988. Aquatic insects as environmental monitors of trace metal contamination: Red River, New Mexico. *Water, Air, and Soil Pollution* 42: 19–31.

Rossaro, B., 1987. Chironomid emergence in the Po river (Italy) near a nuclear power plant. *Entomol. Scand. Suppl.* 29: 331–338.

Zullo, S.J. and J.B. Stahl, 1988. Abundance, distribution and life cycles of midges (Chironomidae: Diptera) in an acid strip-mine lake in southern Illinois. *Am. Midl. Nat.* 119: 353–365.

Chapter 9

INSECTS AND PEOPLE

History of entomology in the United States

> You have never, and can never know what it is to be . . . compelled to
> pursue science as it were by stealth, and to feel all the time, while so
> employed, that you are exposing youself, if discovered, to the ridicule,
> perhaps, at least to the contempt, of those who cannot perceive in
> such pursuits any practical and useful results.
> —T. HARRIS, 1829

THE AMERICAN PROFESSIONAL entomologist is generally accorded respect by the scientific and lay community alike. The past twenty-
five years have seen the extension of the science of entomology
beyond its traditional foundation in natural history into such modern fields as analytical and synthetic chemistry, physiology, systems
analysis, microbiology, and genetics; it is universally recognized, especially in the face of continued insect depredations on health and
economy throughout the world, as a respectable, even vital, scientific pursuit.

Such, however, was not always the case; respectability for the
American entomologist has been a difficult goal to attain. There
was a time, not so long ago, when the study of insects was regarded
at best as a harmless diversion and at worst as an indication of serious mental imbalance. Even after the publication of the decidedly
scientific undertakings of Linnaeus and other early taxonomists,
insect illustrations were benevolently regarded as "prints that
would amuse children and keep them out of mischief" or as challenging models upon which ladies might try out their artistic expertise. Entomology was thought to be the proper pastime for the
idle, the effete, or foppish—for those who could find no better way
to spend their time.

This reputation was hardly merited by the realities of working with insects. The first entomological observations to be made in North America are found in the writings of people who were anything but entomologists, the rough-and-tumble explorers to whom insects were an annoying impediment in their search for lands, treasure, and adventure. In William Wood's 1634 report on the progress of the Massachusetts Bay Colony, the entomological observations are gathered into a chapter titled "Of the evills, and such things are hurtfull in the Plantation." Captain John Smith, of Virginia fame, remarked in his *Generall Historie of Virginia, New England, and the Summer Isles* on the "Musketas and Flies ... and a certaine India Bug, called by the Spaniards a Cacarootch, which creeping into Chests they eat and defile with their ill-scented dung." John Esquemeling, buccaneer and adventurer who boasted of many bloodthirsty deeds, found to his dismay that in America the flies were bloodthirstier than any pirate could ever hope to be: "Flies, which excessively torment all human bodies, but more especially such as never before, or but a little while, were acquainted with these countries."

Despite the discouraging reports, European naturalists soon took an active interest in the wealth of unknown and undescribed insects across the ocean and there began a fruitful exchange of specimens and information. As early as 1587, the first illustration of a North American insect, then loosely described (before Linnaean nomenclature) as *Papilio turnus* (now thought to be *Papilio glaucus australis*) was taken to England by John White, commander of Sir Walter Raleigh's third expedition to Virginia (Fig. 9.1). By 1678,

Figure 9.1
First American insect to be illustrated, *Papilio turnus,* by John White.

John Banister, a Virginia clergyman with entomological leanings, was corresponding with Sir John Ray and other illustrious British entomologists, sending a total of some 52 North American insects across the Atlantic for classification. The first taxonomic publication of H. Fabricius, *Systema Entomologica,* published in 1775, included descriptions of American insects (although the assignation of specific epithets referring to geographic localities in the New World were haphazard at best, due to a limited understanding of American geography). Mark Catesby, perhaps the first extensive illustrator of North American insects, illustrated 26 species in his *Natural History of Carolina, Florida, and the Bahaman Islands,* expressing regret that he was "not able to delineate a great number of them." In two folios published in 1797, John Abbot expanded the number of illustrated North American insects with figures and descriptions of 103 "of the rarer Lepidopterous Insects of Georgia including their systematic characters, the particulars of their several metamorphoses, and the plants on which they feed." In total, he made over 3,000 drawings of American insects.

Abbot was born in 1757 in London and expressed an interest in both insects and drawing early in life. As he wrote in his autobiography, "My peculiar liking for Insects was long before I was acquainted with any method of catching or keeping them. I remember knocking down a Libella [dragonfly]and pinning it, when I was told it would sting as bad as a Rattle Snake bite." He arrived in the colonies in 1773 and was so acclimatized to the New World by the outbreak of the Revolution that he served as a private in Georgia's 3rd Continental Battalion.

Perhaps the first profit-making entomological venture in the United States was the essay written in 1795 by William Dandridge Peck, titled "The description and history of the cankerworm," for which he was awarded $50 by the Massachusetts Society for Promoting Agriculture. (Americans, by the way, were quick to catch on to a good thing; the following year a $25 prize was awarded to the Reverend Noah Atwater for a paper rather unremarkably titled, "Another on the same subject.") Peck won further acclaim (and more prize money) for his observations on slugworms and pests of pear, locust, and cherry trees. In March of 1805, in recognition of this work, he was appointed professor of natural history at Harvard (his alma mater, class of 1782) and, as such, gave the first lectures on entomology in North America. Regrettably, however, he did his share to perpetuate the image of entomologist as eccentric—something of a hermit, he had a self-professed "uncommon fear of exciting public attention" and was prone to serious attacks of mental depression.

A happier, but no less curious, personality was the Reverend Frederick Valentine Melsheimer. A onetime chaplain to the Duke of Brunswick's Hessian Dragoon Regiment, he was captured at the Battle of Bennington and imprisoned; after he was paroled, he settled in Lancaster County, Pennsylvania, to become one of the "ablest of the early Lutheran clergymen" in the country. His entomological exploits were said to "furnish some of his parishioners with mild amusement"; but this "eccentricity" on the part of the reverend led to the publication of the first major entomological work produced in the United States: *A Catalogue of Insects of Pennsylvania.* In these sixty pages are described 1,363 species of insects collected in what Melsheimer referred to as his "Hours of Recreation." The first volume dealt exclusively with beetles; illness prevented the publication of later volumes. The collection Melsheimer amassed in his pursuit of pleasure, faithfully maintained by his son Frederick Ernst, was later purchased by Harvard University (for $150) as the basis of what is now the largest university-owned collection of insects in the United States.

Melsheimer was honored by a successor, Thomas Say, with the appellation "Parent of American Entomology," but later entomologists prefer to accord Say himself with that title. His beginnings were inauspicious to say the least; his formal schooling, which he disliked intensely, ended at the age of 15. His granduncle William Bartram, noted physician and naturalist, got him interested in collecting insects and later in founding the Academy of Natural Sciences in Philadelphia. A need to enlarge the Academy's membership brought about the plan to start a collection of "natural curiosities" for public edification; Say was instrumental in accomplishing this end and wrote that insects being "the great objects of my attention, I hope to be able to renounce everything else and attend to them only." In 1824–28, he published *American Entomology or Descriptions of the Insects of North America,* the culmination of his entomological investigations throughout the Northeast, Southeast, and West. The book met with almost universal praise; one contemporary remarked that "the United States can at last boast of having a learned and enlightened entomologist in Mr. Say." Say died in 1834, at the age of 47, of stomach disorders and general infirmity, perhaps brought on by his irregular habits as a youth; almost always out of funds, he was said to live on bread and milk and to sleep under the skeleton of a horse in the Academy building.

Entomology in the second half of the nineteenth century took its lead from Thomas Say; great strides were made in descriptive taxonomic and systematic work in all the orders. By and large, however, these nineteenth-century productive and dedicated entomologists were entomologists by avocation rather than vocation. Most were medical doctors; Leon Provancher, who worked primarily with the Hymenoptera, was a clergyman; Cyrus Thomas, a Hemiptera systematist, an archeologist, Evangelical Lutheran minister, and lawyer. W.H. Edwards was a lawyer and an actor who turned Shakespearean scholar later in life, perhaps in disappointment over his failure to find financial backing for publication of his monumental work on the butterflies of North America, short of selling his own collection. In the same vein, John William Weidemeyer, another lepidopterist, was an author and playwright. He produced not only a comprehensive catalogue of North American butterflies (1863) but also a well-received New York City play (*The Vagabonds*).

That many interested in entomology did not restrict themselves to that study did not necessarily mean that a living could not be made at it; in that respect, Thaddeus William Harris broke the ground in 1841 with the publication of *Report on Insects Injurious to Vegetation,* a work actually commissioned and funded (at $175) by the Commonwealth of Massachusetts. Harris's steady source of income derived not, however, from sporadic state-financed studies of injurious insects, but from his position as Harvard librarian and lecturer in natural history (which he had studied at Harvard some years earlier with William Peck). Harris, although highly respected in academic circles, nonetheless found himself prey to the lingering prejudices the masses nurtured concerning those who study insects. In a letter to an entomologist friend in England written in 1829, Harris lamented

> You . . . can never know what it is to be . . . compelled to pursue science as it were by stealth, and to feel all the time, while so employed, that you are exposing yourself, if discovered, to the ridicule, perhaps, at least to the contempt, of those who cannot perceive in such pursuits any practical and useful results. (Mallis 1971)

Asa Fitch was probably the first person in North America to be paid on a regular basis to study insects. A reluctant medical doctor who abandoned that career to pursue a more bucolic life, he did not publish his first entomological paper (a thirteen-page essay on the gall midge *Cecidomyia*) until 1845, at the age of 36. Working in the state museum, he was the logical choice of the New York State legislature in 1854 when economic loss in the state due to insect damage prompted the appropriation of $1,000 to study insects injurious to vegetation. Although never formally designated New York State Entomologist, Fitch effectively served as such for the next seventeen years. During that period he published more than fourteen papers in the *Transactions of the New York State Agricultural Society* alone and assembled a collection of approximately 55,000 specimens. Much of what Fitch published, luxuriantly illustrated with plates and woodcuts, is still consulted today by economic entomologists. He was among the first to receive international recognition for entomological work when he was awarded the gold medal from the Central Agricultural Society of France.

At about the same time New York engaged its first entomologist, the federal government, in part spurred on by reports of ever-increasing losses due to insect damage, went about creating the Bureau of Agriculture in 1853; Townsend Glover was appointed as entomologist and special agent, to fill the official position of "expert for collecting statistics, and other information on seeds, fruits, and insects in the United States." Prior to his appointment, Glover had published no conspicuous papers on insects, his chief claim to fame being the construction and exhibition of facsimile models of fruit; he also judged fruit at state fairs. Glover did an admirable job holding up the entomological end of things, considering the trying conditions under which he worked; as entomologist and special agent, his duties included matters relating to fruits, insectivorous birds, and textile materials in addition to insects. A contemporary entomologist, B.D. Walsh, commented:

> This is a good deal like hiring a single cradler to harvest a thousand acres of wheat, and then expecting him, in addition, to cut and fetch in wood, peel and wash the potatoes, and be always on hand ready to wait on the good woman of the house. . . . Will our rulers at Washington never learn that it is bad policy to put a square man in a round hole? And that, whether round or square, no one man can fit a hole that is as wide across as the dome of the Capitol? (Mallis 1971)

Glover was a bit of an odd man out in the estimation of many entomologists; instead of preserving specimens he made elaborate copper etchings (fifty copies only) and distributed them to various institutions of importance around the nation. L.O. Howard, who later became chief entomologist with the U.S. Department of Agriculture, for this reason "thought of him for many years simply as a very eccentric man whose personality and career lent support to the popular idea concerning all entomologists" (Mallis 1971).

The passage of the Morrill Land Grant Act of 1862, setting aside 30,000 acres of land per senator and representative in each state for the construction of an agricultural college, did much to further state interests in entomology. Benjamin Walsh was one who profited from this heightened awareness. Born in England, he was a classmate of Charles Darwin and in later years became a fervent advocate of evolutionary theory. At the age of 30, he emigrated to Illinois, settling at first near Cambridge in Henry County, and then moving to Rock Island to avoid malaria, rampant in other parts of the state. His first entomological foray was a lecture to the Illinois State Horticultural Society at the Bloomington Convention in 1860. He wrote both scientific publications and what would now be regarded as extension pieces for farmers. His prominence led to his appointment in 1867 as the first state entomologist of Illinois. He died from the consequences of a train accident, in which his left foot was mangled and subsequently amputated. Ever of good spirits, he remarked to his wife:

> "Why," he would say to his grieving wife, "don't you see what an advantage a cork foot will be to me when I am hunting bugs in the woods: I can make an excellent pincushion of it, and if perchance I lose the cork from one my bottles, I shall simply have to cut another one out of my foot'"

After his death, the State of Illinois bought Walsh's collection, storing it initially in Springfield but eventually transferring it to the Chicago Academy of Sciences, which, in 1871, burned to the ground along with much of the rest of the city in the legendary Chicago fire. Relatively few of Walsh's enormous numbers of specimens remain in existence today.

Insects, by virtue of numbers and voraciousness, began to put pressure on the United States government for increased recognition. The outbreak of the Rocky Mountain locust plague between 1874 and 1876, so enormous that observers reported seeing trains stopped by the crushed bodies of locusts covering the tracks, led to the appropriation by Congress of $18,000 in 1876 for the establishment of the United States Entomological Commission. Appointed as members were A.S. Packard, Cyrus Thomas, and, as chief, Charles Valentine Riley, whose lobbying had pushed the bill through. Riley had risen from the position of state entomologist of Missouri; his meticulously documented and strikingly illustrated annual reports had gained him a considerable following and made him a logical choice for the newly created position. Upon the death of Townend Glover, Riley assumed the position of USDA Entomologist and contributed considerably to the advance of economic entomology by engineering the first successful application of biological control, introducing the Australian ladybird beetle to California to combat the cottony-cushion scale (see Chapter Five). He also pioneered the use of chemical insecticides. Upon his appointment to the position of curator of insects at the U.S. National Museum in 1885, he donated his entire collection to the institution, thus forming the base for one of the world's largest and most comprehensive collections.

Yet another thing for which C.V. Riley was in part responsible was the founding of the Association of Economic Entomologists. Projected at an American Association for the Advancement of Science meeting in Toronto in early 1889, the Association had its first meeting with the Association of Agricultural Colleges and Experimental Stations at Urbana, Illinois, that November. The participation and membership of Canadian entomologists in the organization made the AEE a uniquely international society for professional entomologists. The society began publication of the *Journal of Economic Entomology* in 1908 and, the following year, changed its name to the American Association of Economic Entomologists.

That there were sufficient numbers of entomologists in 1889 to form a society (and enough of sufficiently different interests to start another national organization, the Entomological Society of America, in 1906) is itself a little surprising, in view of the hitherto prevailing opinions of the public at large toward entomologists; that two acts of Congress actually created this abundance of entomologists almost defies belief. Yet Congress really had little choice in the matter. What with the encroachment of the gypsy moth, the San Jose scale, and the cotton boll weevil in the latter half of the nineteenth century, it was as if the forces of nature were lobbying in Washington on behalf of the entomologists.

Up until the passage of the Morrill Land Grant Act in 1862, the teaching of entomology in U.S. universities was at best a desultory process. It began more or less informally with W.D. Peck's natural history lectures at Harvard. T.W. Harris, one of Peck's students, continued the tradition of informal instruction, but it wasn't until the great zoologist Louis Agassiz appointed Hermann Hagen to the Harvard faculty as professor of entomology that the subject was officially institutionalized. With the passage of the Morrill Act, providing funds, space, and students, academic entomology established itself very quickly. By 1867, A.J. Cook was teaching entomology at the Michigan Agricultural College (although he wasn't appointed professor of zoology and entomology until 1869—he taught the entomology courses as an instructor of mathematics). T.J. Burrill became professor of natural history and entomology at the University of Illinois in 1868. He was only supposed to be an assistant professor; his instant advance through the ranks came only after one Major J.W. Powell, appointed professor of natural history, failed to show up to take charge of his classes. In 1870–71, C.V. Riley, though he himself had never taken a formal college course in entomology, lectured on economic topics at Kansas State Agricultural College and acquired the nickname "Professor." Alpheus Spring Packard, Jr., not only taught but wrote textbooks to assist students beyond his own classroom. Among the books he wrote was the *Guide to the Study of Insects,* published in 1869. Although he enjoyed teaching at Brown University, professorship was not without its travails. According to the great historian of entomology, Arnold Mallis, "The students occasionally hired an organ grinder to play under the windows of the lecture hall, which irritated him so that he would dismiss the class," the object, no doubt, of the students in the first place.

The man remembered best for contributions to the teaching of entomology was, without question, John Henry Comstock. As a young boy earning a living as a ship's cook on the Great Lakes, Comstock stumbled across a copy of T.W. Harris' *Insects Injurious to Vegetation* in a Buffalo bookstore selling at the exorbitant sum of ten dollars; getting an advance on his pay, he bought the book and soon after dedicated his life to the study of insects. After being told by a clergyman that Cornell University in Ithaca, New York, was "the den of the devil himself," he of course immediately sent away for a catalogue and enrolled in the fall of 1869. There was at the time no professional position in entomology at Cornell; by his junior year, by popular demand, Comstock himself was teaching a course. He spent some time at Harvard studying with Hagen and returned to Cornell in 1873 to be appointed instructor of entomology. His syllabus of lectures, first published in 1876, grew to form the nucleus of his textbook *An Introduction to Entomology,* the first of many books that dealt with insects in both a popular and scientific manner. His introduction was sumptuously illustrated by his wife, Anna Botsford Comstock, a onetime student of Comstock's who rose to prominence in the art of wood engraving (she was elected to the American Society of Wood Engravers, an honor bestowed on few women) and eventually became a professor in her own right, of natural history, at Cornell University.

The other piece of Congressional legislation that helped to make entomology a viable profession was the Hatch Act of 1887, which created a nationwide system of state agricultural experiment stations. Prior to the passage of the act, only three states (New York, Illinois, and Missouri) supported entomological research; by 1894, seven years after the act was passed, forty-two states and territories had set up research stations with facilities for rearing and studying insects. Even today, much applied entomological research in the states is conducted at the agricultural experiment stations. One further Congressional endorsement of professional entomology came with the passage of the Smith-Lever Act of 1914, which created the Federal-State Extension Service. The extension staff rose from one (T.H. Parks, hired by the University of Idaho a year before the passage of the act) to over seventy-five in sixty years.

One early entomologist with a particular gift for conveying entomological information to the public was L.O. Howard. Leland Ossian Howard was born in 1857 in Rockford, Illinois, but grew up in Ithaca, New York, where he developed an abiding interest in insects (like Comstock, Howard was inspired initially by T.W. Harris's *Insects Injurious to Vegetation*). He abandoned his plans for a career in civil engineering after enduring a differential calculus class at Cornell and became J.H. Comstock's first student. He graduated in 1872, having completed a study of respiration in hellgrammites, the larval stage of dobsonflies, and shortly thereafter went to work for Riley in Washington. Many of his own writings and research papers were published under Riley's name, as was the custom in those days. He obtained a master's degree under Comstock on chalcid wasp morphology, at which point he began to publish under his own name. After thirteen difficult years under Riley, he was made chief of the Division of Entomology in 1894, in which capacity he served for almost fifty years. During his tenure, he not only discontinued the obnoxious policy of publishing all papers under the name of the chief, he also made a concerted effort to provide entomological information to the general public. He published a book in 1901 on *Mosquitoes, How They Live, How They Are Classified, and How They May Be Destroyed,* followed ten years later by *The Housefly—Disease Carrier.* In the latter book, Howard documented the role of house flies as vectors of many diseases and inspired a "Swat That Fly" campaign nationwide. Other books for the general public included *The Insect Book* (1901), *A History of Applied*

Entomology (1930), *The Insect Menace* (1931), and *Fighting the Insects: The Story of an Entomologist* (1933), all of which served to instruct the public about the impact of insects on the economic status and public health of the nation and to elevate, at long last, the status of the entomologist as a scientist in the eyes of the public.

History of pest control

ONE OF THE earliest acts of civilized humans, which went hand in hand with the invention of agriculture, was to plot the systematic destruction of the innumerable arthropods that infested their lives. The methods of pest control used by prehistoric peoples can only be surmised, since no written record and no graphic representation of anything that can be considered pest control have been found to date. The earliest written records left by our ancestors, however, contain references to pests and pest control. *The Egyptian Book of the Dead,* for example, provides instructions for vermin control. Old Testament references to pests abound and there are even references to specific measures for pest control.

By the Greco-Roman era, a considerable body of literature had arisen around the practices of pest control. In his classic treatise on plants, *Enquiry into Plants,* Theophrastus made a point to provide incidental information on insect pests of plants. The Roman authors Virgil (70–19 B.C.) and Pliny the Elder (A.D. 23–79) also contributed to the early study of pest control. The writings of many of these ancient scholars are available today because they were recorded between A.D. 500 and 600 in a massive compilation called *Geoponika*, published by Cassianus Bassus. This magnum opus summarized 400 years of agricultural writings, spanning the period from 200 B.C. to A.D. 200. Many of the pest control practices of this era were based not so much on science but on religion and magic. Xenophon recommended prayer and divine blessings for bountiful harvests; sacrifices were routinely made to ensure safety from pests. Pliny the Elder made reference to several specific deities, including Robigus, responsible for rusts (plant fungi), and Flora, goddess of mildew. Appeasement of Robigus, during the festival known as the Robigalia, included the sacrifice of a red puppy to guard against rusts. Spiniensis was a god to whom an appeal could be directed to remove thorns from a field. Magical remedies included rubbing trees with green lizard gall to protect against caterpillars and nailing a toad to a barn door to scare away weevils from stored grain. Staking the skull of an ass or a mare in the garden was thought to keep away caterpillars. Pliny was a prolific collector of such choice information. He also mentions that "a menstrual or nubile virgin with bare breast and unbound hair led thrice round a garden hedge caused caterpillars to fall to the ground" (Smith and Secoy 1975).

Although in retrospect it seems difficult to believe that many of these magical practices were widely accepted as practical pest control solutions, at the time there were no workable methods for testing the efficacy of any procedure scientifically. No doubt mystical beliefs arose, at least on occasion, from the association of a particular action with the fortuitous remission of a pest problem. However, some pest control methods may actually have been successful due to an underlying biological activity. Pliny described the practice of storing grain in *siri* (the word from which "silo" derives)—airtight chambers that presumably would fill with carbon dioxide from respiring seeds. In that carbon dioxide induces many insects to shut their spiracles and cease respiring, silo storage may have effectively cut down on stored-insect pest populations. Virgil recommended steeping seeds for storage in houseleek and wild cucumber extracts; such plants have since been found to contain chemicals with insecticidal properties. The encyclopedic *Geoponika* cites numerous plants as effective insect control agents that today are known to contain insecticidal compounds; such plants include bay, asafetida, elder, cumin, hellebore, oak, squill, cedar, absinthe, garlic, and pomegranate. Elemental sulfur, still in use for repelling ectoparasites, was recommended mixed with oil as an insect repellent. Crude fumigation was practiced by boiling together olive oil lees (*amurca*), bitumen, and sulfur to protect vines from caterpillars; *galbanum* (fennel) was burned to repel gnats.

Cultural practices were recommended for reducing agricultural pest problems. The practice of intercropping—planting two or more species together such that the presence of one reduces the probability of insect attack on the other—dates back at least to the times of Pliny, who recommended planting bitter vetch in with turnips, and chickpeas in with cabbage, to protect against caterpillars. Recent studies demonstrate that planting a different crop, such as chickpeas, which are members of the legume family, in amongst cole crops such as cabbage does greatly reduce the ability of cole crop pests to locate and colonize their host plants. The practice today is also known as "companion planting."

The Dark Ages were well-named from the point of view of pest control. For hundreds of years, no new knowledge was gained (nor was misinformation corrected—just dutifully copied and cited over and over). Interest in insect control was reawakened, however, during the general renewal of interest in science in the seventeenth century, spurred by the writings of Francis Bacon and the development of the scientific method. Bacon, among others, introduced the notion that theories could indeed be tested by careful observation and experimental design. The increased emphasis on observation during this era led to a veritable explosion in insect study. Several detailed and profusely illustrated texts on insects were published and, for the first time in centuries, the accuracy of the ancient authorities was brought into question. Ulysses Aldrovandi in 1602 published a book proposing a method of classification and identification of insects and for the first time ever focused attention on systematically distinguishing among insect species. About the same time Thomas Moffet compiled an encyclopedic reference on insect life, called the *Theater of Insects,* which provides many pest control recommendations. Among other things, he described an early form of biological control in which jackdaws were actively encouraged to combat locusts. William Harvey (1578–1657) and Leonardo da Vinci (1452–1519) weren't the only ones with dissecting tools during this era. By dissection and microscopic examination, the seventeeth-century Dutch zoologist Jan Swammerdam (1637–1680) discovered many details of insect physiology and function.

Up until the mid-nineteenth century, chemical control of insects consisted mostly of the use of plant-derived dusts, powders, or extracts, as it had since the days of the Greeks and Romans. Botanical explorations all over the world during the seventeenth and eighteenth centuries had made available to Europeans a tremendous variety of effective plant-derived toxins, including derris, pyrethrum, quassia, and hellebore. Before the recreational use of tobacco became popular in Europe, its prowess as an insecticide was well known, having been described by Nicolas Monardes shortly after its discovery in 1596.

Two major (and radically different) innovations launched the pest control industry by the turn of the century. One was the successful practice on a major scale of biological control, the importation of exotic predators and parasites to control an introduced pest. The introduction of vedalia beetles to control cottony-cushion scale on citrus in California (see Chapter 5) catapulted this approach into public consciousness. The other was the inadvertent discovery of Paris green, the first chemical insecticide. Paris green, aceto-arsenite of copper, was a pigment used in paints and wallpapers that was known to be toxic to people. Based purely on logic,

C.V. Riley reasoned that it might also be toxic to insects and he accordingly recommended it for use against Colorado potato beetle (along with London purple, a waste product from the aniline dye industry). This recommendation languished until Paris green was used by grape growers in France in an effort to discourage passersby from stealing grapes. These grape growers soon recognized that their vines were protected not only from human pests, but from insect pests as well. The successful use of Paris green led to the extensive testing and use of many inorganic compounds, often containing metals. These inorganic compounds are generally referred to as the first generation of insecticidal chemicals. Hydrocyanic acid, one of the most widely used of the inorganic insecticides, came into use as a fumigant in 1886, developed by D.W. Coquillet for use against scale insects on citrus; lead arsenate was introduced as an insecticide in 1892. The redoubtable C.V. Riley pioneered new uses and applications of these chemicals, even to the point of designing equipment for delivering the chemicals to their targets. Riley invented the eddy chamber nozzle, the kind found even in today's pesticide application equipment. This invention earned him an odd kind of immortality in that this nozzle is known in French as *boîte de Riley* or *Riley ordinaire* (Fig. 9.2).

The next era of the so-called second generation of insecticides, was ushered in by the discovery by Paul Muller in Switzerland of the insecticidal properties of a synthetic organic compound he pulled off the shelf in an extensive screening study. First synthesized by Otto Ziedler in 1874, this material—dichloro-diphenyl-trichloro-ethane,

Figure 9.2
Riley nozzle (Riley, 1888).

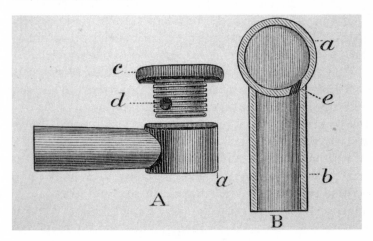

or DDT—proved to be extraordinarily effective against an enormous variety of insects (see Chapter Seven). It was rapidly adopted for use during World War II and was instrumental in heading off the massive epidemics of infectious insect-vectored diseases that had hitherto been an inevitable accompaniment to war (its use contributed to the containment of a typhus epidemic in Naples, Italy, for example, and reduced casualties from malaria in Guadalcanal from 70% to 5%). Muller won a Nobel prize for his finding in 1948.

The discovery of the insecticidal properties of other synthetic organic materials followed quickly. The insecticidal properties of gamma lindane (benzene hydrochloride, or hexachlorocyclohexane—BHC) were discovered simultaneously in England, France, and Spain in 1941–2. The compound itself was first synthesized in 1825 by Michael Faraday and the most active form, the gamma isomer, was discovered by T. van Linden in 1912; it's named lindane in his honor. The German chemist Gerhardt Schrader in 1944 introduced a whole new class of insecticides, the organophosphates, with the discovery of parathion (or schradan, as it was known in Germany); work on parathion actually derived from Nazi chemical warfare work on neurotoxins but the discovery came relatively late during the war and organophosphates found no significant use for chemical warfare purposes.

Postwar development of synthetic organic insecticides proceeded at a frenetic pace. The advantages of these chemicals, particularly those, like DDT, that are chlorinated hydrocarbons (molecules containing only atoms of chlorine, hydrogen, and carbon), over the inorganic insecticides were abundantly clear—they were exceedingly cheap to manufacture, their effects were long-lasting, and their toxicity was considerably higher to insects than to humans. This last feature made the chlorinated hydrocarbons more attractive, for example, than the organophosphates, which are relatively toxic to humans as well as other vertebrates. Aldrin, dieldrin, and chlordane, all chlorinated hydrocarbons, were in wide use by the close of the 1940s. In an unfortunate application of the "more is better" concept, America went spray-happy and fogged, fumigated, and spritzed at the slightest provocation (Fig. 9.3).

And then the inevitable happened. Due to several distinctive insect life-history characteristics, notably, the ability of many species to reproduce several times each year and to produce large numbers of offspring, insects can generally adapt very quickly to changes in their environment. When an insect population

Figure 9.3
At the turn of the century, insecticides in widespread use in households included those of inorganic origin, such as arsenic derivatives (above), and those of botanical origin, such as pyrethrum (right).

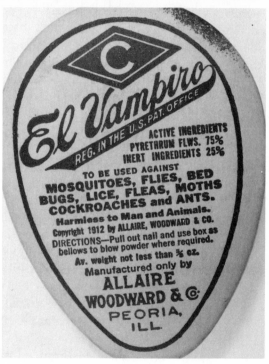

encounters a major mortality-causing agent, natural selection favors those genotypes in a population with an ability to reproduce in the presence of that mortality-causing agent. In every insect population, there is variability with respect to susceptibility to insecticides. When selection pressure from insecticides is intense, the population can rapidly shift from predominantly susceptible individuals to predominantly resistant individuals.

Resistance was a phenomenon well known to entomologists at the start of World War II—the first reported incidence of resistance acquisition was actually in 1908, when a population of San Jose scale *Quadraspidiotus perniciosus* was discovered to be resistant to lime sulfur. In 1916, California red-scale resistance to hydrocyanic acid was reported, as was resistance of codling moths to lead arsenate in 1917. By 1938, seven economically important insect species were known to be resistant to at least one of the first-generation insecticides. However, the rate at which insects developed resistance to synthetic organic insecticides was unprecedented. For one thing, the synthetic organic insecticides are inherently more effective selective agents for promoting insect resistance, because they effect higher rates of mortality in populations. Perhaps more importantly, the way they were applied—indiscriminately, everywhere, and at high concentrations—increased the intensity of the selection pressure.

Resistance to DDT was noted as early as 1942, only three years after the discovery of its insecticidal properties, in a population of Italian house flies. By 1950, almost 20 species were known to be resistant to DDT. Alarmingly, resistance to DDT took on many forms, suggesting that there may be no limit to the types of genetic variation that could lead to resistance. Some species displayed behavioral resistance—that is, they behaved in a manner that protected them from exposure to a toxic dose. For example, mosquitoes in Africa with a proclivity for resting between blood meals on the outside walls of huts were immune from the effects of DDT prophylactically sprayed on the inside walls of huts, where the vast majority of mosquitoes enjoy their postprandial pause. Other species displayed biochemical means of detoxifying the active chemical; J.S. Sternburg at the University of Illinois isolated DDT-dehydrochlorinase, an enzyme that cleaves off chlorine atoms and renders DDT less insecticidal, from the bodies of resistant house flies. Moreover, the acquisition of resistance to one form of insecticide can facilitate the acquisition of resistance to other forms of insecticide. DDT owes its toxicity to the fact that it interferes with sodium transport in nerve cells. Like DDT, pyrethrins also target sodium gates in nerve cells. Many species of insects exposed to DDT that developed resistance to DDT also displayed cross-resistance to pyrethrins, even if they lived in areas never exposed to pyrethrins. By 1960, over 50 species of insects displayed resistance to DDT; by 1976, over 200 species were known to be resistant to DDT, as well as to cyclodienes, pyrethroids, organophosphates, and carbamates.

Resistance was not the only problem that grew with increased use of synthetic organic insecticides. Another aspect of insecticide abuse was its unexpected effect on nontarget organisms. While insecticides had increasingly less impact on the species that prompted their use, they had increasingly more impact on organisms sharing the environment with the target species. Particularly vulnerable were predators and parasites, which, due to the phenomenon of biomagnification, or increase in concentration of an insecticide as it passes up the food chain, were exposed to even higher concentrations of insecticides sequestered in the bodies of the insects they ate. With the devastation of populations of natural enemies, insects that were previously of only minor economic importance grew out of control. Such so-called secondary pests included the corn earworm *Helicoverpa zea,* one of the most important pests of corn in the U.S. today, as well as the spider mite *Tetranychus urticae,* on a variety of crops, including soybeans. In addition to creating new insect problems, biomagnification provoked inquiry into possible health consequences of repeated long-term exposure to pesticides that had relatively low toxicity in the short term; proliferation of dead fish and dead robins in both urban and rural environments started people wondering what the effects of pesticide exposure might be on human physiology.

The publication of one book altered patterns of insecticide use in this country forever. In 1962, the *New Yorker* published a series of articles abstracted from a book by Rachel Carson called *Silent Spring.* The articles appeared in June and the book in July. The thesis of the text was that insecticides should be used intelligently and that satisfactory alternatives, with little or no nontarget impact or adverse health consequences, were available to replace them. Backlash from vested interests followed rapidly. Velsicol of Chicago (makers of chlordane) tried to prevent the publication of the book, claiming it was part of a Communist plot to "reduce American food output to an 'east curtain parity'" (Ordish 1976). Attacks on Carson's scientific credentials abounded. Books were written to counter her arguments, among them *Bugs or People? A Reasoned Answer to*

Opponents of Pesticides, by Wheeler McMillen, and *That We May Live,* by Jamie Whitten, then member of the House Agricultural Appropriations Committee. Articles in respected medical journals (e.g., the *Journal of the American Medical Association,* July 1962) rejected the suggestion that eating food containing pesticide residues that met government standards was in any way harmful.

Despite the scientific and professional backlash, Americans took the message of *Silent Spring* to heart. During the sixties, environmental awareness was tremendously heightened and, along with federal legislation promoting clean air and water, in 1974 the Environmental Protection Agency was created to maintain U.S. environmental quality. After careful and rational data collection and review, the EPA, with the support of the Environmental Defense Fund, initiated hearings in several states on the advisability of continued use of DDT. The outcome of lawsuits and hearings in Wisconsin, California, and Washington was the total ban on domestic use of DDT after 1971–1972.

Proponents of DDT use in particular and chemical pesticide use in general took advantage of a provision in the ban allowing for recall of DDT in times of national health or economic emergency. The EPA, facing severe political pressure, approved a Forest Service request to use DDT to control Douglas-fir tussock moths in forests of the Pacific Northwest in 1974. Over 425,000 acres of forest were then sprayed, despite the fact that the tussock fir moths population was already in decline due to natural causes. The resultant contamination of game species, livestock, and nontarget organisms, as well as the expense and apparent lack of necessity for the spray program, did much to support public (as well as scientific) opposition to the continued use of DDT. Within the next few years, aldrin, dieldrin, heptachlor, and other synthetic organic insecticides were banned or their use severely restricted.

There were two factors that pesticide proponents failed to take into account when predicting nothing short of the total collapse of American agriculture and public health as the aftermath of extensive pesticide bans. The first factor was that the efficacy of synthetic organic insecticides had declined precipitously with overuse; although chemical applications were increasing steadily, yields were often lower than in prepesticide years. The second factor was that effective alternatives to synthetic organic insecticides were available. During the sixties, several innovative pest-control strategies were in development. In 1959, A. Butenandt in Germany succeeded in isolating and characterizing bombykol, the sex pheromone of *Bombyx mori*, the Japanese silk moth. Development of pheromone chemistry techniques led Robert Berger in 1966 to isolate and characterize the sex pheromones of *Trichoplusia ni*, the cabbage looper, an insect that despite its name is a pest on crops in dozens of families. Pheromones have been called the third generation insecticides and can be used for control purposes in several ways: for trapping (as with stored-product insects), for mating disruption, and for monitoring movements and population sizes of pest species.

In 1956, Carroll Williams of Harvard University identified juvenile hormone (JH), a hormone used by a tremendous variety of insects to regulate metamorphosis (see Chapter 2). High hemolymph titers of JH at the time of a molt lead to the retention of juvenile characters, while the molt to the adult stage takes place only when JH titers fall to vanishingly low levels. The improved understanding of hormonal regulation of metamorphosis in insects gave rise to yet another "generation" of insecticides, including synthetic juvenile hormone analogues that prevent insects from reaching sexual maturity and reproducing, and anti-JH compounds (including one isolated from *Ageratum* flowers by William Bowers) that induce precocious metamorphosis of certain bugs into sterile adults. Hormone analogues or antagonists today find a variety of uses, particularly in the control of household pests.

The advantages of third and fourth generation insecticides are abundantly clear; they are so specific in their mode of action that their effects on nontarget organisms are essentially nonexistent, and they are biodegradable and thus unlikely to accumulate through the food chain. Their advantages are to some extent their drawbacks as well. Specificity means that any single chemical is effective only against a very narrow range of pests; development and promotion of such a product are expensive and developers run the risk of never recovering their investments, since they can sell only to a limited market. The fact that these insecticides are so biodegradable contributes to their expense for the consumer in that repeated applications are often necessary to maintain efficacy over the long term. In addition, despite initial optimism that insects would not be able to develop resistance to their own hormones or pheromones, there is increasing evidence that prolonged use of even third- and fourth-generation insecticides can lead to reduced efficacy due to resistance.

The sixties, however, also saw the development of nonchemical alternatives to insect control. In 1960, *Bacillus thuringiensis*, a bacterium that releases a toxic protein in the guts of caterpillars and other soft-bodied insects when its spores are ingested, was registered for use by Nutrilite of Buena Park, California, against cabbage loopers. Various strains of BT, as it is known, are differentially toxic to a wide variety of insects; in its spore stage, it is relatively inert and can be sprayed, like a pesticide, where it is needed. Originally found in Germany infecting stored-product moths, *Bacillus thuringiensis thuringiensis* is relatively specific against caterpillars; *B.t. israelensis* kills mosquitoes and black flies; and *B.t. san diego* is relatively specific for beetle grubs. Because it is relatively inexpensive to produce (it can be grown in large vats) and easy to use, BT has already captured 5% of the world pesticide market. Its market share may expand in the near future. Genetic engineering has allowed the gene for the endotoxin, the crystalline protein that is released in the gut to kill the victim, to be cloned and moved around; among other things, it has been packaged inside *Pseudomonas fluorescens,* a bacterium with a thick cell wall that resists the inactivating properties of sunlight better than BT does. Other bacteria commercially available and in widespread use include *Bacillus popilliae,* a highly specific pathogen of Japanese beetles and other grubs. Known to insect pathologists as milky spore disease (it turns the grubs a milky white color), it has been marketed under the trade name "Doom."

Other pathogens that have proved useful for controlling insects are protozoans, specifically microsporidia. Unlike BT and several other commercial bacteria, protozoan infections can spread throughout a population, at least in part because infected insects live long enough to come into direct contact with other individuals or to produce contaminated frass or webbing. *Nosema locustae* has been used with some success for the control of grasshoppers and locusts in the western part of the U.S. (and has been sold under the catchy name "Hopperstopper"). In addition, in recent years, commercial insectaries have sprung up, where vast numbers of predators and parasitoids are raised and sold for use against specific pests. Tiny parasitic wasps can actually be "sprayed" by aircraft onto crops infested by their hosts.

Even natural enemies and pathogens are not ideal insect control agents. Among other things, in order for populations of natural enemies to become established, low numbers of pest individuals must persist to maintain the populations between outbreaks; otherwise, predators and parasites must be repeatedly ordered and "applied." Pathogens and parasitoids are often slow to kill their hosts, so that crops do sustain some damage while dying pests continue to feed; American consumers (as well as farmers) have grown accustomed to blemish-free commodities. In addition, new concern has arisen over the advisability of importing natural enemies to control native pests. Many of these imported enemies have succeeded in broadening their taste upon arrival in North America and have attacked populations of insects that are not regarded as undesirable. For example, some pathogens imported to control rangeland grasshoppers are now infecting nondamaging native grassshopper species that pose no threat to agriculture.

At present, the most acceptable solution to pest problems in the U.S. (and throughout much of the world) is the practice of integrated pest management (IPM)—a system in which every possible pest-control technique, including chemical pesticide application, is used in an intelligent manner. Pesticide applications can be cut back substantially if they are timed to coincide with the arrival or emergence of a pest population (as determined, for example, by pheromone monitoring traps). IPM practitioners hope that by using every available weapon in the pest control arsenal no single approach will lose its efficacy and that the negative impacts of any single approach will be minimized.

There's actually an official federal definition of integrated pest management, formulated in 1979—it's the "selection, integration, and implementation of pest control based on predicted economic, ecological, and sociological consequences" (Olkowski et al. 1991). Despite initial reluctance to reduce dependency on chemical pesticides, IPM programs for a wide variety of insect pests have been implemented with considerable success. Chemical controls are not abandoned in IPM programs—rather, they are integrated into a comprehensive strategy that makes use of biological, physical, and cultural modifications of the habitat to reduce insects to a population size below that which inflicts economic damage. The synthetic organic insecticides traditionally used are in IPM programs more likely to be replaced by "least toxic" insecticides—repellents, pheromones, hormone analogues, contraceptives, and other chemicals with less negative impact on the environment. In fact, despite dire predictions, IPM practitioners as a rule do not suffer losses in comparison to growers using more traditional chemical controls; according to a comprehensive study examining control programs implemented by 3500 growers in 15 states, those using IPM as a group made $54 million more than did those using chemical controls exclusively (Olkowski et al. 1991).

Fire ants, Congress, and the EPA

"I have been bitten by but one fire ant. Don't send them to me."
—J.P. CAMPBELL, Undersecretary, U.S. Department of Agriculture, before
the Subcommittee on Department Operations, Investigations, and
Oversight, House Committee on Agriculture, June 26, 1975.

PERHAPS NO SINGLE example better illustrates the hazards of indiscriminate pest control than does the war on fire ants that resulted from an act of Congress. Fire ants are not exactly newcomers to the United States. *Solenopsis invicta,* the red imported fire ant, normally resides in western Brazil and Paraguay and is thought to have gained illegal entry into the country around 1940 through the port of Mobile, Alabama. *S. richteri,* the black imported fire ant, probably arrived even earlier, around 1919, from its native region of Uruguay, southern Brazil, and eastern Argentina.

Upon arrival, the ants proceeded to colonize their new home with extraordinary efficiency. By 1975, fire ants had colonized over 52 million hectares of U.S. soil. Biologically, *Solenopsis* species are built for invading continents. Colonies are usually initiated by a single queen but *Solenopsis* species do on occasion form colonies with multiple queens; up to 3,000 queens have been recorded in a single colony. This peculiar bit of reproductive wizardry appears to be unique to North American populations; in their native land, fire ants lead a far more pedestrian existence. The fertilized queen, returning to the ground after her nuptial, or mating, flight, sheds her wings and then starts digging out a burrow, a process that takes the better part of six hours. The burrow is a vertical shaft 1 to 4 inches deep, with small cells about 1/2 inch across near the top and bottom of the shaft. Within two days (sometimes as soon as only a day), the queen begins to lay eggs. The first eggs serve as food for grubs, which hatch from subsequent eggs four to seven days later. Older grubs receive regurgitated food directly from the queen. The grubs undergo four molts in about a week and require another week or so for their pupal stage.

Altogether, egg to adult emergence is a three-week processs. In three months, nests can contain over 300 workers, mostly tiny workers called minima and slightly larger minor workers. In five months, colonies average 1,000 individuals, mostly minors and still larger majors, in a year 11,000, and after three years between 50 and 60 thousand. Along with the enormous population growth comes a pressing demand for nest enlargement. Chambers are added on continuously. After three months, a surface mound forms, with horizontal burrows branching outward. Additional vertical shafts are constructed. A 3-year-old colony features a domed mound up to 3 feet high, with a gallery volume of 40 quarts and tunnels reaching 3 feet or more down to the water table. Ants can move larvae throughout the expansive colony to ensure that they experience optimal temperatures and humidity. Queens are produced continuously as well. The average number of queens produced per year per hectare was, in one area, 462,000; queens departed for nuptial flights at a rate of 100 per minute from some mounds.

Feeding hungry hordes necessitates a broad diet. Fire ants feed primarily on small invertebrates, and rarely on seeds or seedlings. Their habits of consuming large numbers of boll weevils and sugarcane borers would make them appear to some as beneficial (such as Winfield Sterling of Texas A&M, who was prompted to describe this activity in a paper titled "Fire ants wear white hats"). However, they are far more commonly regarded as at least a nuisance and more usually as a serious economic threat to the southern states. They are perceived as agricultural pests not so much because they damage crops directly but because they interfere with the process of growing crops. Their mounds are likely to form in the loose soil of agricultural fields; the hard compacted mounds are formidable obstacles to plows or combines. Disturbing the mounds has the effect of unleashing a massive swarm of fire ants intent on defending the colony, which they do by furiously biting and stinging. Their presence in agricultural fields also interrupts business by promoting a general and completely understandable reluctance among field workers to go into the fields to work.

Fire ants sting humans by sinking their mandibles into the skin and swinging their abdomens around to inject venom. The venom, which is also used by the ants to kill or paralyze prey, contains several toxic alkaloids. The consequences of fire ant stings are dramatic. An immediate burning sensation at the site of the sting is the basis for the name "fire" ant. A swelling soon appears and a blister or vesicle forms at the site. Within a day, the vesicle fills with pus as venom constituents break down cells and tissues. Within a week, the pustules are resorbed, often leaving behind scar tissue. Systemic reactions can also occur in response to the sting and can involve nausea and vomiting, disorientation and dizziness, asthma, and other allergic responses; in some cases, stings can be fatal.

Aside from stinging and nibbling on seedlings, fire ants have the annoying habit of not calling cable companies before they begin excavation and as a result can cause problems by knocking out electrical and telephone service. Not helping matters is the fact that they appear to be attracted to electromagnetic fields, increasing the likelihood that they'll damage electrical insulation or wire connections. The presence of fire ants in the path of the proposed tunnel for the now-defunct national superconducting supercollider near Dallas, Texas, caused concern that construction problems would develop as a result of the ant's predilection for electromagnetic radiation. In agricultural fields, fire ants also tend aphids, which can damage crop plants, and are reported to kill small ground-nesting birds (like quail) and mammals.

So these ants are not well-loved by those who know them well. Since they rose to prominence in the 1940s, Americans, heady with success in insect control by synthetic organic insecticides, demanded a control program to halt the spread of these insects. Early efforts with inorganic insecticides, including sodium cyanide and thallium, met with less than striking success, so the synthetic organic insecticide dieldrin was used in initial control efforts. Control was dramatic and effective, but by 1945 it was apparent that the dieldrin was having a greater impact on quail (and on other vertebrate) populations than on the fire ants, so it was replaced by the synthetic organic insecticide heptachlor. By 1957, 10 to 12 pounds of a granular formulation of heptachlor was recommended per acre for fire ant control.

In 1957, spurred apparently by the success of early control efforts, Congress initiated a cooperative federal/state program aimed at eradicating fire ants from 126 million acres, at a cost of $200 million over a twelve-year period. Heptachlor, in the wake of *Silent Spring,* was banned in 1962, but experimental work in the previous decade provided a compound to fill the breach. Scientists at Auburn University had experimented during the fifties with a material called kepone, formulated in a peanut butter bait, with some success. The USDA Methods Development Laboratory in Gulfport, Mississippi, developed an analogue, mirex, that was equally effective against fire ants, but with lower mammalian toxicity. A toxic bait, based on soy oil and corncob grits, was developed in 1961 and found widespread use, particularly after heptachlor was banned. Initially, excellent control, in excess of 98%, was achieved with increasingly smaller amounts Eradication did indeed seem feasible, at least to some. A successful government initiative aimed at eradicating hog cholera a few years earlier provided a precedent in the public mind that pest problems could actually be eliminated rather than merely controlled.

By 1967, however, the National Academy of Sciences, based on a study it had conducted on the feasibility of eradicating an insect pest, disagreed. Moreover, by 1967 laboratory data had begun to accumulate demonstrating substantial nontarget effects of mirex and kepone on baby rats, chickens, fish, and, most significantly, crabs, crayfish and other crustaceans. Spilloff of insecticide residues from factories in the environmentally sensitive Chesapeake Bay area, home to a thriving seafood industry, created grave concern. In 1970, the Environmental Defense Fund filed for a restraining order against the USDA on the use of mirex.

In 1971, the 3-month-old Environmental Protection Agency canceled registration of mirex over the objections of Allied Chemical Corporation, which held ten of eleven registrations for mirex. By 1972, an amendment to FIFRA, the Federal Insecticide, Fungicide and Rodenticide Act, allowed for public hearings for interested parties. After these hearings were held a year later , the outright ban was lifted but usage was tightly regulated in nonagricultural habitats. On July 14, 1975, for a variety of reasons, Allied Chemical pulled out of the mirex business and transferred its registration to other companies (and eventually all but abandoned agricultural chemicals altogether).

In October 1976, the Mississippi Authority for the Control of Fire Ants successfully petitioned the USDA to end the ban on aerial application of mirex by December 1, 1977. Approval of mirex for ground broadcast was reinstated in July 1976. Mississippi was extremely reluctant to give up on mirex; the Mississippi scientists had developed a new formulation, called ferriamicide, which supposedly had eliminated all or most of the problems previously associated with mirex application. Considerable controversy surrounded these claims and the issue became a political football.

By 1978, when the ban on prior formulations took effect, the U.S. government had spent a total of $91.5 million and the states collectively over $64 million for fire ant control, prompting noted Harvard entomologist E.O. Wilson, a world authority on ant biology, to dub the eradication effort the "Vietnam" of insect control efforts. There is still concern over the spread of fire ants. Only two natural forces seem to limit their potential: their intolerance for cold weather and competition from *Solenopsis gemmata* and other native fire ant species. However, both of these barriers are fragile at best; among other things, *S. invicta* is interbreeding with native ant species and may by hybridization acquire the capacity to withstand cold winters. Current research efforts into new control measures include insect growth regulators, natural enemies, pathogens, and pheromones.

Entomophobia and delusory parasitosis

> From ghoulies and ghosties and long-leggety beasties
> And thing that go bump in the night
> Good Lord, deliver us.
> SCOTTISH PRAYER

FIRE ANTS AND their ilk are but one of the many reasons why the vast majority of people regard insects either with indifference or, more often, with contempt. For example, in a telephone survey of over 1,000 households in Arizona, Byrne et al. (1984) found that 37.6% of respondents "dislike" arthropods encountered in their yards and a whopping 83.5% "dislike" arthropods encountered inside the home. In contrast, less than 1% of respondents admitted to enjoying the presence of arthropods in the home. Of the arthropods reported in these homes, few posed any serious threat to human health or to the structural integrity of the homes. Nonetheless, when respondents were asked to rank-order animals based on likeability, four of seven arthropods earned scores lower than the skunk.

People dislike insects for obvious reasons. Insects, more than any other recognizable type of organism, can gain access not only to the homes of humans but even to their bodies, and invasions of privacy are not generally welcomed irrespective of the taxonomic identity of the interloper. In addition, insects, being in general so small, are difficult to keep track of; an inability to monitor their movements and their population growth tends to discomfit home owners. Finally, there really are arthropods that pose a serious threat to health and well-being, both by direct injury (as is the case for venomous biting or stinging arthropods) and by indirect injury (as is the case for disease vectors).

For a relatively small percentage of the population, however, the sight of an insect provokes blind panic or sheer terror. Entomophobia is properly defined as an irrational fear of insects (phobia, from the Greek *phobos,* for "terror" or panic-fear). Closely related is arachnophobia, an unreasonable fear of spiders; these phobias may overlap due to an inability of the sufferer to distinguish between classes of arthropods. According to one authority, "Among the common fears reported by Western societies that involve animals, insects and spiders rank relatively high as fear-inducing organisms, running close behind rats and snakes" (Hardy 1988). To some extent, fear and loathing of rats, snakes, insects and spiders is not all that unreasonable—all are capable of inflicting nasty painful bites that on occasion may even prove fatal. But a reasonable fear becomes a phobia when an individual is no longer capable of carrying out a normal daily routine because of that fear.

True phobic responses to insects and spiders are probably rare—on the order of 3% or less in the population at large. Entomophobia is most likely to develop in young children, usually between the ages of 2 and 7 years; at these ages, both sexes are equally prone to display symptoms of entomophobia. Generally, children outgrow entomophobia at least by their twelfth year. On occasion, however, entomophobia persists or even develops de novo in adults. In adults, there are more female entomophobes; estimates of a biased sex ratio range from 2:1 to 9:1 in favor of women. Education doesn't necessarily affect the incidence of entomophobia. A survey of college freshmen conducted in 1984 revealed that, while 53% of those who had taken biology were afraid of spiders, only 35% of those who had not taken biology acknowledged being afraid of spiders (Hardy, 1988).

Phobias can be classified into several different categories. Entomophobia is considered a simple phobia, fear of a specific object or situation; other familiar simple phobias include claustrophobia (fear of closed places), acrophobia (fear of heights), ailurophobia (fear of cats), and the like. Symptoms of simple phobias resemble any sort of panic attack and can be divided into three elements. One, subjective reactions involve perceptions, and a phobic response is characterized by intense anxiety, "feelings of impending doom" (Hardy 1988), and a fear of loss of control. Two, physiological reactions are typical of fright responses in general: rapid heartbeat, cold sweat, difficulty breathing, shaking, and light-headedness. Finally, there are motor responses (voluntary muscular movements), which generally involve either running away or freezing motionless.

One aspect of serious entomophobia is that the fear elicits avoidance behaviors, not only to the objects responsible for the fear (i.e., the insects or spiders) but also the places or activities with which the objects are associated. This avoidance behavior is a big

problem in entomophobia in particular because insects are so ubiquitous—avoiding them may be an exercise in futility. Phobics also
tend to generalize their fears—a fear of cockroaches may escalate
into a generalized fear of basements or dark places. A generalized
fear of insects may reduce an entomophobe to avoiding all forms of
outdoor activities.

How insect phobias begin is a much-debated question. Traditional folk wisdom maintains that such fears result from the transfer of a maternal experience to a child in the womb; a shock
experienced by an expectant mother manifests itself as a phobia in
the child years later. Freudian theory, in contrast, holds that phobias result from unresolved mental conflicts, usually over repressed
sexual desires. Freudian theory revolves around conflicts between
the ego or self, the superego (a sort of "conscience") and the id, the
seat of hungers or animal drives. According to film critic N. Carroll
(1981), "The majority of phobias . . . are traceable to such a fear on
the ego's part of the demands of the libido. Giant insects are a case
in point. The giant spider, for instance appears . . . as an explicit
symbol of desire. Perhaps insects, especially spiders [*sic*], can perform this role not only because of their resemblance to hands—the
hairy hands of masturbation—but also because of their cultural association with impurity. At the same time, their identification as
poisonous and predatory—devouring—can be mobilized to express anxious fantasies over sexuality." Others have suggested that
arachnophobia results from a subconscious association between
predaceous spiders and all-consuming mothers.

A seemingly more reasonable theory is that phobias develop
from innate fears of potentially dangerous objects or situations.
That such phobias exist may be due to the fact that those individuals with a genetic predisposition for avoiding dangerous situations
may have enjoyed a selective advantage early in human evolution. A
fear of high places, for example, may have led to avoidance behaviors in primitive peoples that reduced mortality from accidental
falls. In support of the biological basis for phobias is the fact that
virtually all children as a matter of course go through an entomo-
phobic phase that gradually wanes. Moreover, there is evidence that
there is an underlying genetic component to phobic behavior; there
are, for example, familial tendencies toward certain sorts of phobic
behaviors. It can be argued, if fear of danger underlies animal phobias, that lions, tigers, sharks, or other vicious predators should
evoke greater fears than do, say, maggots or caterpillars, Fear of
things that are disgusting, rather than dangerous per se, though,

may nonetheless be adaptive in that such things are often associated with filth and contamination, and hence disease, which can of course be as deadly as a conventional predator. On the opposite side of the argument are behaviorists who maintain that all phobias derive from conditioned responses to aversive stimuli. The aversive stimulus may be experienced directly or it may be provided vicariously, as by a parent who repeatedly cautions against some particular object, action, or situation.

Given that there are people suffering from incapacitating cases of entomophobia, there is considerable debate as to what can be done to eliminate or alleviate these fears. Freudian psychoanalysts try to unearth repressed desires and direct the patient to confront them. Behavior therapists have on the whole been more successful using a variety of techniques. Systematic desensitization is one such approach—relaxation followed by increasingly more terrifying mental images of the object. In vivo therapies are similar but involve confrontations with the actual object rather than a mental image. There's also implosive therapy, in which a phobic must envision a terrifying situation that culminates disastrously ("caught in a giant spider web . . ."). Flooding therapy spares the phobic's imagination and is the most aggressive—it actually places him or her in a catastrophic situation (for example, strapped down in a room overrun with cockroaches). Finally, modeling is a system in which a role model (therapist, parent or friend) touches the feared object in the presence of the phobic. There's also drug therapy, but such treatment tends to mask, rather than change, the condition.

Quite distinct from entomophobia is a psychosis known as delusory parasitosis. In this condition, a person labors under the delusion that he or she is being attacked, externally or internally, by insects or other small parasites. It's not a new phenomenon (one French description dates back to 1896). Delusory parasitosis is also known as Ekbom's syndrome, since it was thoroughly documented and described by the German psychiatrist K.A. Ekbom in 1938 ("*der präsenile Dermatozenwahn*"). Hundreds of case histories can be found in the literature and are reported in a variety of journals, including not only psychiatric journals but also entomological and dermatological journals as well. Again, this affliction is more common in females than in males but unlike entomophobia it is almost exclusively found in adults (in one study, 39% of fifty-one cases were ages 50 to 59 and 14% were 60 to 69, so over half were over 50).

Whereas phobics may seek out psychiatrists, the first professional usually contacted by a person suffering from delusory parasitosis is a pest-control operator or entomologist. There is a set of observations that can be used to distinguish delusory parasitosis from a routine pest-control request (Waldron, 1961):

1. the "bugs" are often black or white initially and can change color later

2. the "bugs" jump (as do particles charged with static electricity)

3. "bugs" frequently infest hair and can be seen on a comb or shaken onto a sheet

4. the "bugs" (unlike the vast majority of ectoparastic arthropods) are associated with "common household items" like toothpaste tubes or cosmetics

5. "bites" are almost invariably visible (which are usually the result of secondary infection from frenzied scratching)

6. the "bugs" can be so numerous as to drive the occupants out of their home into extraordinarily uncomfortable accommodations (e.g., sleeping in their car, or sleeping on a kitchen counter)

7. the "bugs" are highly "transmissible"—the afflicted person is so convincing that friends and relatives not only support his or her story but also manifest symptoms themselves (such as paresthesia— an itching, crawling sensation)

8. the "infestation" continues for two or three months or longer—most arthropod infestations are usually self-limiting over a shorter period of time

9. there may once have been an actual arthropod infestation at the beginning of the episode.

The very serious side of delusory parasitosis is the length to which the person goes to rid himself or herself of an imaginary infestation. Bathing in gasoline, "dog dip," cleaning compounds, and the like can cause serious skin damage; others resort to self-mutilation with sharp instruments such as razors and knives. Luis Bunuel wrote in his autobiography of a visit with Salvador Dali in the great surrealist painter's hotel room in Montmarte:

> I found him stripped to the waist, an enormous bandage on his back. Apparently he thought he'd felt a "flea" or some other strange beast and had attacked his back with a razor blade. Bleeding profusely, he got the hotel manager to call a doctor, only to discover that the "flea" was in reality a pimple.

Most cases of delusory parasitosis are symptomatic of psychic disorders such as involutionary melancholia (induced by menopause or male "change of life" and possibly precipitated by actual urticaria or pruritis) or full-blown paranoid disorders.

The first thing a physician, a pest-control operator, an entomologist, or even a concerned relative should do in a case of suspected delusory parasitosis is to be sure that the infestation is indeed imaginary. There are several species of ectoparasitic mites that are effectively invisible to the naked eye (scabies, for example); persistent infestation can cause intense itching and discomfort. In addition, such infestations are indeed highly contagious upon contact. If, however, scrupulous investigation reveals no arthropods on the premises, it's generally considered unethical at worst and unhelpful at best to treat the nonexistent pest problem. For that matter, it's even illegal, according to the amended FIFRA; even well-meaning pest-control operators risk violating the law by treating a nonexistent pest problem. There is no sure-fire entomological solution to delusory parasitosis; the most qualified professional to refer to in such cases is a psychologist or psychiatrist.

Insects and the law

> Laws are like cobwebs, which may catch small flies,
> but let wasps and hornets break through.
> —JONATHAN SWIFT, *A Critical Essay Upon the Faculties of the Mind*

SINCE INSECTS ARE renowned for their ability to establish themselves in the unlikeliest of places, it should be no surprise to anyone that, in addition to kitchens, bedrooms, classrooms, greenhouses, restaurants, and bank president's offices, an occasional insect wanders into a courtroom. It may be surprising, however, that insects are on occasion expressly invited into courtrooms—as material witnesses, evidence, or even defendants. Cases involving insects are as much a part of legal history in general as are lawyers' fees and motions for mistrials. In early days, court proceedings involving insects were strictly ecclesiasticial since, in those days, it was reasoned that insects were not answerable to civil authorities (see Chapter Four). At least as early as the seventeenth century, one result of the scientific revolution was that insects began to lose their divine protection and to figure prominently in civil court cases.

Today, insects are involved in court cases from many branches of the law. In business law, for example, most insect cases involve the liability of an owner or operator of a business for injuries to customers caused by insects or small animals. Injury is usually direct, as by bite or sting, and negligence on the part of the owner must be proved. For example, in *Flippo v Mode O'Day Frock Shops,* 248 Ark. 1, 449 S.W.2d 692 (1970), the owner of a women's clothing store was ruled not liable for injuries caused by a spider bite (note that, according to the legal profession, spiders are classified as insects). While trying on a pair of stretch slacks inside a clothing store, a woman was bitten by a spider and claimed a breach of implied warranty (i.e., the slacks were unfit for the use intended). By the way, although she lost the case, the woman did buy the slacks.

In another such case, *CeBuzz, Inc. v Sniderman* (1970), a grocery store owner was negligent and therefore liable for injuries caused by a spider bite, since it was determined that the "store owner had prior knowledge of the existence of such spiders on the premises." These spiders were lurking in a bunch of bananas where spiders could be expected to lurk. In *Cunningham v Neil House Hotel* (1940), an innkeeper was found not liable for injuries to a guest who had been stung by an insect "where, among other things, there was nothing to indicate either the nature of the insect, where it came from, or how long it had been in the room." Mere occurrence of an unusual accident does not warrant application of the *res ipsa loquitur* ("things speak for themselves") doctrine.

Business owners have to worry not only about actual injuries inflicted by insects but also about injuries resulting from the fear brought about by insects on their premises. For example, in *Stuckey's Carriage Inn v Philipps* (1970), a hotel owner was sued by a hotel guest "who, while attempting to get rid of a cockroach which crawled up her thigh, got her legs entangled in a bedspread and stumbled and fell over a chair." The owners were ruled not negligent with respect to the presence of a cockroach in the room, since they treated the premises regularly to eliminate cockroaches. However, the point in question was whether the hotel was negligent in allowing the bed to be made up in a dangerous way. The guest, in "a moment of stress of sufficient magnitude" was correcting a patent defect and therefore could recover damages.

In innkeeping law, insects figure prominently in cases of infestation of leased dwellings in which their presence represents a violation of an implied warranty of habitability on the part of the landlord. Their presence therefore justifies nonpayment of rent by tenants or serves as grounds for constructive eviction by the landlord (a situation in which a place becomes so uninhabitable that the tenant is effectively driven out). For example, in *Buckner v Azulai*

(1967), an infestation of psocids, or book lice, in a building was sufficient grounds for constructive eviction "in a state with housing regulation to keep leased premises free from vermin." The tenant couldn't keep his place vermin-free because his neighbors in the same building didn't make any effort to rid their place of vermin. In *Delamater v Foreman* (1931), an infestation of bed bugs constituted constructive eviction since the court decided that, in a modern apartment, an individual tenant cannot be expected to eliminate vermin, so it's the landlord's responsibility. In *Ray Realty Co. v Holtzman* (1938), the case was construed as constructive eviction because the landlord didn't incinerate the garbage properly, and as a result the building was infested with "rats, mice, roaches, and bugs." Another successful tenant, Bassinger, was unsuccessfully sued for back rent by Hancock Constr. Co. in *Hancock Constr Co. v Bassinger* (1923); bed bugs "frequently interrupted the sleep of the tenant and . . . caused [him] embarrassment in his business by appearing in his clothes."

Sometimes, tenants just have to live with vermin. In *Fisher v Lighthall* (1885), even though the tenant couldn't stand ants "from the basement to the garret," he still was liable for rent because the lease didn't indicate that the landlord would keep the dwelling free of vermin; there was "no implied contract or condition . . . that the premises shall be habitable." Another case of *caveat emptor,* or "let the buyer (or tenant) beware," was *Pomeroy v Tyler* (1887), in which the court ruled that bed bugs, cockroaches, water bugs, and red ants should be no surprise to anyone renting an apartment in New York. And, in *Griffin v Freeborn* (1914), a kitchen infestation of cockroaches was deemed not inconvenient enough to constitute constructive eviction. A factor in the decision was that the tenants waited too long to complain and to vacate the premises.

The distinctions made by legal minds are often puzzling from an entomological point of view. In many cases, the type of insect isn't even specified and, even when it is, there is often a lack of appreciation of insect behavior or ecology. Doutt (1959) lamented, "Entomologists, although they may be saddened, will not be surprised to learn that in many insect cases that reach the courts the judge cannot resist ponderous attempts at judicious humor." One such case was *Ben Hur Holding Corp. v Rox* (1933), in which a small number of crickets was deemed insufficient cause for withholding rent. The judge reasoned "while a cricket is technically an insect and a bug, it would appear from the study of his life, that instead of being obnoxious, he is an intellectual little fellow, with certain attainments of refinement and an indefatigable musician par excellence." Contrary to legal opinion, however, in an apartment situation, crickets and cockroaches are not appreciably different except for the fact that cockroaches at least have the decency to remain silent at night while they are chewing through the garbage.

Insects enter into insurance law usually in the form of an act of God by causing an accident, usually involving a motor vehicle. In *Hanford v Omaha and CB Street R. CB* (1925), a streetcar motorman who was strangled by an insect that blew into his mouth was ruled negligent, not for attempting to wipe his eyes but "in allowing his car to drift while doing so." Another streetcar incident prompted *Bradley v Seattle* (1930). A passenger in a streetcar that skidded because of caterpillars on the tracks sued for damages (an inbound car collided with an outbound car on the same track). Since the motorman had reported 45 minutes before the accident that "the caterpillars were thicker than ever before," the defense of *vis major* or act of God was withdrawn since he knew about the caterpillars and should have known that they presented a hazard.

Often insects play a role in the application of the emergency rule to automobile accidents. These are cases in which insects fly into a car window, bite or sting, and thereby cause an accident. There are dozens of such cases (*Vassia v Highland Dairy Farms Co.* (1937) involved a horse-drawn milk truck and a car). Typical is *Heerman v Burke* (1959). A wasp flew up the left sleeve of the short-sleeved sport shirt of the driver of an automobile; when the wasp stung him, he stopped quickly and veered, which threw the plaintiff guest forward, causing injury. The plaintiff lost the case, since the presence of the wasp in the vehicle constituted a sudden peril not created by negligence.

Along the same lines, people run into insect law when they run into insects and attempt to collect from insurance companies. In *Tracey v Standard Accident Insurance Co.*, "injury to the insured's eye by coming in contact with an insect while he was riding a motorcycle was held to be accidental within the meaning of an accident insurance policy." However, in *Preferred Mutual Accident Assoc. v Beidelman* (1889), a jury decided that a sting by venomous insect did not constitute accidental poisoning.

Certain insects are also regarded as a health hazard and public nuisance under the law. Even power companies are obliged to correct conditions that are conducive to mosquito growth (*Yaffe v Ft Smith* ante 1138). In *Towaliga Falls Power Co. v Sims* (1909), although a public service corporation can exercise right of eminent domain in damming a body of water to provide power, it is not permitted to create "stagnant ponds or polluted pools of water which endanger the health of neighboring communities." Flies and maggots are also, understandably, considered undesirable and potential health hazards; in *Coker v Birge* (1851), a tavern owner was granted an injunction to stop the construction of a stable near his tavern for just that reason.

Of the million or more arthropod species, honey bees occupy a very special place in the law. Although they are recognized as insects, because of their economic importance they have been recognized as property as well. From a legal point of view, bees are difficult to classify. The law recognizes two types of animals: domesticated, or *ferae domitia,* and wild, or *ferae naturae.* Bees are considered *ferae naturae.* However, there are two kinds of wild animals, "those which are free to roam at will and those which have been subjected to man's dominion" (Loring 1981). Bees, of course, can fall into either category depending on the situation. When bees are and are not property depends on finer points of law. Moreover, liability on the part of property owners depends on the extent to which they have exercised dominion over the bees. The whole situation makes for some interesting legal wrangles.

This legal wrangling has been going on literally for centuries. First- and second-century Roman law differentiated, for example, between doves, which return to their place of origin after flying about, and bees, which do not. Representing *animus revertendi,* or *mansuetae,* doves are wild by nature but can be individually tamed. Contemporaries disputed finer points of the law, stating (as did Celsus) that bees do return to their hive (as foragers do after gathering nectar or pollen); when they cease returning to a particular hive, they cease being owned. This discussion was not as esoteric as it may seem; the same doctrines were discussed in the context of slaves and their ownership by Roman citizens as well.

The precedent that bees are ownable and thus stealable items really goes back to a Hittite code found in the ruins of Boghazkoy in 1330 B.C., which describes punishments for theft of bee swarms. The Roman jurisconsults also stated that stealing bees is a punishable offense. As early as the seventeenth century, bees were considered not only property but taxable property subject to church tithes (which the bees' owner could pay in honey and wax).

A landmark case of sorts involved the exclusion of bees from the common-law case that ownership of wild animals ceases when the animal returns to the wild state. According to most traditions, swarming bees belong to the owner as long as he or she is in prompt pursuit—the owner can see and recognize his or her bees. The traditions diverge, however, when bees swarm onto someone else's property. Scottish law (*Harris v Elder* 1893) holds that an owner has no legal right to trespass to pursue his bees; if the bees settle, the owner loses out. In general, however, American law, picking up on a tradition as old as Justinian, dictates that temporary escape does not deprive the owner of his ownership rights *per industriam*.

Ownership of bee products has also proved to be a thorny legal issue. A 1977 legal decision in Pennsylvania nicely sums up the dilemma regarding honey ownership (Loring, 1981):

> If the bees are his property (landowner) the honey is his: for it is the manufacture of his bees. It is the production of what may be called his flock. But this requires proof, in the first instance, that the bees are his property. The treeing on his land may form some presumption of it. . . . It cannot necessarily be inferred, therefore, that the honey made in a tree, on another's land, is made by the bees of the owner of the land. Here, however, a difficulty occurs; it may not have been the herbage of the owner of the land (*herbagium terrae*) from whence the liquid was extracted that was made into honey. Bees do not confine themselves to the fruits or flowers growing near, but move to a distance.

The question of whether bees are wild or domestic animals (or whether, if kept, they can be considered livestock) has repercussions with respect to zoning ordinances. The case of *People v Kasold* (1957) California involves a beekeeper accused of violating a zoning ordinance that prevents certain activities in residential areas other than "the raising of poultry, rabbits, and chinchillas, and the keeping of domestic animals, in conjunction with the residential use of a lot." At issue was whether or not bees could be considered domestic animals. The court effectively decided that "bees are not rabbits"—that they are not domesticated enough to fall within zoning restrictions.

The right of people to keep bees is often restricted not by zoning ordinances but by bee behavior. Bees can constitute a nuisance in the sense that they can interfere with a neighbor's rightful enjoyment of his or her property. In the case of *Olmstead v Rich* (1889), a neighbor's bees, according to the plaintiff, were "driving him, his servants and guests from his garden and grounds and stinging them and interfering with the enjoyment of his home and . . . family, while engaged in the performance of their domestic duties, soiling articles of clothing when exposed on his premises and making his dwelling and premises unfit for habitation and unsafe, constituting a nuisance to plaintiff's damages of $1,500." Although the nuisance issue was conceded, the plaintiff received only six cents in damages. In other more recent cases, bees are generally not regarded a priori as a nuisance, although they can become a nuisance depending upon how they are housed and kept. Whenever beekeeping legitimately interferes with a neighbor's ability to use his or her property, bees constitute a nuisance.

Liability for bee sting is also a tricky legal issue. In 1904, an action of *trespass quare clausum fregit* ("common law remedy for the recovery of damages for the wrong of intruding upon the real property of another") was brought by a mule owner whose mules were attacked and stung ("bit") by bees (*Delaware, Petey Mfg. Co. v Dryden* 1904). He lost—generally, there is no absolute liability of beekeepers for the actions of their bees. Liability occurs only if there is demonstrable negligence: "In order to be held liable, the owner of the bees must know of their vicious tendencies and . . . such owner is under a reasonable duty to locate the bees in a place where they will not be in contact with persons traveling roads and similar places." A plaintiff also has to demonstrate conclusively that the bee or bees doing injury belong to a particular beekeeper.

In a curious case of turnabout is fair play, beekeepers of late have been suing their neighbors for negligence, in particular with respect to pesticide application and drift. In 1949, *Jeanes v Holtz*, a beekeeper could not recover for his bees, which died on a neighbor's property as a result of pesticide residues on his neighbor's land:

> A beekeeper may not recover for losses occasioned when his bees contacted poisonous dust on defendant's and neighboring fields. Some mention of "trespassing bees" appears but such reasoning is questionable. Certainly one could not sue the beekeeper for trespass. If superficially related legal categories must be applied to this new problem, the rules on "business guests" would seem more applicable in light of the value of bees in pollinating crops. (Doutt, 1959)

Beekeepers can recover damages, however, when pesticides drift onto their own property and cause damage. Damage is reckoned in terms of how it affects the colony. In the case of *S.A. Gerard and Co. v Fricker* (1933), figuring damages from insecticide drift involved estimating not only the value of lost honey, but the value of lost seed or fruit crop due to inadequate pollination. As recently as 1985, in *Tutton v A.D. Walter Ltd*, a crop-dusting farmer was held potentially liable due to negligence. He was growing oilseed rape, which is not insect-pollinated, and sprayed his crop, contrary to recommendations of both the Ministry of Agriculture and the manufacturer of the insecticide, while it was flowering. Not only did he know his neighbor's bees were at risk (as they had been in previous years) but he was spraying at the wrong time to protect his crop against seed weevils and midges. The farmer tried to argue that the bees were trespassers and he owed them nothing. The court ruled that he "still owed the 'duty of common humanity,' the limited duty to avoid carelessly harming trespassers." In a commentary on this case in the *Cambridge Law Journal,* J. Spencer wrote, "The decision is obviously a famous victory for country parsons, village schoolmistresses, and everyone else who indulges in the agreeably dotty pursuit of keeping bees, but it is of wider significance than this," again amply illustrating the general ignorance in the legal profession of the importance of bees in particular (which are responsible for over $9 billion worth of agricultural production in the U.S. annually through pollination) and the biology of insects in general.

References

History of entomology in the United States

Holland, W.J., 1929. The first picture of an American butterfly. *Scientific Monthly* 29: 45–48.

Kirby, W. and W. Spence, 1857. *An Introduction to Entomology.* London: Longman, Brown, Green, Longmans, and Roberts.

Mallis, A., 1971. *American Entomologists.* New Brunswick: Rutgers University Press.

Messersmith, D.H., 1976. Long live the entomologist! *Insect World Digest* 3: 21.

Musgrave, C. and D. Bennett, 1976. Bicentennial review of early American entomology. *Fla. Entomol.* 59: 329–333.

Smith, J., 1966. *Generall Historie of Virginia, New England, and the Summer Isles; with the Names of the Adventurers, Planters, and Governours from their First Beginning Anno: 1584 to this Present 1624.* Cleveland, Ohio: World Publishing Co.

Weiss, H.B., 1945. Some early entomological ideas and practices in America. *J. New York Entomol. Soc.* 53: 301–308.

Wilkinson, R.S., 1973. George Starkey, an early seventeenth century American entomologist. *Great Lakes Entomol.* 6: 59–64.

History of pest control

Brown, W.L., 1961. Mass insect control programs: four case histories. *Psyche* 68: 785–789.

Carson, R., 1962. *Silent Spring.* Boston: Houghton Mifflin.

Lockwood, J.A., 1993. Benefits and costs of controlling rangeland grasshoppers (Orthoptera: Acrididae) with exotic organisms: search for a null hypothesis and regulatory compromise. *Environ. Entomol.* 22: 904–913.

McMillen, W., 1965. *Bugs or People? A Reasoned Answer to Opponents of Pesticides.* New York: Appleton-Century.

1964. Obituaries. *Time.* 83 (April 24): 73.

Olkowski, W., H. Olkowski, and S. Daar, 1991. What is integrated pest management? *IPM Practictioner* 13: 1–7.

Ordish, G., 1976. *The Constant Pest.* New York: Scribner.

Riley, C.V., 1888. Insecticide appliances. Modifications of the Riley or cyclone nozzle. *Insect Life* 1: 243–249.

Smith, A.E. and D.M. Secoy, 1975. Forerunners of pesticides in classical Greece and Rome. *J. Ag. Food Chem.* 23: 1050–1056.

Smith, A.E. and D.M. Secoy, 1981. Plants used for agricultural pest control in western Europe before 1850. *Chem. Ind.* 1981 (3 January): 12–17.

Snetsinger, R., 1983. *The Ratcatcher's Child: The History of the Pest Control Industry.* Cleveland: Franzak and Foster.

Whitten, J., 1966. *That We May Live.* Princeton: Van Nostrand.

Fire ants, Congress, and the EPA

Coniff, R., 1990. Fire ants: too hot to handle? *Smithsonian* 21: 48–57.

Daniel, P., 1990. A rogue bureaucracy: the USDA fire ant campaign of the late 1950s. *Agricultural History* 64: 99–114.

1975. Hearing before the subcommittee on Department operations, investigations and oversight of the Committee on Agriculture, House of Representatives, 94th Congress, First Session, June 26, 1975, Serial No. 94-T. Washington, D.C.: Government Printing Office.

Kaiser, K., 1978. The rise and fall of mirex. *Env. Sci. Tech.* 12: 520–528.

Komarek, E., 1978. *Proceedings of the Tall Timbers Conference on Ecological Animal Control by Habitat Management.* Tallahassee: Tall Timbers Research Station.

Lofgren, C., W. Banks and B. Glancy, 1975. Biology and control of imported fire ants. *Annu. Rev. Entomol.* 20: 1–30.

Revkin, A.C., 1989. March of the fire ants. *Discover* 10: 71–76.

Shapley, D., 1971. Mirex and the fire ant; decline in fortunes of a "perfect" pesticide. *Science* 172: 358–360.

Entomophobia and delusory parasitosis

Bunuel, L., 1982. *My Last Sigh.* New York: Vintage.

Byrne, D., E. Carpenter, E.Thoms, S. Cotty, 1984. Public attitudes toward urban arthropods. *Bull. ESA* 30: 40–44.

Davey, G.C.L., L. Forster, and G. Mayhew, 1993. Familial resemblances in disgust sensitivity and animal phobias. *Behav. Res. Ther.* 31: 41–50.

Ekbom, K.A., 1938. Der präsenile Dermatozenwahn. *Acta Psychiat. Neurol.* 3: 227–259.

Halprin, K.M., 1966. The art of self-mutilation II. Delusions of parasitosis. *J. Am. Med. Assn.* 198: 1207.

Hardy, T.N., 1988. Entomophobia: the case for Miss Muffet. *Bull. Entomol. Soc. Am.* 34: 64–69.

Miller, L.A., 1954. An account of insect hallucinations affecting an elderly couple. *Can. Entomol.* 86: 455–456.

Olkowski, H. and W. Olkowski, 1976. Entomophobia in the urban ecosystem: some observations and suggestions. *Bull. Entomol. Soc. Am.* 22: 313–317.

Pomerantz, C., 1959. Arthropods and psychic disturbances. *Bull. Entomol. Soc. Am.* 5: 65–67.

Smith, R.C., 1934. Hallucinations of insect infestation causing annoyance to man. *Bull. Brooklyn Entomol. Soc.* 29: 208–212.

Waldron, W.G., 1961. The role of the entomologist in delusory parasitosis (entomophobia). *Bull. Entomol. Soc. Am.* 8: 81–83.

Wykoff, R.F., 1987. Delusions of parasitosis: a review. *Rev. Inf. Dis.* 9: 433–437.

Insects and the law

M.H.B., Annotation—law of bees. *Annotated American Law Reports,* 39: 352–364.

Daube, D., 1959. Doves and bees. Drois de l'Antiquité et Sociologie Juridique. Melanges Henri Levy-Bruhl. *Publ. de l'Institut de Droit Romain* 17: 63–75.

Doutt, R.L, 1959. The case of the trespassing bee. *Ann. Entomol. Soc. Am.:* 93–97.

Loring, M., 1981. *Bees and the Law.* Hamilton, IL: Dadant.

Spencer, J.R., 1986. A duty of common humanity to bees. *Cambridge Law J.:* 15–17.

Chapter 10

APPRECIATING INSECTS

Insects as symbols

> The butterfly the ancient Grecians made
> The soul's fair emblem, and its only name.
> —S. T. COLERIDGE

INSECTS HAVE BEEN important throughout history not only for what they are and what they can do but also for what they represent to people. People have for thousands of years attributed all kinds of symbolic significance to different sorts of insects for different kinds of reasons. Often the symbolism obviously derives from morphological or behavioral attributes of the insects themselves but it can also be quite mysterious, at least from the perspective of Western twentieth-century civilization. Insects as symbols, however, are common to virtually all cultures.

Totemism is perhaps the most overt form of human identification with insects. Totemism, derived from an Ojibwa Indian word *ototeman,* meaning "his brother sister kin," is the association of an animal or plant with a group of blood-related individuals. Anthropologically, totemism is recognized by several attributes. Generally, the clan is associated by name with an animal, plant, or natural phenomenon; along with that association is the tacitly understood relationship of every member to the namesake either mystically or genetically—every example of the namesake, conversely, is a clan member. This relationship leads to the veneration or adoration of the namesake species with attendant religious taboos and proscriptions.

Many anthropologists believe that totemism arose as a means of guaranteeing a stable and predictable food resource. With a clan

315

dedicated to the protection and conservation of their usually edible totem (i.e., ancestor), there will theoretically be a steady supply of that food item for members of different clans within the tribe, not restricted by the same food taboos. There are, however, other explanations for the origin of totemism. According to another theory, small groups of people tend to live in a restricted range of habitats; within each range, different plant or animal species predominate and are used disproportionately for food or trade. The name, then, would be imposed from the outside by clans that dealt in one way or another with the totem group (e.g., "the guys who eat witchetty grubs," or "the guys who make honey beer"). As for other interpretations, there are many—it's an area ripe for speculation.

Totems themselves represent a wide variety of organisms—everything from the fearsome wild aurochs bull of the ancient Near East to the less-than-awe-inspiring vegetation god of Neolithic Eurasia. Relationship to the totem varies according to the attributes of the totem animal, ranging from fear to admiration to desire (as in the case of totems that are edible). Insects enter into totemic systems mostly as staple food items. In central Australia, where totemism exists in its most unchanged and complete form, the Arunta tribe contains no less than six groups associated with insect totems: these include *wutnimmera* (cicada), *idnimity* (cerambycid or longhorn beetle), *inchalka* (grub), *ilpirla* (eucalyptus or acacia manna psyllid), *yarumpa* (honey ant), and *udnirringitta* (witchetty caterpillars, in the families Cossidae and Hepialidae). As in many such groups, the totem animal is forbidden as food; to eat one is tantamount to eating your own ancestor. The one exception is at fertility-inducing ceremonies called *intichiuma*. Ceremonies are held just before the insect breeding season (the best time to hold a fertility ceremony if you want results). The headman of each totem group is the only member permitted to eat the totem; he does so to keep power over the animal.

In India and Africa, bees and honey appear as tribe totems (in India, for the Bhramada and Juang clans, and in Africa for the Suk). Membership in such a clan supposedly confers the power and ability to control bees. The Nandi of Africa have as totems the jackal and the cockroach, for reasons that are not immediately apparent. In Africa, the mantis plays an important role in the religion of many tribes. The Northern Swahili call it *kukuwazuka* or "fowl of the ghosts" and the Thonga believe it to be an emissary of ancestor gods (which is not quite totemism in the strict sense). It also appears, although not as an object of worship, in the religious doctrine of Zulus and the Kalahari bushpeople. The fact that at least several totemic species that were not food items were predaceous may indicate early recognition of the importance of insect predators in protecting plant sources of nutrition.

In preliterate cultures, insects are important not only as totemic symbols but as participants in creation myths. Worldwide, cultures provide insects with a role in the creation of the world. There are at least two possible explanations for this widespread phenomenon. One is that the creative activities of insects are readily visible: webs are spun, tunnels are dug, nests are constructed, cases are built, galls are formed, and so on. There may also be an innate recognition of the relative evolutionary age of organisms. Virtually all cultures create a hierarchy of sorts (in which humans are almost invariably at the pinnacle) and the relatively ancient origin of arthropods may be implicitly acknowledged in this hierarchy.

Written records make understanding insect symbolism that much easier. Among the earliest are Egyptian hieroglyphs, going back to around 3100 B.C. The founder of the First Dynasty, King Menes, chose the hornet *Vespa orientalis* as the symbol for his kingdom due to its reputed fierce and dangerous nature. Other cults of the era that perhaps arose in awe of the destructive power of insects included locust worshippers in ancient Iraq, and the sect in the Philistine city of Ekron, where Beelzebub, lord of the flies, was worshipped. Among the many powers attributed to this god was the ability to foretell disease, an interesting observation in view of the fact that flies of one form or another have been implicated in the transmission of a number of epidemic diseases. Of course, no discussion of insects as symbols would be complete without reference to the role of the sacred scarab, *Scarabaeus sacer,* in the spiritual lives of ancient Egyptians (see Chapter 8).

Insects appear frequently in Greek writing, usually in figurative language. One recurrent metaphorical reference is to the soul as an insect; *psyche* means both "butterfly" and "soul." The original concept, however, was not *psyche* but *phalaene* or "moth" (derived from *phallus*). Metamorphosis provided an apt metaphor for life, death, and resurrection (as in the Egyptian scarab cult). The butterfly in Egypt was the emblem of Osiris, who was supposed to have been confined after his death in an oak coffin until he arose again, with renewed life. The butterfly symbol also appears in Oriental and Finno-Ugric mythology. The Finns believed that even vegetables have butterfly souls, interesting with respect to the fact that many vegetables are food for developing lepidopteran larvae. The

Kwakiutl of northwest North America also depicted the soul as a butterfly, as did the ancient Slavs, although they also claimed that the soul could leave the body in the form of a fly, a dove, a duck, a nightingale, a swallow, cuckoo, eagle, raven, mouse, snake, hare, or even a small flame.

The bee also appears fairly frequently as a symbol of the soul. In Orphic teaching in Greece, the bee was the emblem of the soul not only because of its honey-producing capabilities but because bees tend to form swarms and to migrate; the Orphic adherents likened this behavior to the swarming of souls toward the divine unity. The Buriats of Siberia also depicted the soul as a bee or wasp visible when issuing from the mouth of a sleeping person; such soul-animals must not be disturbed (a good way, incidentally, to avoid being stung). Indo-Malaysian and Moslem traditions also speak of the soul as a bee. Grasshoppers were symbols of nobility; nobles wore golden grasshoppers in their hair to indicate descent of noble ancestors and local breeding "for such is the natural property of the grasshopper, that in what soil he is bred in the same he will live and die; for they change not their place, nor hunt after new habitations" (Whittick 1960).

Greek writers made prolific use of insects in literature for their symbolic value. Aristophanes the playwright made entomological allusions throughout his plays—fleas, flies and the like. Far better remembered today for his use of insect symbolism was Aesop, a Greek slave who wrote a series of fables. Probably the most famous of his many fables is the account of the ant and the grasshopper, in which ants are industrious workers with foresight and grasshoppers lazy and concerned only with the passing pleasures of the moment. This particular fable has been rendered again and again in the context of contemporary times; a short animated film from the 1920s depicts the lazy grasshoppers as Bolsheviks, and in another, from the 1960s, the grasshopper appears as a wastrel hippie.

Roman writings were on the practical and prosaic side; most entomological references were made in the context of taxonomy or control. Figurative references to insects are curiously few and far between. The Roman empire was partitioned in A.D. 395 and for several hundred years Christianity (more specifically, the Catholic Church) dominated all writing (which consisted for many years simply of copying ancient Greek and Roman writings). Early Christianity was characterized by mysticism, and symbolism was rife. *The Physiologus,* an anonymous zoological text popular until it was placed by Pope Gregory the Great in A.D. 600 on the index of heretical writings, was filled with moralistic fables illustrated with animal symbols. Flies symbolized the devil, as they had in previous centuries. Ants symbolized industry (as in the Biblical injunction "Go to the ant, thou sluggard, and consider her ways"). As described in the fifty-fifth chapter of the work, ants gather food for themselves, protect themselves from foul weather, and distinguish wheat from barley, gathering only wheat; "Barley is food for animals and comparable to the teachings of the heterodox. Man should abstain from barley and take only wheat which represents the true faith in Christ" (Morge 1973).

In early Christian mystical writings in general, scarabs represent sinners, moths the temptations of the flesh, and so on. Bees came to symbolize virginity, perhaps because nobody ever saw them mating (the act involving only one queen and a handful of drones consorting 100 feet in the air or higher). Dung beetles were catapulted from their deified status of ancient days to symbolize for early Christians sinfulness and heresy; "the dung beetle is declared to be a heretic, sullied by the stench of heresy, and the balls of dung are explained as evil thoughts and heresies" (Morge 1973). Many of the encyclopedias and compilations of the period had this sort of tone.

A major advance in insect symbolism arose during the eleventh century not in the context of art or literature but in the context of armed combat. The art of war had progressed to the point that soldiers were covered with heavy metal plates (the fabled "knights in shining armor"), which attire had the decided advantage of protecting the wearer from mortal blows but had the disadvantage of rendering friend and foe indistinguishable on the battlefield. Knights began to depict their identity with symbols on their shields so as to become more easily recognizable. Thus was born the art of heraldry. Initially, heraldic symbols were simple geometric patterns but they gradually became more elaborate and more explicit, particularly as the whole art of combat became less practical and more ritualized (as in tournaments). Inspirational figures, called charges, were soon superimposed on the geometric patterns. Many were of animals, usually fierce and destructive ones such as lions, eagles, stags, and a number of mythical monsters. However, a number of knights adopted insects for their shields. Some insects might have been selected because of their instant familiarity to a large number of people. Bees are quite common as heraldic symbols, sometimes in dorsal view with wings stretched, sometimes perched on a flower, sometimes in a swarm around a skep. Bees were also popular as elements of a pun or rebus on a family name. Bee, Beebee, Beeston, Beston, Bie, Combe, and Humby are among the families with bees on their coats of arms. Two families in Germany named Hummel converged on their heraldic use of bees, and the Abella family in Spain did the same.

Other heraldic symbols included a punning reference to a cockchafer or dorbeetle in the shield of the Doore family of Cornwall. The first recorded use of a butterfly, according to the College of Heralds in London, was in 1633, requested by a musical family named Bassano. Probably these are actually silk moths (as Glover's *History of Derbyshire* records 200 years later) depicted on mulberry branches rather than on laurel as originally recorded; the branch of the Bassano family remaining in Italy may have been involved in the silk industry. Butterflies appear in the arms of Baron du Bois Geffroy et de la Galissonnière (*azure a trois papillons* or, blue with three gold butterflies) for no known reason. The grasshopper was depicted in the arms of Sir Thomas Gresham, who founded the Royal Exchange in 1566. It's an odd symbol for a banker—an insect that Aesop considered improvident. However, the use of the symbol goes back before Sir Thomas—his great grandfather James used it on his seal and it appears in correspondences going back to the middle of the fifteenth century. Again, the use of the grasshopper appears to be as part of a pun or rebus involving the name Gresham. In the fifteenth century, the insects were also known as greshey, a close enough resemblance to Gresham to suit the College of Heralds.

In Japan, heraldic symbols were also in common use during the Tokugawa era. Every daimyo had three such symbols or *mon*—a *jo-mon* for ceremonial garments and battle gear (in which centipedes figure prominently), the *kai-mon* that adorned regular clothing, and the third *mon* on doorways. In the eighteenth century, Kabuki actors were identifiable on stage by their *mon*. At least six wore butterflies, one of whom, Sanjo Kantaro, was renowned for playing women's roles.

The scientific revolution took a lot of the romance and mystery out of the animal world. Insects suffered as symbols as a result, although more knowledge of their anatomy and habits allowed them to play a more varied role in literature. About the only contemporary use of insects to identify membership in a clan, family, or group with a common mission is in sports. Among professional sports teams, insect symbols are few and far between—the Charlotte Hornets grace the National Basketball Association and the

Sting play soccer for Chicago—but no fewer than thirty-five college sports teams have adopted mascots and symbols with six legs. The vast majority of these are hymenopterans, including yellow jackets (Georgia Tech), wasps (Emory and Henry), hornets (Sacramento State in California and Lyndon State College in Lyndonville, Vermont, where female athletes are called the Hornettes), and bees (Saint Ambrose University in Davenport, Iowa, where, interestingly, female athletes are the Queen Bees and male athletes are the Fighting Bees—the latter vastly more inspiring than, say, the Fighting Drones). Some team names are references to species whose distributions are associated with a particular locality—nothing is more southern than boll weevils (University of Arkansas at Monticello), and nothing says Texas like scorpions (Texas Southmost). Why the Salem Community College teams play as the Mosquitoes is not immediately apparent.

Insects in art

> Lastly his shinie wings as silver bright,
> Painted with thousand colors passing farre
> All painters' skill, he did about him dight:
> Not halfe so manie sundrie colors arre
> In iris bowe. . .
> —E. SPENSER

INSECTS HAVE BEEN artistically rendered by people since the beginning of recorded time not only to exploit their symbolic value but also to appreciate their esthetic value. While their small size may have presented a technical challenge, their almost limitless diversity in terms of shapes, colors, and textures undoubtedly has served as a source of inspiration. Insect images early on were by and large utilitarian; as food items, they appear in records of successful hunting/gathering expeditions or of feast days. Scenes of honey hunting dating back 6,000 years grace the walls of the Cave of the Spiders near Valencia, Spain; locusts on skewers are part of a banquet scene in an Egyptian tomb near Ninevah.

In ancient Greece, insects began to appear in a broader range of artistic contexts. Sculptors enjoyed the technical challenge of rendering tiny creatures in stone. Such celebrated figures as Phidias, Callicrates, and Myrmecides (in the fifth century B.C.) playfully depicted miniature scenes of chariots pulled by flies. Insects abundantly adorned decorative pins, finger rings, and coins. Their small size may have made them particularly well-suited for decorating small objects. The trend toward using insects in decorative motifs continued even through the Dark Ages. Monks faithfully copied manuscripts of various descriptions for posterity, "illuminating" them with brilliantly colored illustrations. One such cleric, the Master of the Brussels Initials, favored aposematic insects such as bright-red fire bugs (Pyrrhocoridae) for his marginalia. Many religious texts of the period were enlivened by naturalistic figures, including many plants and insects.

Secular art came into its own in the fourteenth and fifteenth centuries and insects were frequently included in works of art to showcase the artist's talent. Art throughout Europe was at the time starkly representational, accurate to the last detail. Due to their small size and anatomical complexity, insects, as in previous centuries, presented a formidable technical challenge to an artist. Some artists took the challenge a step further and drew insect subjects, particularly flies, in such a way as to create the illusion that they had alighted on an otherwise unrelated painting, generally a portrait; they appear, lifesize and out of scale with other images, almost anywhere in the painting, as a live fly would appear. This trompe l'oeil effect suggests that artists were not averse to having a joke at the expense of their audience. Among the fifteenth century masters of this trick were Petrus Christus, the Master of Frankfurt, Carlo Crivelli, and Il Vecchietto. This is not to say that insect images of the period were devoid of symbolic meaning. Insects were common elements of religious works of the day, possibly symbolizing human frailty and insignificance.

Scientific investigations and geographic explorations of the sixteenth and seventeenth centuries if anything served to increase the artistic emphasis on accuracy in detail. Painters of the Dutch and French schools produced innumerable still-life studies of flowers, excruciating in their precision, and these flowers were frequently accompanied by insect visitors. Maria Sibylla Merian (1647–1717) completed the masterful *Neues Blumenbuch*, a series of spectacular floral and insect images that she modestly suggested might serve well as models for embroideries. Among the Dutch realists who painted insects were Jan van Huijsum (1682–1749) and Paulus Theodorus van Brussel (1754–1795); Pierre Joseph Redouté (1759–1840) won acclaim throughout Europe for his painstaking but elegant flower portraits. Acknowledgment of the decorative value of insect images continued well into the twentieth century. E.A. Seguy, architect of the French decorative arts movement of this century, published books of "visual ideas" for artists and designers. Of these, two dealt with insects, *Papillons,* devoted exclusively to butterflies, and *Insectes,* which covered a staggering array of species ranging from dragonflies to beetles.

By the late nineteenth century, realistic images had lost much of their appeal. With the advent of impressionism, insect images became more important for their symbolic significance than for their ability to lend an air of authenticity to a bucolic scene— hence the 1889 painting *Death's Head Moth,* by Vincent Van Gogh. From impressionism to surrealism, insects gained in popularity. Salvador Dali, a fixture in the surrealist movement, was positively fixated on insects. He wrote on several occasions about his own insect experiences and frequently incorporated insects into his works. Ants appear in *Accommodations of Desire* (1929), *Great Masturbator* (1929), and *Portrait of Gala* (1931); stylized grasshoppers in *Myself at the Age of Ten When I Was the Grasshopper Child* (1933); and daddy longlegs in *Daddy Longlegs of the Evening—Hope!* (1940).

Even artists who create their work using computers have succeeded in incorporating insect images and themes in their work. Michael Newman's 1984 *Metamorphosis* not only uses a familiar insect image—that of a butterfly—but makes reference to a time-honored and distinctively insect-related theme. It's pretty much a certainty that, as new media develop and new styles become popular, insects will continue to be an integral part of artistic expression worldwide.

Eastern and western traditions vary greatly in their approach to life, expecially with respect to insects as subjects for artistic interpretation. Perhaps one of the most important cultural influences on art in general is religion, and differences in religious teachings could underlie the differences in attitudes between East and West. Eastern religions are generally more compatible with nature. In Japan, for example, the Shinto religion recognizes deified spirits in rivers, trees and mountains. Buddhism as well, the other major religious presence throughout much of Japanese history, emphasizes spiritual unity with nature. This is particularly the case for the contemplative sects, such as Zen (called *Ch'an* in China). Epitomizing this sentiment is perhaps the most oft-quoted line of Chinese philosophy in entomological circles, the statement by philosopher Chuang-tze around 330 B.C. that he dreamed he was a butterfly and awoke to wonder whether he was actually a butterfly dreaming it was a man.

This kind of unity with nature is seldom encountered in Western civilization where, at least since the Renaissance, people have felt that nature exists to fulfill human needs and that, in order to understand, one is almost obligated to dissect, analyze, and objectify. Transmigration of souls never sat well with the Christian Church, so early Europeans could be confident of not jeopardizing the health of some ancestor by stepping on a cockroach. The Christian religion, with its heavy emphasis on man as ultimate perfection (after all, who was created in God's image?) made it easy for people to feel superior to the so-called lower life-forms. Nature in European art is on the whole highly objective and representational.

Not so for much of Asian art. In Chinese, as well as Japanese, painting, the mood is the thing, not the mundane morphology. Basic shapes are conventionalized and artists mold these to fit their needs. Rather than reflect what can be seen, Asian art tends to reflect what can be sensed or experienced. In a way, Asian artists were typologists, concentrating on essence rather than on individual variation.

Kano Kagenobu, in the early nineteenth century, prepared a hand scroll of sketches in the styles of Japanese and Chinese artists spanning almost a thousand years; this scroll, *Wakan Hissha Churui,* or *Insects by Japanese and Chinese Artists,* is probably the most complete single compilation of Eastern insect art through the ages. Subjects are copies from the work of artists ranging from Chinese masters from the Northern Sung (960–1127) through Japanese contemporaries of Kagenobu. Among the Japanese artists copied is Motonobu (1457–1559), referred to as *Ko Hogen.* Motonobu was one of the founders of the Kano school of painting during the Muromachi period; screen painting (which expanded the canvas for the artist) and the use of colors came into prominence. Sesshu (1420–1506) was noted for bird, flower, and landscape screen painting. This is the period during which Zen Buddhism became prominent in Japan; with its emphasis on nature and purity, Zen influence was manifested by simplicity of both subject and style. The works of Sotatsu, the founder of the *Rimpa* movement, are represented by the grasshopper on a pink flower (*Rimpa* being a decorative style that blended elements of both Chinese and Japanese traditions).

While it is true that insects are depicted in paintings as an element of the natural world, they also appear in Asian, particularly Japanese, art for rather unique reasons as well. One long-standing tradition in Japanese art is to use insects for allegorical purposes.

This use of insects transcends religious and political boundaries (e.g., Walt Disney, creator of the animated Jiminy Cricket, wasn't exactly a Zen master). One possible reason insects are such good substitutes for humans is that they are generally perceived as being so different from humans; what better way to express a universal truth than by having it uttered by creatures with little or no resemblance to humans? One of the earliest examples of animal allegory (including insects) goes back to the eleventh century, the Heian era, in Japan. The handscrolls titled *Choju Giga* (*Frolicking Animals*) are attributed to the priest-artist Toba Sojo Kakuya (1053–1140). Animals played human roles while the priest made dire predictions about the fate of humanity based on contemporary standards of human behavior. This work inspired a whole series of caricature paintings called, after the originator, *Toba-e,* in which humor was a crucial element.

Insects appear frequently in allegorical context perhaps as a result of the watchful censorship of a powerful centralized government; it's more difficult, one expects, to take offense or express righteous indignation when roosters, monkeys, or caterpillars express revolutionary ideas. One of the earliest forms of this type of endeavor appeared in the Muromachi period (1392–1573) and was titled "Competitive Verses on Four Living Creatures," which included a section on insects. In the Edo Period *irui-monogatari* (tales of nonhuman beings) became popular. One excellent example is the *Mushi Uta-Awase* ("Poetry Match of Insects"), a blatant send-up of the "Poetry Match of 36 Important Poets." Basically, a traveler stumbles upon a garden in which insects are staging a poetry contest (to be judged by a toad). Thirty species pair up, beginning with the cricket that proposes the match and his opponent, a wasp. Among the other participants are a bell-cricket, a firefly, a cicada, a gold beetle, a fly, an ant, a mosquito, a flea, a louse and, in the broadest sense of the word "mushi" (which means "small crawling animal" and not strictly "insect"), an earthworm. The poet-author, Kinoshita Katsutoshi (1569–1649), or Choshoshi, works both outrageous puns and learned literary references into the competition and commentary. For those keeping score, the match was declared a draw.

A later work, *Ehon mushi erabi,* the *Picture Book of Selected Insects* (translated for Westerners as *Songs of the Garden*), published in 1788, served more as a natural history text than a thinly disguised political statement. The editor, Yadoya no Meshimori, was one of the preeminent members of the *kyoya* or comic verse movement, a movement that rejected classical formalism and tradition (hence, devoting an entire poetry anthology to insects, not one of your more traditional topics for poetic discourse). The artist, Kitagawa Utamaro, was perhaps best known for his woodblock prints of beautiful women, perhaps not an inappropriate choice, since the volume was originally conceived as a collection of poems about love. As in Choshoshi's work, a total of thirty are paired off. Again, the use of insects in expressing lofty romantic sentiments may have been a way around excessive sentimentality during the morally conservative Edo period. One stylistic advance (or at least direction) evident from Utamaro's work is an altogether new emphasis on realism. Utamaro's insects are often, if not always, recognizable with respect to species and are depicted behaving in a manner becoming to their taxon. Some art historians attribute this emphasis on scientific precision and realism to Western influence and, indeed, the appearance of this and other works does coincide with an increase in contact with Western culture. Striving for realism was not exactly new to Eastern art; in third-century China, one Tsao Fu-shing converted a small spot on a screen into a fly so realistic that the emperor attempted to brush it off with his sleeve. However, realistic depiction as an end in and of itself was not a common theme until comparatively recently.

Insects were instrumental in creating an illusion of reality in one particular art form in the early nineteenth century: that is, in the production of *netsuke* and *inro*. Kimonos not being equipped with pockets, Japanese gentlemen devised an assortment of devices for transporting small necessaries such as money, tobacco, or medicine. *Inro* are medicine cases, *kinchaku* money pouches and *netsuke* little toggles on the end of the cord (*ojime*) that attached the case or pouch to the belt. *Netsuke* became objects of industrial art from about the mid-seventeenth to mid-nineteenth centuries. These articles are of necessity very small; thus it became a matter of pride to demonstrate one's technical skills by creating works of art in miniature. The Edo period was characterized by general economic well-being, when more material possessions were within the grasp of merchants and traders; industrial art gained a certain amount of respectability due to the growth of this class (pride in one's work and all). Of the various sorts of *netsuke*, insects appear most often in *katabori* (figure carving). Insects may have been favored by *netsuke* carvers because they provided realistic scale—a life-sized insect could be convincingly carved and still be functional as a *netsuke* (in contrast with, say, a horse or tiger). A tremendous diversity of arthropods appear either on *inro* or on *netsuke,* ranging from spiders through hymenopterans (including bee larvae that can actually move in their combs). As early proponents of the twentieth-century philosophy "small is beautiful," Japanese *netsuke* artists can find no rivals in the West and no more appropriate subjects than insects.

Bugs on the big screen

AS VISUAL MEDIA come and go, insects rarely go unrepresented. Insects have even had a long and illustrious career in the movies—a somewhat surprising state of affairs considering how small they are and how difficult to photograph. In the early days, insect appearances in film were limited to animated films, since drawings or puppets, with crude early equipment, were more easily photographed than were the real things. Insects may actually even have contributed to the invention of animated film. Around the turn of the century, Segundo de Chomon, a Spanish filmmaker, was shooting intertitles for a silent film and noticed that a fly, accidentally included on the footage exposed a frame at a time, appeared to move in a jerky fashion. This is the essence of animation; one or two frames are exposed at a time and between exposures small changes are made (one drawing could be substituted for another slightly different drawing, or a mechanical model could be slightly repositioned); when the film is projected at normal speed, the illusion of motion is created. Insects are in some ways ideally suited for animated films. To get close, all that is necessary is to draw a bigger picture: no fiddling with f-stops, macro lenses, or lighting is needed. But there are some anatomical features of insects that pose some hefty challenges to an animator. Specifically, in drawing insects, it's difficult to keep track of six moving legs, two moving antennae, and from two to four moving wings. Thus, in animated films, insects often appear missing any number of appendages. (Nevertheless, while insects in cartoons usually have four, instead of six, legs, spiders have six, instead of eight, thereby preserving a basic appreciation of the greater legginess of spiders).

One of the earliest uses of insects before the camera was by entomologist-turned-animator Wladislaw Starewicz, who, in trying to film insect behaviors, discovered that it was easier to manipulate them when they were dead. He carefully wired several stag beetle carcasses and painstakingly repositioned them for shot after shot in the short film *The Fight of the Stag Beetles*; that film and its sequel,

Beautiful Lucanida or the Bloody Fight of the Horned and the Whiskered were enormously popular with audiences. Starewicz took his act on the road, showing these films to packed houses. He continued to expand this new genre with larger-than-life models and progressively more complex plots (as in *Revenge of the Kinematograph Cameraman,* a story of love and betrayal among a variety of insect species, Fig. 10.1).

The first appearance of insects in cartoons involving line drawings was in a film by pioneer animator Winsor McCay in 1910—*How a Mosquito Works*—featuring a vintage plot (mosquito finds man, mosquito bites man). Even this early film contains many of the conventions typically found in insect cartoons through the decades: an adversarial relationship between humans and insects as well as the depiction of insect mouthparts as tools. The success of his first film inspired others (such as *Bug Vaudeville* in 1921). McCay used to tour the country and show his films in vaudeville houses to large crowds.

Perhaps the crowning achievement in arthropod animation was Jiminy Cricket, who appeared in a supporting role in the 1940 feature *Pinocchio* from the studio of Walt Disney. Originally, the film didn't include the cricket character but Disney animators turned to the cricket to unify disparate elements within the film.

Figure 10.1
Scene from *Revenge of the Kinematograph Cameraman,* by W. Starewicz, the first puppet animation film.

The idea comes from the original Pinocchio story by Carlo Collodi; a "talking cricket" appears in an early chapter and chastises Pinocchio for bad behavior, whereupon Pinocchio smashes him with a mallet. The Disney studio, however, allowed him to survive not only through the feature film but through several subsequent short subjects and educational films. Although Jiminy had a more central role than do most animated arthropods, he embodies virtually all the liberties animators take with insects. He has virtually no insect attributes at all. He lacks wings, he has two arms and legs instead of six legs and he has humanlike facial features, with eyes equipped with pupils and mouthparts unlike any in the phylum Arthropoda.

In live action films, insects kept a relatively low profile until mid-century. In an early effort, Willis O'Brien, the special effects wizard, shot some footage of giant spiders for use in *King Kong* (1933) but it ended up on the cutting room floor. Insects made it big (as it were) in the movies only in the fifties. In the watershed year 1954, *Them* was released by Universal Studios and featured giant mutant ants produced by atomic testing in the Arizona desert. The film proved to be the biggest moneymaker for Universal studio for the year. Its success in retrospect is understandable. The majority of people don't care for insects to begin with and would grudgingly grant that the only thing in their favor is that they're small (and hence easily squashed, swatted, or otherwise disposed of). Were this one tolerable feature of insects to be taken away, many people would have serious problems coping with them. Insect gigantism had been done before *Them*—for example, *Mesa of Lost Women* (1953) features a giant spider created by the evil Dr. Aranya through ill-conceived hormonal transplants between normal-sized spiders and beautiful women (Fig. 10.2). But *Them* was actually a good movie (by contemporary standards); well plotted, well directed, and featuring several big-name actors of the era.

The sincerest form of flattery is imitation, and *Them* was flattered extensively through the next few decades. One of the most prolific practitioners of this genre was director/producer Bert I. Gordon, who made so many big-animal films that he was known as "Mr. Big" (a reference as well to his initials: B.I.G.). Gordon made several of the most inept big bug films in the fifties (e.g., *Beginning of the End*, a 1957 film featuring giant radiation-induced grasshoppers threatening to destroy Chicago) and even cranked one out in 1977 (*Empire of the Ants*). Other notable titles of the fifties in the "big bug" genre include *Tarantula* (1955), and *Deadly Mantis, Black Scorpion, Monster from Green Hell,* and *The Spider* (all 1957). Japan was a relative latecomer to this genre, but more than made up for its late entry with the longevity of its principal big bug star, Mothra, a giant moth. The first film, *Mothra,* was released in 1962; Mothra was featured in four subsequent feature films, co-starring with other "big" science fiction stars such as Godzilla and Rodan.

A possible spin-off of the idea that humans have no business tampering with the powerful and destructive forces of nature (e.g., atomic energy) is the idea that the men and women who do such tampering are doomed to destroy themselves. The fifties also saw a long string of misguided-scientist films, many of which featured insects. Among these misguided souls were brilliant scientist Andre Delambre (portrayed by Al "David" Hedison), who in the 1958 film *The Fly* attempted a matter-transport experiment that went awry, leaving him with the head and foreleg of a common house fly; and cosmetics magnate Janice Starlin (played by Susan Cabot), who in the 1959 film *Wasp Woman* attempted to regain her lost youth by undergoing a series of injections of hormones extracted from wasps; although initially she grows more youthful in appearance, she, too, ends up with the head and front appendages of an insect.

The 1969 film *The Hellstrom Chronicles* kicked off a new trend in insect fear films—the use of real footage of real-sized insects made threatening by purposeful action and social organization. The cinematography in this film was of such high quality that the

Figure 10.2
Poster promoting *Mesa of Lost Women,* 1953 film featuring giant spiders.

film actually won a cinematography award at the Cannes festival the year of its release. Films began to appear using footage of real insects engaging in more or less normal insect behaviors such as *Phase IV* (1974) and *Bug* (1975), (although in this film cockroaches end up committing some very unnatural acts, including spelling out death threats with their bodies on the wall of a house). The appearance of so-called "killer bees" on a container ship in San Francisco harbor in 1974 may have inspired filmmakers to capitalize on a real threat: invasion by Africanized killer bees. These films were appealing for two reasons. First, filmmakers could capitalize on the fact that the audience had at least a passing familiarity with the film's antagonists even before entering the theater. The other advantage from the filmmaker's perspective is that bees, of all insects, are probably easiest to manipulate (while still alive). Raised by the hundreds of thousands by hobbyists, bees could be obtained in huge numbers on relatively short notice, Moreover, bees can be controlled chemically—by pheromones—to cluster or land in a particular spot and so are more easily manipulated for special effects. At least five films were made about killer bee invasions between 1974 and 1978, although none of them was particularly successful at the box office (particularly surprising in the case of the 1978 film *The Swarm* with a screenplay by Arthur Herzog and a cast including such Academy Award-caliber actors as Henry Fonda, Fred MacMurray, Olivia de Haviland, and Michael Caine).

Another common theme in seventies' fear films is the fact that tampering with the environment, usually by contaminating it with toxic wastes, is bound to come to no good. Remember, the Environmental Protection Agency was founded in 1972 and environmental awareness became pervasive as EPA prosecution hit the papers. Hence, the 1977 film *Empire of the Ants* featuring giant ants, created by toxic *and* radioactive waste in a Florida swamp, who take over a local sugar factory and enslave the locals with pheromones. The following year, *Kingdom of the Spiders* was released, featuring rampaging tarantulas (whose normal food supply was destroyed by indiscriminate use of DDT), attacking and consuming dogs, cattle, and, in the final reel, humans.

In the eighties, insect fear films acquired a new life with the release of David Cronenberg's *The Fly*. Although as ludicrous as earlier efforts by entomological standards (6-foot-tall flies can't walk on ceilings), it was generally regarded by critics as an artistic success, thematically along the lines of Kafka's *Metamorphosis*—physical transformation leading to mental and emotional change, stress on relationships, and so on. Although *The Fly II* (directed not by Cronenberg but by Chris Walas, special effects artist on the earlier film) wasn't as welcomed by critics, it nonetheless was perceived as more than just a horror film, with allegorical elements relating the physical and emotional changes of adolescence with the metamorphic transformation of the protagonist. Despite the presence of redeeming intellectual content in these films, they attracted considerable attention for their starkly graphic and gorily repulsive special effects, surpassing anything even imaginable in the 1950s.

The year 1990 saw the release of what may well have been the biggest budget arthropod film in history—*Arachnophobia*—with movie mogul Steven Spielberg serving as executive producer. Its overall budget included location shooting in cloud forests in Venezuela as well as support for importing and screen-testing dozens of species of spiders from as far away as Australia. It was tremendously successful at the box office and the genre shows no signs of declining through the 1990s. Real scientific discoveries and developments continue to be mined for plotlines. A case in point: 1983 saw the publication of the first report of successful genetic transformation of an insect (which happened to be *Drosophila melanogaster*, the common laboratory fruit fly) and by 1989 genetically engineered insects (specifically, mutant killer cockroaches) made their first appearance in an otherwise forgettable science fiction film (*The Nest*). Entomologists thus, by conducting their research, perform a useful service function for the novelty-starved, moviegoing public.

Collecting insects for fun and profit

> There . . . he found amusement in chasing butterflies. . . .
> The chase of butterflies was an apt emblem of the ideal pursuit in
> which he had spent so many golden hours. But, would the Beautiful Idea
> ever be yielded to his hand, like the butterfly that symbolized it?
> —Nathanial Hawthorne, *Artist of the Beautiful*

ONE OF THE nice things about insects is that their ubiquity invites entrepreneurial enterprise; many people have discovered ways to, if not make a living, at least supplement a salary by exploiting insects. Such enterprises don't require decades of training at institutions of higher learning. Rather, most simply require an appreciation of the ecology and behavior of a few select species. One of the most straightforward ways to make a living out of insects is to collect them. There are a surprising number of potential markets for insects, but it's no surprise that the largest market is collectors who appreciate insects for aesthetic reasons. The most popular insects with collectors by far are the Lepidoptera, and butterfly collecting is in the popular conception almost synonymous with insect collecting in general. Beetles, though, have their fans, particularly the showier families (including the scarabs, the leaf beetles, and the longhorn beetles). Early collecting efforts date back to the seventeenth and eighteenth centuries, when European imperialism and expanding economies sent explorers all over the globe in search of new products to develop and monopolize and new markets for old products already manufactured. The early equipment was awkward and heavy; the clap net or bat-folder was standard issue (as depicted in Ingpen's *Instructions for Collecting,* published in 1827). The now universally used round ring-net was developed sometime during the nineteenth century. The history of its invention is lost and is much disputed (according to one probably apocryphal account, it was invented by the Russian scientist Alexander Netsky Popov and named in his honor).

Famous names among collectors include H.W. Bates who, in 1848, went on an expedition to the Amazon (accompanied by Alfred Russell Wallace) and collected about 8,000 new species of animals and plants. Alfred Russell Wallace himself left in 1854 on an eight-year trip to the Malay Archipelago; during this extended collecting trip, he made observations that contributed to his elaboration of a theory of evolution, published simultaneously with Charles Darwin's. Yet another conspicuous collector of the era was A.S. Meek, who discovered the largest species of butterfly in the world, Alexandra's birdwing, *Ornithoptera alexandrae*. He was accompanied on his collecting trips in New Guinea by members of the Papua tribe who would shoot down the large species of butterflies with small arrows. Andre Avinoff, gentleman-in-waiting to Czar Nicholas, was a lawyer (son of a Russian general) who financed forty collecting expeditions for butterflies throughout Asia; his collection, which included over 10,000 *Colias* butterflies alone, was nationalized after the Revolution and now makes up part of the Leningrad Academy of Science collection.

The Rothschilds were tremendously important figures in the history of entomology by virtue of their enormous energy and enthusiasm for collecting. Walter Rothschild was a member of the House of Lords, and Charles, his younger brother, a banker and partner in the banking firm N.M. Rothschild and Sons. Both were avidly interested in collecting—when still a child, Walter started his own natural history museum at Tring, a family estate. Charles was enamored of fleas and amassed a collection of world fleas unequaled anywhere in its breadth, naming in the process *Xenopsylla cheopis,* the Oriental rat flea, principal vector of bubonic plague. Walter was less discriminating in his collecting habits (although he was particularly fond of butterflies and moths and co-authored monographs of swallowtail butterflies and hawk moths with his museum curator Karl Jordan) and either collected himself or had collected all manner of wildlife from all over the world. He worked with over 400 collectors to amass the largest natural history collection assembled by a single individual. He had in his collection over 2,250,000 butterflies alone, from even the remotest regions of the world; thus the collection includes many one-of-a-kind type specimens (specimens from which original species descriptions were made). It is a collection unlikely to be equaled ever again.

Despite popular notions that butterfly collecting was for the effete and foppish, nineteenth and early twentieth century naturalists had to undertake long and physically arduous journeys through uncharted territories and face disease, natural disasters, and even death on numerous occasions. The most successful became intimately familiar with the life history of the objects of their expeditions; in order to capture perfect specimens of rare species it was often necessary to locate the immature stages and rear them through on their host plants to adult eclosion, necessitating a thorough knowledge of botany as well as entomology.

Today it's almost impossible to imagine the interest generated by specimens of butterflies and other insects brought back from exotic places. Auctions were held regularly in Covent Garden in London from 1818 to the 1960s. Enormous prices have been paid for single specimens. An unusal marbled white (an albino) was sold for £20 in 1843; a century later, it was again sold for £49 to a buyer who had purchased another unusual marbled white and wanted to make a pair, which was subsequently sold for over £100 (about $450) shortly after World War II. Even relatively recently, unusual specimens have fetched spectacular prices. In 1966, the type specimen (the individual from which a new species was described) of *Ornithoptera allotei* was sold at auction for £750. Although most specimens today sell for under $100, about 1 to 2% sell for over $200 and a few are advertised at prices greater than $2,000. Even today, collecting is not necessarily a poor person's pastime, although it certainly has become easier; approximately 80% of all of the species of swallowtail butterflies in the world, for example, can be bought through the mail from dealers.

Demand for rare and unusual specimens has been so great that collecting has led on occasion to overexploitation. One butterfly in England, *Lycaena dispar dispar,* the large copper, attracted the interest of collectors at the end of the eighteenth century due to its bright copper-red wings. The species lived only in the Marshy Fens area around Huntingdon. Residents of the area learned quickly that profits could be made from collecting both adult and larval stages; by 1849, the last large copper in the wild state was collected and killed and the species went locally extinct. In the way of a happy ending, however, another race of the butterfly, *Lycaena dispar batava* , was discovered in Holland in 1915 and British butterfly enthusiasts imported thirty-eight individuals into the Wood Walton Fen in 1927, where they were successfully established (to the extent that, in 1931, 100 pupae were shipped back to Holland to restock a reserve that had been created to protect the butterfly from overcollecting there).

The danger of overcollecting still exists throughout the world. One major market for butterfly wings is tropical factories in which folk-art items in bulk quantities are manufactured out of the wings. These are abundant in Taiwan and other places in Asia, where landscape scenes are painstakingly created out of butterfly wing scales. About fifteen years ago, the trade in such deadstock was worth about $20 to $30 million. In Mexico and Brazil, butterfly wings are removed and then glued to cardboard or paper bodies. Morpho butterflies, prized for their brilliant blue iridescent wings, are now farmed in some parts of Brazil rather than collected from the wild. In Papua New Guinea, where many of the more spectacular species of birdwing swallowtails are now protected by law, butterfly farming enterprises have sprung up as well. Beetles, too, have found wide use in art and artifacts, especially the brilliantly metallic leaf beetles and buprestids, which frequently substitute for gemstones in pins, necklaces, and other ornaments. The use of buprestids in particular has a long history—in Japan, the Buprestid Beetle Shrine (*tamamushi-no-zushi*), erected by the Empress Suiko in A.D. 600, features Buddhist scenes and symbols rendered with the wing cases of these iridescent beetles. During the Mughal era in India (17th century), elaborate garments and accessories were frequently ornamented with bits of beetle wings. Even today, the Shaur of the Amazon forests of South America use buprestid beetle wings to adorn necklaces, headgear, and armbands.

One other more recent demand for strange and exotic insects comes from Hollywood, where film directors interested in particular effects often hire people to supply them with large numbers of exotic (or not so exotic) species. These people are often credited in the movies as "insect wranglers." Insect wranglers are not only responsible for providing insects but also for manipulating them on the set. This calls for some basic understanding of insect behavior (e.g., phototaxis or pheromone chemistry). Entomologist Steve Kutcher provided 40,000 carpenter ants for an episode of the television program "Wonder Woman," 3,000 locusts for the feature film *Exorcist II: The Heretic*, and 18,000 ladybugs for a television commercial. David Brody, of the American Museum of Natural History, traveled to Trinidad and other parts of the Caribbean to provide 20,000 cockroaches for the feature film *Creepshow*. Bee biologist Norman Gary not only provided bees and expertise to filmmakers throughout the 1970s, he also acquired a Screen Actors Guild membership and appeared in at least one of these films (*Terror Out of the Sky*) as a farmer who is attacked by an angry swarm.

For the less adventurous, money can be made by selling insects without ever leaving home. A wide variety of insects are reared and sold commercially for a variety of purposes. One large market is hobbyists who are either looking for pet food for predators that require live prey and anglers who are looking for bait. The Lazy H Bait Co., which sells between 10 and 12 million crickets every year, was started up by Leroy and Elaine Henderson after they quit their jobs at a Florida manufacturing plant and bought 10 acres, on which they constructed cricket rearing facilities. The crickets are raised on chicken mash; 15,000 crickets consume 5 pounds of food and drink half a gallon of water daily. The 15,000 crickets lay 75,000 eggs, which hatch in eleven days; in six weeks, they are shipped out in lots of 1,000, retailing at $2.50 to $3.00 per 100. Most go to bait shops but others to pet shops where owners of lizards, snakes, turtles, frogs, and tarantulas buy them (*Champaign-Urbana News Gazette* 6 September 1985).

Another major market for reared insects is in biological control. Classical biological control involves the use of natural enemies to reduce numbers of pest species. There are two basic approaches to biocontrol by natural enemies. Inoculative releases consist of introducing a control agent with the hope of establishing it in an area on a permanent basis. Inundative releases are basically the biological equivalent of a chemical spray—biocontrol agents are released only when a pest problem arises and continue to be released until the problem is under control. In inundative release programs, establishment of the natural enemy is not necessarily an objective. For both strategies, large numbers of biocontrol insects must be mass-reared. It's getting to be big business. Many companies have sprung up to provide biocontrol agents for home gardeners. Various species of ladybugs, predatory mites, and praying mantids are offered for sale through seed catalogues and garden suppliers. Typical 1990 prices as advertised by Gardens Alive! of Lawrenceburg, Indiana include: fifty immature spined soldier bugs (for use against Mexican bean beetles) for $6.95 (good for 10 feet of garden row); one package of 100 greenhouse whitefly parasites *Encarsia formosa* for $6.95; a half pint of ladybugs (3,500 individuals) for $9.95 (good for 2,500 square feet of aphid-infested garden); and 1,000 lacewing larvae or eggs for $6.95. (Interestingly, *Trichogramma* parasitic wasps were available for sale as far back as 1931 for as little as 50¢ for 1,000 parasites, for the control of codling moth—demonstrating that even the price of insects isn't what it used to be).

A relatively new and increasingly popular way of making money from raising insects is to build butterfly zoos and voliaries, where butterflies fly free and where tourists walk through admiringly. This is not simply a business for altruistic nature-lovers—as the *Wall Street Journal* phrased it in a January 10, 1989, account, it's a potential source of "big net profits." Worldwide, there are about sixty or seventy such operations, over thirty in England alone. The London Butterfly House in Syon Park is a 10-thousand square foot glasshouse; Clive Farrell, the owner, also operates butterfly houses in Edinburgh, Weymouth, and Stratford-upon-Avon. A fifty-room mansion in Sherborne is the home of Worldwide Butterflies, run by Robert Goodden, who sells them for weddings and movies; and Lullingstone Silk Farm raises *Bombyx mori* for royal silk. Collectively, the British butterfly houses pulled in a gross gate of about £5 million in 1986. In the United States, there are at least a half a dozen such enterprises. Perhaps the most successful to date is Butterfly World, near Fort Lauderdale, Florida. It is a 3-acre, $1.5 million park, equipped with 3,000 butterflies, with an admission price of $6 per person. Other parks include Callaway Gardens in Georgia and Marine World/Africa U.S.A. in Vallejo, California. In Tokyo, Tama Zoological Park contains a butterfly house, and in Sri Lanka there is a voliary in a 0.8-acre greenhouse, on the grounds of the Dehiwela Zoological Garden, that houses at least 50 native species. Australia's Melbourne Zoo spent almost half a million dollars

Figure 10.3
Insect artifacts: (above) Mexican jumping bean, actually the larva of *Carpocapsa saltitans*, a small moth, inside the fruit of a shrub (*Sesbatiana* S. Passoa); (right) Gag "ice cubes," plastic encasing honey bees.

building a butterfly house for over 300 native species. Lest anyone think that butterfly houses are suited only to warm climes, it is worth mentioning that the Fjärilhuset, the most northerly butterfly house in the world, is situated in one of the most popular city parks in Stockholm, Sweden.

Perhaps the greatest insect attraction, historically, has not been the butterfly house but the more prosaic flea circus. Flea circuses were popular attractions up until relatively recent times (one stood in Times Square until the early fifties). Typically, flea performers are *Pulex irritans,* the human flea—females are preferred at least partly because, being larger, they're more visible. Traditionally, would-be flea circus operators purchase or otherwise collect their fleas and begin their training by enclosing them in bottles underneath a light. Sensitive to both light and heat, the enclosed fleas tend to remain in the bottom of the bottle. After a while, the cork in the bottle is replaced by a glass cover; fleas that jump and hit the glass cover gradually habituate to jumping only a particular height (which height can be determined by controlling the height of the container in which they're confined). After about a two-week training period, the fleas are ready for the stage. They can be hooked up with fine tiny wires to wagons or coaches, dressed up as historical personalities, or equipped with tiny swords for mock battles. Some audiences were invited to view the performances under a magnifying glass. In the 1930s, the price of admission at one show was 15¢, slightly more expensive than a double feature at a movie house at the time. The end of the performance is traditionally marked by the owner offering the performers a hearty meal—on his arm—in full view of the audience, a nice touch with few parallels in the world of show business. Even when they're dead, fleas can provide entertainment; a longtime staple in folk art exhibits in Mexico are dressed fleas, actually flea corpses decked out in elaborate, although tiny, handsewn costumes. Dead bees find their way into novelty items as well (Fig. 10.3).

Artists have found ways to use insects to create artwork as well. Character-marked wood is what comes of making the best of a bad situation. Woodboring insects, by tunneling through wood, cause millions of dollars' worth of damage annually to the lumber industry. Insects cause around $90 million worth of damage to 2 billion boardfeet of oak alone each year. Some damaged lumber goes into railroad ties, pallets, pulp chips and the like, but other such lumber is enterprisingly marketed as "character-marked" and is sold for large sums of money as wall paneling, touted as unique and "never duplicated." The Hardwood Plywood Manufacturers Association has sponsored showings of such woods. In one survey, 62% of 4,398 families queried expressed a preference for furniture constructed with character-marked wood. Examples include wormy chestnut, marked by *Melittoma sericeum,* the chestnut timberworm; spot-worm white oak, marked by *Agrilus acutipennis,* flatheaded borer; pinhole oak, marked by ambrosia beetles (Scolytidae); and timber-worm oak, marked by oak timberworm *Arrhenodes minutus.*

Finally, there is a small cadre of artists that has actually gone into partnerships of sorts with insects for esthetic purposes. Garnett Puett, for example, has made quite a reputation as the world's first apisculptor. To create a work of art, he prepares a plaster form and encloses it in a glass case along with about 100,000 bees. The bees proceed to build comb to form an eerie, wax-encased sculpture. The sculptures typically sell for $5,000 or more; "Apiscaryatid", commissioned by the Laumeier Sculpture Park in St. Louis, is one of the most popular attractions in the park. Puett actually comes from a long line of beekeepers. Another enterprising artist is Kazuo Kadonaga, who sets up wooden grids for silkworms to spin cocoons in. He has created (with some six-legged assistance) over ninety such pieces. After the sculptures are complete, the silkworm pupae are gently heated and killed, the ultimate proof that artistic accomplishment requires great personal suffering and sacrifice. And, Richard Boscarino is the artist who created the medium of Roachart. He poses dead cockroaches in lifelike backdrops equipped with tiny props (among his tableaux are a beauty parlor, a sales counter at Bloomingdale's, a public restroom, a bowling alley, a diner, and a reenacted Last Supper, among others). These little dioramas sell for between $500 and $900 and have inspired a line of greeting cards. He buys most of his cockroaches already dead from a biological supply company rather than collect them himself. He also makes traditional jewelry and regards roachart as just a sideline. Among other things, it is labor-intensive—his diner diorama required 100 hours to create. Boscarino once aspired to become an entomologist but a summer job at Princeton University pinning specimens convinced him that it was not the profession for him.

References

Insects as symbols

Anonymous, 1987. *Blue Book of College Athletics.* Akron, Ohio: Rohrich Corp.

Berenbaum, M., 1978. Bees as symbols in heraldry. *Am. Bee J.* 118: 178–180.

Bodenheim, F., 1950. *Insects as Human Food.* The Hague: Junk.

Cherry, R.H., 1993. Insects in the mythology of Native Americans. *Am. Entomol.* 39: 16–21.

Cirlot, J.E., 1962. *Dictionary of Symbols.* New York: Philosophical Library.

Cowan, F., 1865. *Curious Facts in the History of Insects.* Philadelphia: Lippincott.

Faulkner, P., 1931. Insects in English poetry. *Sci. Monthly* 33: 53–73.

Harpaz, I., 1973. Early entomology in the Middle East. Pages 21–36 in *History of Entomology* (R.F. Smith, T.E. Mittler, and C.N. Smith, eds.). Palo Alto: Annual Reviews Inc.

Kritsky, G., 1985. Tombs, mummies, and flies. *Bull. Entomol. Soc. Am.* 31: 18–19.

Morge, G., 1973. Entomology in the western world in antiquity and in medieval times. Pages 37–80 in *History of Entomology* (R.F. Smith, T.E. Mittler, and C.N. Smith, eds.). Palo Alto: Annual Reviews Inc.

Phipson, E., 1883. Insects. In: *The Animal-Lore of Shakespeare's Time.* London: Trench, pp. 389–433.

Robbins, R.H., 1960. *Encyclopedia of Witchcraft and Demonology.* New York: Crown.

Sawicki, J. A. and H. G. Scott, 1983. Depiction of invertebrates on U. S. army field artillery and infantry insignia. *Bull. Entomol. Soc. Am.* 29: 18–19.

Walton, W.R., 1922. The entomology of English poetry. *Proc. Entomol. Soc. Wash.* 24: 159–203.

Weigle, M., 1982. *Spiders and Spinsters: Women and Mythology.* Albuquerque: University of New Mexico Press.

Whittick, A., 1960. *Symbols, Signs and Their Meaning.* Newton, CT: Branford.

Insects in art

Berliner, N.Z., 1986. *Chinese Folk Art: The Small Skills of Carving Insects.* Boston: Little Brown.

Betchaku, Y. and J.B. Mirviss, 1984. *Utamaro: Songs of the Garden.* New York: Viking Press.

Davis, M. and J. Kathirithamby, 1986. *Greek Insects.* New York: Oxford University Press.

Gaunt, W., 1972. *The Surrealists.* London: Thoms and Hudson.

Hutchinson, G. E. 1974. Aposematic insects and the Master of Brussels initials. *Am. Sci.* 62: 161–171.

Pressly, W. L., 1973. The praying mantis in surrealist art. *Art Bull.* 55: 600–615.

Priest, A., 1952. Insects: the philosopher and the butterfly. *Metropolitan Museum of Art Bull.* 10: 172–182.

Wilson, S., 1980. *Salvador Dali.* New York: Tate Gallery.

Bugs on the big screen

Bonansinga, J., 1987. The Fly et al. From Neumann to Kafka to Cronenberg, fear of flying finds new insectual imagery. *Filmfax* 5: 39–42.

Carroll, N., 1981. Nightmare and the horror film: the symbolic biology of fantastic beings. *Film Quarterly* 34: 16–25.

Leskosky, R. and M. Berenbaum, 1988. Insects in animated films; or, not all "bugs" are bunnies. *Bull. Entomol. Soc. Am.* 34: 55–63.

Mertins, J., 1986. Arthropods on the screen. *Bull. Entomol. Soc. Am.* 32: 85–90.

Stein, M., 1987. "Them!" *Filmfax* 5: 44–47.

Warren, B., 1982. *Keep Watching the Skies!* Jefferson, North Carolina: McFarland.

Collecting insects for fun and profit

Anonymous, 1990, Steve Kutcher may have butterflies in his stomach, but his showbiz bugs never miss a cue. *People Weekly* 33 (April 16): 111.

Eidmann, H., 1928. Insekten als Haarschmuck in China. *Entomolog. Mitteil.* 17: 46–49.

Emmel, T.C., 1975, *Butterflies.* New York: Knopf.

Howard, L. O, 1931. *The Insect Menace.* New York: Century.

Hubbell, S., 1987. Onward and upward with the arts: bugs. *New Yorker* 63 (28 December): 79–89.

Kenly, J.C., 1938. *Little Lives: The Story of the World of Insects.* New York: Appleton-Century.

Karges, J., 1985. *Nabokov's Lepidoptera: Genres and Genera.* Ann Arbor: Ardis.

Klausnitzer, B., 1981. *Beetles.* New York: Exeter.

Morris, M. G., N. M. Collins, R. I. Vane-Wright, and J. Waage, 1991. The utilization and value of nondomesticated insects. In: *The Conservation of Insects in Their Habitats.* N.M. Collins and J.A. Thomas, eds. London: Academic, pp. 319–347.

Rives, V.Z., 1994. An overview of beetle elytra in textile and ornaments. *Io Vison: Cultural Entomology Digest* 1: 2–9.

Rothschild, M., 1985. *Dear Lord Rothschild: Birds, Butterflies, and History.* Philadelphia: ISI.

Russell, H., 1913. *The Flea.* Cambridge: Cambridge University Press.

Smart, P., 1985. *The Illustrated Encyclopedia of the Butterfly World.* Essex: Salamander.

Simpson, G., 1948. Can fleas really be trained? *Sci. Digest* 24: 24–25.

Solomon, J.D., 1986. *Using hardwood lumber with insect, fungus and bird defects as character-marked woods.* USDA Forest Service Res. Paper SO 226.

Tekulsky, M., 1985. The *Butterfly Garden.* Boston: Harvard Common Press.

Chapter **11**

EQUAL TIME
(THE INSECT PERSPECTIVE)

MUCH HAS BEEN said of the tremendous impact that insects have had on human affairs, so it is only fair to mention, at least in passing, that the interaction is not just a one-way street—humans have likely had a greater impact on insects than any other life form on the planet. In many cases, that impact was benign, if not downright beneficial. By growing crops, domesticating livestock, storing food, making clothing, building houses, and just by being themselves, humans have provided food, drink, and shelter to incalculable numbers of arthropods. But there's a darker side to the relationship as well.

In some cases, the negative impact of human activities on insect welfare is deliberate. Insects are routinely swatted, sprayed, or otherwise targeted for destruction. But oftentimes, the destruction far exceeds the designs of even the most ill-intentioned person. The indiscriminate use of insecticides has brought about the destruction of untold billions of insect lives, the vast majority of which were not the intended targets of the spray campaign. Even collectors and hobbyists, professed admirers of insects, have on occasion collected insects into extinction (see Chapter Ten). Several new laws have been enacted that exact stiff penalties on collectors who attempt to take species that are considered in jeopardy.

In the course of their comings and goings through history, humans have managed to shuffle things up pretty comprehensively. Humans rarely travel unaccompanied—whenever people change locations they almost invariably bring along arthropod hitchhikers. It's been estimated that over 2,000 insect and arachnid species currently residing in the United States—about 2% of the total number of U.S. insect residents—are actually immigrants from foreign shores. In many cases, these aliens have been decidedly undesirable.

Among their ranks are yellow fever mosquitoes, gypsy moths, fire ants, and most of the cockroaches that infest hearth and home. All told, nonindigenous insect species wreak over $90 billion worth of havoc on the U.S. economy.

The dollar figure, though, is not the whole story. Resident American insects often suffer terrible consequences as the result of the introduction of nonnative species. These foreign invaders can usurp the homes, the sustenance, and even the loved ones of native species. A small parasitic wasp, *Pleolophus basizonus,* was introduced into Canada over thirty years ago to control jackpine sawflies; as a result, a native parasitic wasp, which used to parasitize the same sawflies, is all but extinct, outcompeted by the aggressive newcomer. And imported fire ants can interbreed with native fire ants (which don't exhibit the aggressive tendencies of the South American émigrés) and in the process prevent them from mating with their own kind.

These exotic imports have an impact beyond the insect community as well. Although most of the human residents of South Africa undoubtedly did not welcome the accidental introduction of the Argentine ant, an aggressive stinger resembling the imported fire ant in its habits, the introduction has had implications far beyond the inconvenience and irritation of the few people that run across the ants. The Argentine ants imported accidentally into South Africa are, like many of the native South African ants, voracious seed-eaters. The ability of ants to nibble certain seed structures and then to transport the uneaten portion, usually to a "garbage dump" outside the ants' nest, is critical to a wide variety of plants, which rely on these ants to disperse their seeds to suitable germination sites. Unlike the native ants, though, the Argentine ants are far more careless about the way in which they dispose of the seeds they don't finish eating and as a result some plants in areas where Argentine ants have displaced the local ants are on the verge of disappearing.

From the insect perspective, even more catastrophic than human disruption of the biological community has been the impact of human activities on the physical environment. Intensive urbanization and development have all but eliminated certain kinds of habitats—and, along with them, their insect inhabitants. Many insects have extremely specific habitat requirements and thus are highly vulnerable to disturbance. The Karner blue butterfly, *Lycaeides melissa samuelis,* for example, feeds as a caterpillar only on wild blue lupine, a plant restricted to sand and pine barrens. Because sand and pine barrens, large stretches of sandy soil with little standing vegetation, have proved especially amenable to transformation into airports, shopping malls, and the like, the Karner blue has undergone a drastic reduction in range. Originally discovered near Center, New York, in the Karner pine bush, a large inland pine barren, the Karner blue is now known only from about fifty localities in New York, New England, and parts of the Midwest, almost invariably in tiny populations where patches of sand or pine barren persist.

Even species that appear to be abundant may be vulnerable. Monarch butterflies, for example, are known from virtually every state east of the Rocky Mountains and are frequently spotted sipping nectar along roadsides in towns, cities, and rural areas everywhere throughout their range. Monarchs are rather unusual among insects, however, in that every year the entire population undergoes a mass migration to specific localities for passing the winter. Whereas American retirees may head south to spend the winter in just about any city in Florida, Arizona, or California with good shopping and shuffleboard, when monarchs head south they invariably end up in only one of two specific locations: a mountainous region just west of Mexico City, Mexico, or a narrow strip of land along the Monterey Peninsula near Pacific Grove, California. In these overwintering sites, monarchs rest on trees for up to five months, wings folded and relatively immobile, by the millions. Come spring, they mobilize, move out, and head north again, to lay eggs on the emerging milkweeds in southern states. Encroaching development in these two, relatively speaking, tiny spots can have an untold impact on the entire monarch population. In California, the main concern is that further commercialization of the sites will so disturb the butterflies that they will abandon the area; in Mexico, extensive logging, even in sites protected by law by the Mexican government, threatens to eliminate the very existence of the overwintering area, leaving millions of monarchs without a protected winter haven.

Even more insidious than outright elimination of habitats is the fact that human activity can so profoundly alter habitats as to make them uninhabitable for insects. Human activity has been affecting insect lifestyles ever since they put in an appearance on the planet—merely by walking a path regularly, people can compact the soil and reduce its suitability as a home for arthropods. But technology has expanded the ability of humans to alter the environment in ways that are unprecedented in the history of life on earth. Manufacturing and industrialization have profoundly altered the ways in which materials cycle through ecosystems, in most cases to the detriment of other organisms. Virtually every form of environmental pollution—carbon dioxide emissions, acid rain, ozone, lead, road salt, and the like—reduces the growth and vigor of plants, which in turn reduces the suitability of plant foliage as food for herbivorous insects. Insects consuming plants adversely affected by these pollutants grow more slowly and produce fewer offspring than insects consuming healthy, robust plants.

Patterns of energy flow through ecosystems have also undergone profound change as a result of technological advances. Thermal pollution from nuclear power plant discharges interferes with aquatic insect life cycles, often initiating premature (and fatal) spring emergence. Even artificial lighting can disrupt the life cycles of insects. The introduced ailanthus silk moth, for example, failed to flourish in Philadelphia because light from street lamps shining at night disturbed its breeding cycle. Electrical discharge from power lines can also cause problems for insects. Bees in colonies situated directly under power lines can perceive both electric shock and electromagnetic fields. Honey bee colonies under power lines have electric fields and hive currents 100 to 300 times higher than do shielded colonies. If intensities are high enough, developmental abnormalities and even death can result.

That insects can die, or even go extinct as species, as a consequence of human activities is not likely to inspire people to change their ways—most people simply don't feel that strongly about insects to react in such a manner. But the fact that insects—by all rights and purposes the world's most adaptable creatures—are having troubles coping with the world as it has been restructured by humans should serve as a warning. Because they live everywhere, insects are excellent monitors of environmental conditions—canaries in every conceivable kind of coal mine. When they're in trouble, it means that every other form of life is potentially in trouble. For no other reason, then, it's worth caring about the health and well-being of insects because, as long as their health and well-being are safeguarded, the planet can be reckoned as habitable for humans.

References

Alstad, D.N., G.F. Edmunds, and L. Weinstein, 1982. Effects of air pollutants on insect populations. *Annu. Rev. Entomol.* 27: 369–384.

Beyer, W.N. and J. Moore, 1980. Lead residues in eastern tent caterpillars (*Malacosoma americanum*) and their host plant (*Prunus serotina*) close to a major highway. *Env. Entomol.* 9: 10–12.

Bindokas, V., J. Gauger, and B. Greenberg, 1988. Mechanism of biological effects observed in honey bees (*Apis mellifera* L.) hived under extra-high voltage transmission lines: implications derived from bee exposure to simulated intense electric fields and shocks. *Bioelectromagnetism* 9: 285–302.

Bolsinger, M. and W. Fluckiger, 1987. Enhanced aphid infestation at motorways: the role of ambient air pollution. *Great Lakes Entomol.* 20: 237.

Bond, W. and P. Slingsby, 1984. Collapse of an ant-plant mutualism: the Argentine ant (*Iridomyrmex humilis*) and myrmecochorous Proteaceae. *Ecology* 65: 1031–1037.

Butt, S.M., J. Beley, T. Ditsworth, C.D. Johnson, R. Balda, 1980. Arthropods, plants and transmission lines in Arizona—community dynamics during secondary successsion in a desert grassland. *J. Environ. Management* 11: 267–284.

de Viedma, M.G., R. Escribano, M.R. Gomez-Bustillo and R.H.T. Mattoni, 1985. The first attempt to establish a nature reserve for the conservation of Lepidoptera in Spain. *Biol. Conserv.* 32: 255.

Dindal, D., R. Norton, 1979. Influence of human activities on community structure of soil Prostigmata. *Rec. Adv. Acarol.* 1: 619.

Dirig, R., 1988. Nabokov's blue snowflakes. *Nat. Hist.* 97: 64–68.

Drewett, J., 1988. Never mind the whale, save the insects. *New Scientist.* 120: 32–36.

Elkins, N. and W. Whitford, 1984. The effects of high salt concentration on desert soil microarthropod density and diversity. *Southwestern Naturalist* 12: 239–241.

Flogaitis, E. and P. Blandin, 1985. Trampling effects on soil macroarthropods in suburban forests: experimental study. *Acta. Oecol.* 6: 129–142.

Frank, K.D., 1986. History of ailanthus silk moth in Phila. Pa: a case study in urban ecology (Lepidoptera: Saturniidae). *Entomol. News* 97: 41–51.

Gall, L., 1984. Population structure and recommendations for conservation of the narrowly endemic alpine butterfly, *Boloria acrocnema* (Lepidoptera: Nymphalidae). *Biol. Conserv.* 28: 111–138.

Gersich, F., M. Brusven, 1981. Insect colonization rates in near-shore regions subjected to hydroelectric power peaking flows. *J. Freshwater Ecol.* 1: 231–236.

Jahn, E., 1986. Physical power zones and insects. A review. *Anzeig. Schädlings. Pflanzenschutz Umweltschutz* 59: 8–13.

Lemasson, M. and E. Bruneau, 1986. Ecotoxicology of a chlorophenoxyacetic weedkiller (MCPA) on honeybee (*Apis mellifica* L.). Study in glassroom. *J. Appl. Entomol.* 102: 263–272.

Lincoln, D., D. Couvet, N. Sionit, 1986. Response of an insect herbivore to host plants grown in carbon dioxide enriched atmospheres. *Oecologia* 69: 556–560.

Lockwood, J., 1987. The moral standing of insects and the ethics of extinction. *Fla. Entomol.* 70: 70–89.

Lockwood, J.A., 1993. Environmental issues involved in biological control of rangeland grasshoppers (Orthoptera: Acrididae) with exotic agents. *Environ. Entomol.* 22: 503–518.

Luckenbach, R., R. Bury, 1983. Effects of off-road vehicles on the biota of the Algodones dunes, Imperial County, California. *J. Appl. Ecol.* 20: 265–286.

Meyer, W., G. Harisch, and A.N. Sagredos, 1986. Biochemical and histochemical aspects of lead exposure in dragonfly larvae (Odonata: Anisoptera). *Ecotox. Env. Safety* 11: 308–319.

Morton, A.C., 1983. Butterfly conservation: the need for a captive breeding institute. *Biol. Conserv.* 25: 19–34.

Newhouse, V.F., L. D'Angelo and R.C. Holman, 1979. DDT use and the incidence of Rocky Mountain spotted fever: a hypothesis. *Env. Entomol.* 8: 777–781.

Office of Technology Assessment, 1993. *Harmful Non-indigenous Species in the United States.* Washington, D.C.: U.S. Congress.

Piart, J. and D. Duviard, 1985. Study of the epigeic collembolan community in a heathland of central Brittany: changes in the number of springtail catches after spreading of crushed household refuse. *Rev. d'Ecol. Biol. Sol.* 22: 97–120.

Price, P.W., 1970. Characteristics permitting coexistence among parasitoids of a sawfly in Quebec. *Ecology* 51: 445–454.

Read, H.J., C. Wheater, and M. Martin, 1987. Aspects of the ecology of Carabidae (Coleoptera) from woodlands polluted by heavy metals. *Environ. Pollution.* 48: 61–76.

Robel, R.J., C. Howard, M. Udevitz, and B. Cornutte, 1981. Lead contamination in vegetation, cattle dung, and dung beetles near an interstate highway. Kansas *Environ. Entomol.* 10: 262–263.

Robertson, P., M. Woodburn, and D. Hill, 1988. The effects of woodland management for pheasants on the abundance of butterflies in Dorset, England. *Biol. Conserv.* 45: 159.

Robin, A.M., J. Geoffroy, 1985. Experimental trampling in a suburban forest—description of the experiment and pedological investigations. *Rev. d'Ecol. Biol. Sol.* 22: 21–34.

Rossaro, B., 1987. Chironomid emergence in the Po River (Italy) near a nuclear power plant. *Entomol. Scan.* S29: 331–338.

Scott, D.C., 1993. Logging squeezes butterflies' winter home. *Christian Science Monitor* 13 April, 1993.

Seastedt, T. and D. Crossley, 1981. Microarthropod response following cable logging and clear-cutting in the southern Appalachians. *Ecology* 62: 126–135.

Shuey, J., J. Calhoun, and D. Iftner, 1987. Butterflies that are endangered, threatened, and of special concern in Ohio. *Ohio J. Sci.* 87: 98–106.

Thome, J.P., M.H. Debouge, and M. Louvet, 1987. Carnivorous insects as bioindicators of environmental contamination: organochlorine insecticide residues related to insect distributions in terrestrial ecosystems. *Int. J. Env. Analyt. Chem.* 30: 219–232.

Udevitz, M., C. Howard, R. Robel, and B. Cournutte, 1980. Lead contamination in insects and birds near an interstate highway, Kansas *Environ. Entomol.* 9: 35–36.

Wolcott, T.G., and D.L. Wolcott, 1984. Impact of off-road vehicles on macroinvertebrates of a mid-Atlantic beach. *Biol. Conserv.* 29: 217–240.

Wolf, P., 1989. Planners get conflicting reports on Monarchs. *Pacific Grove Monarch* 11 August, pp 1, 6.

Appendix

Introduction to the Insect Orders

come come come
come in your billions
tiny small feet
and humming little wings
crawlers and creepers
wigglers and stingers
scratchers borers slitherers
—DON MARQUIS, archy and mehitabel

THE CLASS INSECTA (*in*, in, *secta*, cut) is the largest class of organisms in the world, with upwards of 850,000 described species. In order to impose some sort of order on this bewildering array of living things, entomologists have divided the class into smaller more manageable groups, called, logically enough, orders. Although there are over two dozen orders recognized by professionals, several of these groups are so obscure that most people can live their entire lives blissfully unaware of their existence. This survey is intended to introduce the conspicuous orders, the ones that are most likely to have an impact on human affairs. Thus, rockcrawlers (order Grylloblattodea), webspinners (Embioptera), zorapterans (Zoraptera, not even common enough to have a common name), and their ilk must dwell a little longer in obscurity. For convenience, orders are listed alphabetically.

Order Coleoptera (*koleos,* sheath, *ptera,* wings) =beetles (370,000 species). In terms of species, there are more beetles than there are any other kind of insect (or any other kind of animal, for that matter). As a group, they are amazingly variable; among the few features in common are the two pairs of wings, with the forewings, or elytra, hardened to form a kind of sheath for the flightworthy hind wings. Generally, mouthparts are of the chewing variety, although there are exceptions to the rule, particularly among the larvae (often called grubs). Whatever plant or animal material you might think of, there's probably a beetle that eats it; among the diet choices include fruits, flowers, leaves, seeds, wood, fungi, mold,

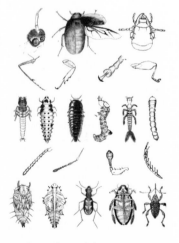

other arthropods, stored products, blood, dead bodies, and dung. Beetles can be found in a staggering range of habitats on land as well as in the water (they are, for example, among the only insects that are aquatic as adults as well as larvae).

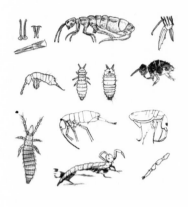

Order Collembola (*colla,* glue, *embolon,* bar)= springtails (6,000 species). These are tiny wingless six-legged creatures that molt throughout their adult life. Whether or not they can be considered insects has been subject to question, mostly because of their steadfast winglessness as well as certain peculiarities of mouthpart structure. Equipped with chewing or sucking mouthparts, they feed primarily on organic matter. Their most prominent characteristic feature is the "spring," a locomotory device that folds up under the abdomen and is slapped against the ground to propel the springtails skyward (or at least six inches in that direction). The scientific name of the order derives from another structure on the abdomen, a ventral tube, or "glue peg," which helps the springtails regulate humidity.

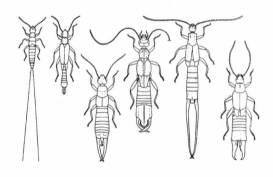

Order Dermaptera (*derma,* skin, *ptera,* wings) = earwigs (1,200 species). In earwigs, the first pair of wings is very short and thickened while the second pair, vaguely ear-shaped, is folded fanlike underneath; contrary to popular opinion, earwigs do not crawl into auditory orifices of unsuspecting sleepers. Although most are omnivorous, some can damage plants and still others are ectoparasites of vertebrates, generally bats or rodents. The most distinctive morphological trait of the order (missing, though, in parasitic forms) is the pair of forceps or pincers on the tip of the abdomen; these are used for wingfolding and for pinching potential predators. Their most endearing trait, however, is the fact that they exhibit maternal care, the mother residing with her offspring and protecting them from danger.

Order Dictyoptera (*dictyo,* net, *ptera,* wings)=cockroaches (4,000 species), and mantids (2,000 species). Dictyopterans have two pairs of wings, the first pair of which is thickened and the second pair folded and membranous. The wings fold flat over the body, an arrangement that allows cockroaches to scuttle into tight spots (not the least of which are spaces under sofas and behind refrigerators, although many

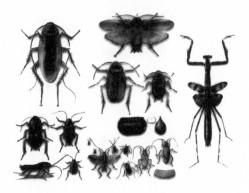

cockroaches live peaceful and obscure lives ensconced in leaf litter). Cockroaches of all ages are omnivorous, which means they eat anything and everything. Mantids, on the other hand, are voracious predators, mostly of other insects, for their entire lives; they grab their prey with their powerful spined grasping (raptorial) forelegs. Entomologists place these unlikely partners in the same order based on certain obscure characteristics—wing musculature, an unusual twist to the male genitalia, and the habit of females of packing eggs into an egg case or ootheca.

Order Diptera (*di,* two, *ptera,* wings)=flies (120,000 species). Although many insects are known as flies (e.g., fireflies, stoneflies, dragonflies, alderflies, and the like), only members of the order Diptera are true flies, from the entomological standpoint. True flies have only a single pair of membranous wings; their hind wings are reduced to knob-like structures called halteres, which act as balancing organs. The mouthparts of adults and larvae are basically of the sucking/lapping type but they are tremendously variable, as is the diet; it is difficult to think of anything organic that is not consumed by some fly or other. They suck

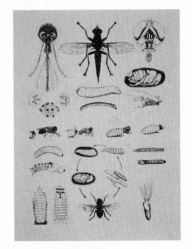

blood of both vertebrates (e.g., mosquitoes, black flies, stable flies, horse flies) and invertebrates (punkies), they eat dung (house flies, latrine flies), they eat plants (gall flies, leafminer flies) and fungi (fungus gnats, phorid flies), they invade the bodies of vertebrates (bot flies, screwworm flies, warble flies) and invertebrates (tachinid flies, humpbacked flies), and they prey on insects and other arthropods (robber flies, hover flies, stiletto flies). Even after they are dead, animals and plants still serve as food for flies; blow flies and flesh flies eat dead animals and fruit flies and crane flies eat plant debris. Needless to say, they live in a wide range of terrestrial and aquatic habitats. The headless, legless larvae of some flies are called maggots.

Order Ephemeroptera (*ephemera,* short-lived, *ptera,* wings)= mayflies (2,000 species).Mayflies have two pairs of wings, held above their body; they are more or less forced to hold them upright because they lack any sort of mechanism for folding flat. While the immature stages, mostly aquatic vegetarians, have chewing mouthparts, adults lack functional mouthparts and live only a day or so (hence *ephemera* or short-lived). Characteristically they possess two or three long tails in both adult and immature stages. Mayflies are unique in the insect world in that, as subadults (or subimagoes), they molt with wings on. Mayflies are important components of freshwater food chains and are relished by vertebrate and invertebrate predators alike.

Order Hemiptera (*hemi,* half, *ptera,* wings)= true bugs (50,000 species). The hemipterans are the only insects regarded by entomologists as "bugs"—properly speaking, that name belongs to insects with two pairs of wings, the first pair thickened at the base, and sucking mouthparts. True bugs can be found in a variety of aquatic and terrestrial habitats, where their sucking mouthparts imbibe a wide variety of foods; some (e.g., leaf bugs, seed bugs, lace bugs, stainers, leaf-footed bugs, and some stink bugs) are plantfeeders, some (e.g., ambush bugs, assassin bugs, other stink bugs, minute pirate bugs, waterscorpions, and giant water bugs) are predators, some (e.g., the flat bugs) eat fungus, and some (e.g., bed bugs and bat bugs) are even ectoparasites.

Order Homoptera (*homo,* same, *ptera,* wings)=
aphids, scale, mealybug, whiteflies (32,000
species). Like hemipterans, homopterans have
sucking mouthparts, but their forewings are uni-
form in texture, either thickened or membranous and are generally held rooflike over the
body. All homopterans feed on plant fluids.
Whereas the cicadas and hoppers (tree-, frog-,
and leaf-) are fairly active, aphids and scale insects are sedentary (often wingless) for much of
their lives. Reproduction in these sedentary
forms is often parthenogenetic (asexual).

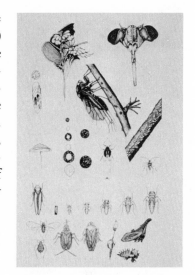

Order Hymenoptera (*hymen,* membrane, *ptera,*
wings)=bees, ants, wasps, sawflies (108,000
species). Although members of this large order
are diverse, they generally possess two pairs of
wings, both membranous, linked by a series of
hooks, or hamuli. In all hymenopterans other
than sawflies, the first segment of the abdomen
is fused to the thorax, giving them a distinctive
"wasp-waisted" appearance. The mouthparts
can be modified for both chewing and sucking
(particularly nectar) in adults; while larval
sawflies are caterpillar-like, other hymenopter-
ous larvae are grub-like. Habits are extremely
variable throughout the order; many are parasitic as larvae on other arthropods, a number are
predaceous, still others are herbivorous, and

many are eusocial. In the parasitic and social species, the ovipositor
has been modified to serve as a venom injection device.

Order Isoptera (*isos,* equal, *ptera,* wings)=termites (1,900 species). Termites have two pairs of wings, at least for a while, that are similar in size and venation; adults generally shed their wings shortly after mating. Equipped (with few exceptions) with chewing mouthparts, they consume wood, soil, fungus and the like; their ability to eat these unappetizing things is due in part to the presence of symbiotic cellulose-digesting microbes in their guts. Every species in the order is eusocial, that is, displays complex social behavior, with kings and queens to produce offspring, workers (often the immatures or nymphs) to care for the nest, and soldiers (who assume a variety of bizarre forms) to defend the nest against invaders.

Order Lepidoptera (*lepidos,* scale, *ptera,* wings) =butterflies, moths, skippers (140,000 species). The most distinctive feature of lepidopterans are their scaly wings, held roof-like by moths and upright by butterflies. The adults have sucking mouthparts, a coiled tube used for sucking nectar or occasionally other fluids (even in a few rare instances blood or even tears). The larvae, called caterpillars, for the most part eat plants, although a small number are carnivorous on soft-bodied, slow-moving fellow insects. Caterpillars can spin silk, and some species spin themselves a cocoon in which to pupate.

Order Neuroptera (*neuron,* nerve, *ptera,* wings)=dobsonflies, lacewings, antlions, alderflies, snakeflies, fishflies, owlflies (4,700 species). The four membranous wings of neuropterans, generally held rooflike over the body, are networked with a complex system of fine veins. Adults have chewing mouthparts and are mostly carnivorous, although some (such as the lacewings) specialize in slow-moving soft-bodied prey and some don't feed at all. Immature stages (larvae) have biting or sucking mouthparts, variously sickle-shaped or short, and are all carnivores, mostly on other arthropods but spongillafly larvae feed on freshwater sponges.

Order Odonata (*odontos,* tooth)=dragonflies, damselflies (5,000 species). Like mayflies, dragonflies are equipped with wings that do not fold flat over the body—they're held above the body, in the case of damselflies, or out to the side, in the case of dragonflies. The aquatic immature stages and aerial adult stages are all exclusively predaceous. Anatomical adaptations to facilitate these carnivorous habits include the "labial mask" of the immatures (a grappling-hook arrangement that flaps forward, snags prey, and hauls it inward) and the angled thorax of the adults, which allows the legs to swing forward to form sort of a "shopping basket" for scooping-

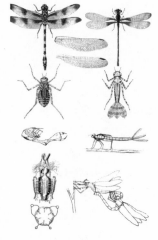

up prey on its wing. The thin, elongated bodies of the odonates has inspired the common name "darner," or darning needle.

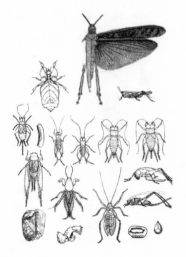

Order Orthoptera (*orthos,* straight, *ptera,* wings)=grasshoppers, crickets, katydids (17,000 species). Orthopterans have two pairs of wings, with the first pair thickened and the second pair folded. Whereas grasshoppers and locusts have short antennae, the crickets and katydids have long threadlike antennae. Grasshoppers and locusts use their chewing mouthparts to consume plants; crickets and their long-horned relatives are omnivorous. Eggs are produced in egg capsules. Species in this order are well known for their saltatorial (jumping) habits and their ability to produce sound by rubbing various body parts together.

Order Phasmida (*phasma,* ghost)=walkingsticks (2,000 species). Walkingsticks are generally wingless or equipped with very short wings (particularly the forewing). All are herbivorous and chew plants at all life stages. Eggs are scattered rather cavalierly in amongst the vegetation. They are perhaps best known for their cryptic coloration and their resemblance not only to sticks but to twigs, leaves, stems, and other plant parts.

Order Phthiraptera (*phthirius,* louse, *aptera,* wingless)-chewing lice (2,600 species) and sucking lice (230 species). Lice are wingless permanent ectoparasites of vertebrates; all are drab-colored, flattened top-to-bottom, and equipped with strong claws for clinging to fur or feathers. Chewing lice, with chewing mouthparts, tend to have heads that are wider than their thorax; sucking lice, with piercing/sucking mouthparts that fold up into a little pouch, have heads narrower than their thorax. Most lice spend their entire lives (including the egg, or nit, stage) on their vertebrate hosts.

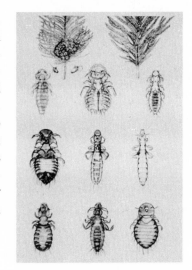

Order Plecoptera (*plekein,* braided, *ptera,* wings)=stoneflies (1,700 species). The two pairs of wings possessed by stoneflies are membranous and fold flat over the back, the hindwing in many species pleated into accordion-like folds. As immatures (nymphs), they are aquatic vegetarians or carnivores and often can be found clinging to the undersides of stones. Adults of many species don't feed at all; since they're weak fliers, they're often found near streams, lakes, or other bodies of water. Both nymphs and adults have long paired cerci, or tails, projecting from the tip of their abdomens.

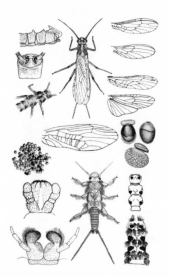

Order Psocoptera (*psoco,* rub, *ptera,* wings)= psocids, bark or book lice (2,400 species). The membranous wings of psocopterans, when present, are held rooflike over the body; the wingless species often encountered in books and other printed material are called booklice. These small, soft-bodied insects consume all manner of debris, including mold, fungi, and even dead insects.

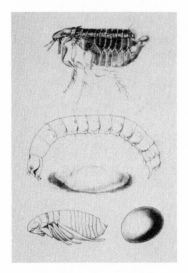

Order Siphonaptera (*siphon*, tube, *aptera*, wingless)=fleas (2,300 species). Fleas are a fairly homogeneous group; all are wingless permanent and intermittent ectoparasites of vertebrates. Their body is laterally flattened and exoskeleton thickened, to facilitate movement in and around host fur and feathers. The distinctive piercing/ sucking mouthparts of the adults are used to consume blood; larvae live in nests and use their mouthparts to chew on nest debris, including their parents' excrement. Apart from their diet, their most distinctive feature is their remarkable hopping ability.

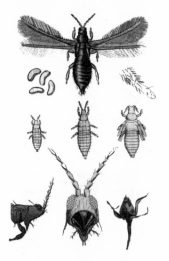

Order Thysanoptera (*thysano*, fringe, *ptera*, wings)=thrips (5,000 species). Thrips are tiny, cigar-shaped insects, generally equipped with two pairs of narrow, fringed wings. Their un-usual rasping/ sucking mouthparts are asym-metrical (the right mandible is missing). A diverse group, their diet variously includes plant tissue, pollen, and even other arthropods. In ad-dition to their peculiar wings and their peculiar mouthparts, they have a peculiar bladderlike structure on the tip of their feet and a peculiar form of metamorphosis that includes a winged "prepupa" and winged "pupa"- like stage.

Order Thysanura (*thysanos,* tassel, *ura,* tail)= silverfish, firebrats (350 species). Like springtails, thysanurans are wingless and molt throughout their adult life. They use their chewing mouthparts to feed on debris—some species of silverfish damage books and wallpaper in houses and firebrats are known to frequent bakeries and other warm spots. The "tassels" (*thysanos*) are the three long "tails" at the tip of the abdomen. They're called silver "fish" because their bodies are frequently ornamented with shiny scales.

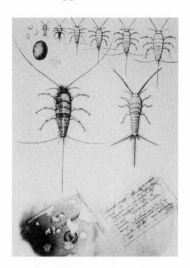

Order Trichoptera (*trichos,* hair, *ptera,* wings)= caddisflies (7,000 species). Trichopterans owe their name to their hairy wings, held roof-like over the body. Although the terrestrial adults have reduced mouthparts and probably don't feed, the aquatic larvae are variously herbivorous or carnivorous. Some larvae are free-living but the majority construct little cases for themselves out of sand, stones, twigs, or other debris.

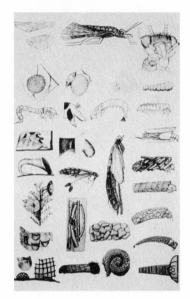

Index

363